ECOLOGICAL ASSEMB1

Perspectives, advances,

It has been more than 20 years since Jared Diamond focused attention on the possible existence of assembly rules for communities. Since then, there has been a proliferation of studies trying to promote, refute or test the idea that there are sets of constraints (rules) on community formation and maintenance (assembly). This timely volume brings together carefully selected contributions which examine the question of the existence and nature of assembly rules with some rigor and in some detail, using both theoretical and empirical approaches in a variety of systems. The result is a balanced treatment, which encompasses a wide range of topics within ecology, including competition and coexistence, conservation and biodiversity, niche theory, and biogeography. As such, it provides much to interest a broad audience of ecologists, while also making an important contribution to the study of community ecology in particular.

EVAN WEIHER is an Assistant Professor in the Biological Sciences Department, University of Winsconsin, Eau Claire, Winsconsin, USA.

PAUL KEDDY holds the Edward G. Schlieder Endowed Chair for Environmental Studies at Southeastern Louisiana University, Hammond, Louisiana, USA.

ECOLOGICAL ASSEMBLY RULES

Perspectives, advances, retreats

Edited by

EVAN WEIHER AND PAUL KEDDY

CAMBRIDGE
UNIVERSITY PRESS

PUBLISHED BY THE PRESS SYNDICATE OF THE UNIVERSITY OF CAMBRIDGE
The Pitt Building, Trumpington Street, Cambridge, United Kingdom

CAMBRIDGE UNIVERSITY PRESS
The Edinburgh Building, Cambridge CB2 2RU, UK
40 West 20th Street, New York, NY 10011–4211, USA
10 Stamford Road, Oakleigh, VIC 3166, Australia
Ruiz de Alarcón 13, 28014 Madrid, Spain
Dock House, The Waterfront, Cape Town 8001, South Africa

http://www.cambridge.org

First published 1999
First paperback edition, with corrections, 2001

Printed in the United Kingdom at the University Press, Cambridge

Typeset in Times 10/13pt, in QuarkXPress™ [WV]

A catalogue record for this book is available from the British Library

Library of Congress Cataloguing in Publication data

Ecological assembly rules : perspectives, advances, retreats /
edited by Evan Weiher and Paul Keddy.
p. cm.
Includes index.
ISBN 0 521 65235 9 (hc.)
1. Biotic communities. I. Weiher, Evan, 1961– II. Keddy, Paul, 1953–
QH541 .E3165 1999
577.8′2–ddc21   98-44324   CIP

ISBN 0 521 65235 9 hardback
ISBN 0 521 65533 1 paperback

Contents

Contributors

Karla L. Balent
Department of Biological Sciences
SUNY College at Brockport
NY 14420
USA

Barbara D. Booth
Department of Botany
University of Guelph
Guelph
Ontario N1G 2W1
Canada

James H. Brown
Department of Biology
University of New Mexico
Albuquerque
NM 87131
USA

Marcelo Cabido
Instituto Multidisciplinario de Biologia Vegetal
(CONICET – Universidad Nacional de Córdoba)
Casilla de Correo 495
5000 Córdoba
Argentina

Fernando Casanoves
Grupo de Estadistica y Biometria
Departmento de Desarrollo Rural
Facultad de Ciencias Agropecuarias
Unioversidad de Córdoba
Casilla de Correo 509
5000 Córdoba
Argentina

Martin L. Cody
Department of Biology
University of California
Los Angeles
CA 90095–1606
USA

Tamar Dayan
Department of Zoology
Tel Aviv University
Ramat Aviv
69978 Tel Aviv
Israel

Sandra Díaz
Instituto Multidisciplinario de Biologia Vegetal
(CONICET – Universidad Nacional de Córdoba)
Casilla de Correo 495
5000 Córdoba
Argentina

James A. Drake
Complex Systems Group
Ecology and Evolutionary Biology
University of Tennessee
Knoxville
TN 37996
USA

Theodore C. Foin
Division of Environmental Studies
University of California
Davis
CA 95616
USA

Barry J. Fox
School of Biological Science
University of New South Wales
Sydney NSW 2052
Australia

Paul Keddy
University of Ottawa
30 Marie Curie Street
PO Box 450 Stn. A
Ottawa
Ontario K1N 6N5
Canada

Douglas A. Kelt
Department of Wildlife, Fish & Conservation Biology
University of California
Davis
CA 95616
USA

Douglas W. Larson
Department of Botany
University of Guelph
Ontario N1G 2W1
Canada

Julie L. Lockwood
Department of Ecology and Evolutionary Biology
The University of Tennessee
Knoxville
TN 37996
USA

Mark V. Lomolino
Department of Zoology and Oklahoma Natural Heritage Inventory
Oklahoma Biological Survey
University of Oklahoma
Norman
OK 73019
USA

Michael P. Moulton
Department of Wildlife Ecology and Conservation
The University of Florida
PO Box 110430
Gainesville
FL 32611
USA

Stuart L. Pimm
Department of Ecology and Evolutionary Biology
The University of Tennessee
Knoxville
TN 37996
USA

Tom Purucker
Complex Systems Group
Ecology and Evolutionary Biology
University of Tennessee
Knoxville
TN 37996
USA

Carmen Rojo
Complex Systems Group
Ecology and Evolutionary Biology
University of Tennessee
Knoxville
TN 37996
USA

Daniel Simberloff
Department of Ecology and Evolutionary Biology
University of Tennessee
Knoxville
TN 37996
USA

Lewi Stone
Department of Zoology
Tel Aviv University
Ramat Aviv
69978 Tel Aviv
Israel

Elizabeth M. Strange
Department of Fishery and Wildlife Biology
Colorado State University
Fort Collins
CO 80523
USA

Evan Weiher
Department of Biological Sciences
PO Drawer GY
Mississippi State University
MS 39762
USA

J. Bastow Wilson
Botany Department
University of Otago
PO Box 56
Dunedin
New Zealand

Craig R. Zimmerman
Complex Systems Group
Ecology and Evolutionary Biology
University of Tennessee
Knoxville
TN 37996
USA

Introduction: The scope and goals of research on assembly rules

Paul Keddy and Evan Weiher

Why should assembly rules be studied, why should a symposium be organized, why should a volume on the topic be published, and why should anyone bother to read it? These are all perfectly reasonable questions, and it is our task to briefly address them by way of introducing this volume.

For some time, 'community ecology' has been the name loosely applied to a collection of studies and methods that apply to more than one organism, but that apply at scales below the landscape. Many books on community ecology appear to offer little more than a disparate hodgepodge of studies that are unified solely by the above vague restrictions. This may seem a harsh criticism, and a peculiar way to open a book written for community ecologists. But, these sort of criticisms form the ground of this volume (some might suggest charnel ground), and it is not an original observation by any means. Indeed, Pianka (1992), a prominent member of our discipline, has felt obliged to apologize on our behalf in a paper titled 'The state of the art in community ecology', observing therein that 'community ecology has for too long been perceived as repugnant and intractably complex'. He apologizes to a world symposium that 'the discipline has been neglected and now lags far behind the rest of ecology'.

He is not alone in his survey, opinion and prognosis. Nearly two decades earlier Lewontin (1974) wrote of the 'agony' of community ecology. More recently, the late Rob Peters (1991) annoyed many when he decried the lack of apparent progress in our discipline. Lawton (1992) then attacked Peters; Keddy (1992) criticized his criticisms. Scheiner (1993) then criticized Keddy for criticizing Lawton, and Keddy replied (Keddy, 1993), and then Scheiner replied to him (Scheiner, 1994). Lacking tangible progress, people turn upon one another. If there is agony in community ecology, as Lewontin suggested, much of it appears to be self-inflicted. Meanwhile, thick compendia under the name of community ecology arise with frustrating regularity and repetitive content. This situation is what led Keddy (1993) to observe that, without far more emphasis upon measurable properties, critical tests and rational decision-making, community ecologists run the risk of becoming more like the humanities than the sciences, prone to political and emotional conflicts rather than debates using rational criteria. In Camille Paglia's (1992) essay 'Junk bonds and corporate raiders: academe in the hour of the wolf' one can read her view that 'the self-made inferno of the academic junk bond era is the conferences, where the din of ambition is as deafening as on the floor of the stock exchange. The huge, post-1960s' proliferation of conferences ... produced a diversion of professional energy away from study and toward performance, networking, advertisement, cruising, hustling, glad-handing, back-scratching, chitchat, groupthink'. Mercifully, community ecologists (and this volume) are completely immune to this risk because we are doing science.

Why a symposium on assembly rules, and why a book you may still be asking? Plans for the symposium were rooted in the above circumstances, combined with two common-sense observations. These were:

(a) if there are not some common goals for community ecology, then they are unlikely to be achieved;
(b) a prominent theme in the discipline is the attempt to predict the composition of ecological communities from species pools.

The first observation appears self-evident, but apparently in some circles there is still a suspicion of research 'agendas'. There appears to be a naive belief that the way to build a spaceship and land a human on the moon is to trust that, if everyone indulges themselves in an idiosyncratic and self-indulgent pastime at the taxpayer's expense, the outcome will be positive. Like Voltaire's Professor Pangloss, there is the insistence that this must be the best of all possible worlds, however inefficient or painful it may appear on the surface. If progress is forgotten about entirely, and our discipline is seen as a mere pastime for the tenured and well to do, there is no particular need to be concerned about goals, progress and social contribution. As long as ecologists have jobs and the chance at a large grant, why worry? Volumes with this kind of philosophy are too frequent as it is, and given that it has been said in print (Keddy, 1991), as editors we were careful to insist upon a common purpose. Within this common purpose, a diversity of views about the details and the strategies and tactics for achieving it was welcomed.

The second observation may be less evident, but if the many topics that have arisen in community ecology over the years are considered, the common thread may not be in level of organization, methodology or number of species, but in the underlying problem that is being addressed. Moreover,

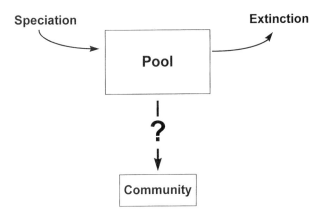

Fig. 1. Assembly rules address a central theme in community ecology: how are communities assembled from species pools. (Evolutionary ecology, by contrast, deals with the formation of the pool.)

adopting this goal clarifies a common source of confusion among ecologists themselves: community ecology is different from evolutionary ecology. Figure 1 shows a possible framework for community ecology. Community ecologists are concerned with the question mark: how does one get from the pool to the community? Evolutionary ecologists are concerned more with the top box and arrows: how do speciation and extinction produce the pool? From the perspective of community ecology, the pool is just the source of raw materials, and the process that creates the pool, while of interest, generally occurs at longer time scales than are normally considered relevant.

A new name could be invented to describe the study of how communities are built from pools, but if the literature is looked back on, there is a good deal of terminology that can already be borrowed. There is always a risk in using pre-existing terminology, because it all comes with baggage. The baggage includes assumptions about the kinds of organisms worthy of study, past controversies that are actually tangential to the issues at hand, and past confusions that entangled ecologists. It is for this reason that the Roman armies called baggage *impedimenta*. Ecologists do not need more impediments. But, neither does there seem to be any point in inventing new terms when perfectly good ones are already there. To do so would be to throw out the wisdom of past work because the baggage is feared. Thus the term 'assembly rules' has been adopted to describe the problem of assembly communities from pools; this accords rather well with Diamond's original (1975) usage of the word. There may be doubts about using birds as a model system, about descriptive as opposed to experimental studies, about inferences about mechanisms that may not be justified, about controversies that have generated more heat than light, and about habitual ways of trying to solve these problems that appear self-defeating. All of these objections (and more,) were raised either by participants in the symposium, or by other practising ecologists. The term assembly rules, however, still captures the essence of the problem in Fig. 1. Moreover, it nicely fits in with Pirsig's (1974) observation that assembling a rotisserie is not unlike fixing a motorcycle, that the challenge of putting something together properly from the pieces (assembling it) is a challenge with worthy mechanical and philosophical dimensions.

And so, the symposium was called 'assembly rules', and researchers were sought out who were studying how communities were assembled out of pools. In spite of ourselves, the perceptive reader will discern certain biases. For example, Fig. 2 gives one perspective upon the composition of ecological communities upon Earth's surface; more recent calculations would expand the invertebrate and fungal component. This would seem to be a common-sense starting point in designing the discipline of community ecology. In spite of ourselves, we have ended up with a disproportionate representation of vertebrate examples. Our defence is that, while trying to collect a representative set of studies on community assembly, a highly biased and artificial pool from which to make the draw was being dealt with, and so the distortions of our literature have been included. Our only plea is that we consciously tried to avoid the worst distortions. Further, to the extent that

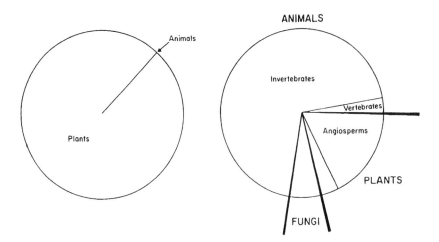

Fig. 2. The importance of different life forms in the biosphere, measured according to biomass (*left*) and number of species (*right*). (From Keddy, 1989.)

they have been reproduced, others can only be encouraged to rectify the situation.

Two existing paradigms

Within the literature, and within this volume there are at least two developing paradigms for the assembly of communities (Fig. 3). The first we call the island paradigm because it deals with mainlands, islands, immigration, and coexistence. The second we call the trait–environment paradigm because it begins with pools, habitats as filters, and convergence. This is not to suggest that there are only two ways forward, or that either of these is the best. These simply happen to be two themes which will be evident in this concert. The challenge for a musician is to build upon a theme in an entertaining way without being repetitive.

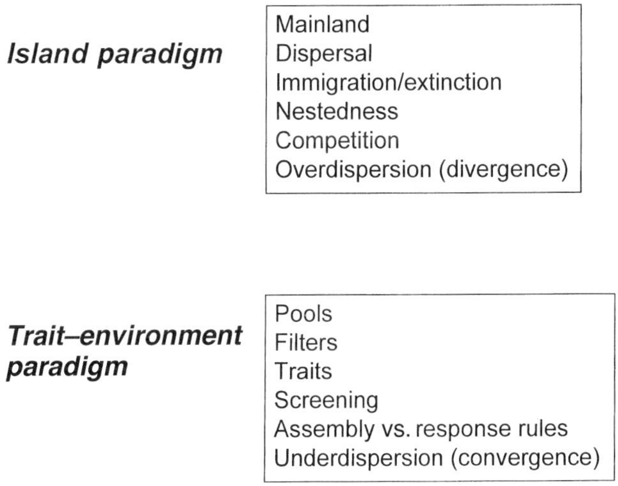

Island paradigm

Mainland
Dispersal
Immigration/extinction
Nestedness
Competition
Overdispersion (divergence)

Trait–environment paradigm

Pools
Filters
Traits
Screening
Assembly vs. response rules
Underdispersion (convergence)

Fig. 3. The two most common paradigms for community assembly are the island paradigm (*top*) and the trait–environment paradigm (*bottom*).

Type 1: Island models

Many studies are built upon the raw data lists of species on islands. In terms of Fig. 1, the pool is the adjoining mainland, and the list of species from the island is considered to be the community. The basic series of steps is as follows:

(a) make lists of organisms in each habitat;
(b) create one or more null models for possible patterns;
(c) test for patterns in these lists;
(d) offer explanations for these patterns;
(e) state the explicit rules for community assembly.

A good example of this sort of study comes from Diamond's (1975) work on the avifauna of New Guinea (Fig. 4). There is now a large literature on this topic, and a growing body of literature on null models, but a good deal of controversy about the costs and benefits of null models and the kinds of data appropriate to them (Gotelli & Graves, 1996). Moreover, while there have been many searches for evidence of pattern, few brave souls have reached step (e).

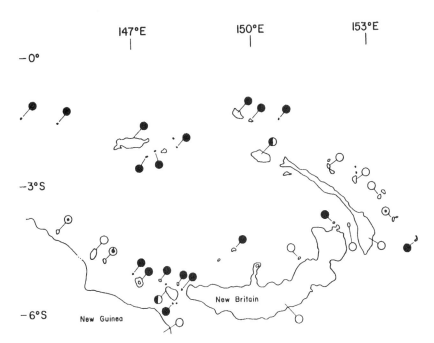

Fig. 4. The island paradigm for assembly is built upon studies of bird distribution on offshore islands. Are there patterns, do they differ from those predicted by null models, and are there rules that predict them? This example shows the distribution of two species of *Ptilinopus* fruit pigeons, where split circles are co-occurrences and dots are co-absences. (From Keddy, 1989 after Diamond, 1975.)

Type 2: Trait–environment models

One can also approach community assembly not by using lists of organisms, but by focusing upon their traits. The environmental factors are then viewed as filters acting upon these traits. In this case the procedure is as follows:

(a) determine the key traits the organisms possess;
(b) relate the traits to key environmental factors;
(c) specify how trait composition will change with specific changes in environment;
(d) relate this back to the particular organisms possessing those traits.

We are not interested in general properties of the traits themselves, but in the relative abundances of the organisms that possess them. A good example of this sort of study is the work on prairie potholes by van der Valk (1981). The water level in the pothole acts as a filter determining the kinds of plant species that will occur there; the two key states are drained v. flooded (Fig. 5).

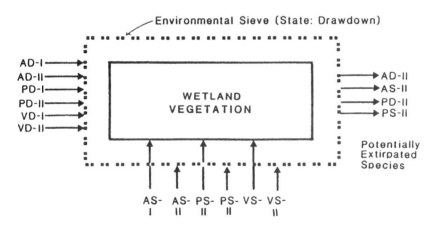

Fig. 5. The trait environment paradigm for community assembly focuses attention upon the pool of species, the traits they possess, and the environmental filters operating in a particular situation. The example shows that the composition of vegetation in a pothole wetland depends upon whether the pool of buried seeds is exposed to flooded or drained conditions. (From van der Valk, 1981.)

What is an assembly rule?

What would an assembly rule look like if one were found? A goal cannot be attainable unless some criteria are set up to tell when it has been achieved. An

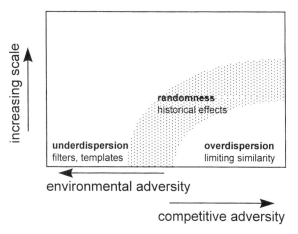

Fig. 6. A first step in the search for assembly rules is the search for pattern. Island modes have tended to search for overdispersion (*right*) whereas trait–environment models have tended to focus upon trait underdispersion (*left*). A larger view suggest that both are possible, depending upon the scale of enquiry. (From Weiher & Keddy, 1995.)

assembly rule specifies the values and domain of factors that either structure or constrain the properties of ecological assemblages.

Overall, there are four parts in the procedure of finding assembly rules:

(a) defining and measuring a property of assemblages;
(b) describing patterns in this property;
(c) explicitly stating the rules that govern the expression of the property; and
(d) determining the mechanism that causes the patterns.

Contrary to common practice, merely documenting a pattern is not the study of community assembly. Plant ecologists have described plant patterns for more than century, and appear ready to continue doing so for yet another; simply describing patterns is not the study of assembly rules. Nor is an improvement, the added demonstration that pattern exists against a null model, sufficient to qualify. Asking if there is pattern in nature is akin to asking if bears shit in the woods. Null models provide a valuable and more rigorous way of demonstrating pattern, but they still do not specify assembly rules. Within this realm of pattern (step 2), one, of course, needs to ask what kinds of patterns might occur and at what scales (Fig. 6). But actual assembly rules, step 3, require further effort yet. They might include statements such as the following:

'If an assemblage of plants is flooded, the subset of species that survives will all have aerenchyma.'

'In the absence of predators, a pond in the temperate zone can be expected to have between 5 and 10 amphibian species. If a predatory fish species is introduced to the pond, this will fall to between 0 and 2 species.'

'The ratio of insectivorous to granivorous birds in deciduous forests is between 0.25 and 0.33, whereas in boreal forests the ratio falls between 0.45 and 0.55.'

'If a herbaceous plant community with biomass of 500 g m^{-2} is fertilized with NPK fertilizer, the mean number of species per m^2 will decline by 10% with each x g m^{-2} of fertilizer.'

'There is a linear relationship between the number of beetles in deciduous forests and the volume of coarse woody debris. The equation relating the two is as follows ...'

These statements all are expressed in terms of measurable properties and their range or variation in relation to another factor. The following statements would not qualify as assembly rules.

'Similar organisms tend not to coexist.'

'Competition controls the distribution of birds on islands.'

'Copepod communities in ponds are not randomly assembled from a species pool.'

'Tree species diversity increases with decreasing latitude.'

'The distribution of lizards upon islands deviates significantly from null models.'

Such statements certainly belong within community ecology, and may have within them concepts or models that increase our understanding of certain assemblages. But they are not assembly rules. Rules themselves must be explicit and quantitative if they are to qualify. The other statements are, perhaps, steps on the way to the goal. There seems to be some current confusion on this matter. For example, the existing literature suggest that merely finding a pattern in properties is an assembly rule. Confusing a step on the path with the attainment of a goal only generates confusion.

Obstacles on the path

Good generals study the failures of other generals so that they do not repeat their mistakes. One of the consequences of accepting a goal is the recognition and study of obstacles and past errors. This could be considered annoying, something to be avoided at all costs, perhaps because no one wants to admit to having fallen prey to an obstacle. Perhaps our early days in Sunday school are remembered, being told of our committing a sin. But, the delightful side of this is that if obstacles cannot be recognized, they cannot be avoided. That is why road signs warning 'detour ahead' are so useful; without them we could damage our car. Returning to generals, in his chapter Doctrine of Command, General Montgomery (1958) wrote 'I hold the view that the leader must know what he himself wants. He must see his objective clearly and then strive to attain it; he must let everyone else know what he wants and what are the basic fundamentals of his policy.' (p. 81)

The frequent reluctance to admit that there are both goals and obstacles to our work is certainly unmilitary, and perhaps unprofessional. It may reflect a desire to remain child-like, innocent and naive, with no responsibility for one's actions. Once, like General Patton, our intention to be in Berlin next year is announced, everyone will know if it is not achieved. It takes some bravery to announce our goals, and to suggest that society should care whether Berlin or Paris is achieved. Are there some obstacles that have interfered with past campaigns in community ecology. What are some of these errors that might have led to Pianka's despair? At the symposium the participants were specifically warned against some pitfalls. For those who were not in attendance, they are briefly listed below.

Before the list is presented, one more clarification is necessary. The list offers styles of research which are obstacles to advancement. Elements of such styles are contained in everyone, but in different relative proportions. In an exactly analogous way, all humans have anger, negativity, arrogance, envy, ignorance, greed, suspicion in their psychological make-ups. One of the reasons humans have lists such as the seven deadly sins, or the three poisons, is to be warned to watch out for these states as they arise within own minds. Tradition has taught that these states create confusion for ourselves, and problems for our human communities. In the same way, the following list of styles in intended to illustrate approaches that can be slid into if one remains unaware of one's own behavior. The intention is not to list obstacles in order to imply that these flaws are found in only a few bad people (see the exchange between Scheiner, 1993 and Keddy, 1993), nor so readers can try to guess who falls into which category, but rather to acknowledge that all humans are subject to such tendencies. Such a list, then, provides gauges for an instrument panel that will warn if one wanders too far into unproductive terrain. Five obstacles are considered.

(a) Bigger is better ('Mine is bigger than yours')
Sometimes it is thought that if someone does not know where they are going, then at least the neighbours can be impressed by seeing them drive a bigger car while they try to get there. This is common in the animal and plant kingdoms. Large insects tend to be dominant over smaller ones (Lawton & Hassell, 1981) just as large plants tend to be dominant over smaller ones (Gaudet &

Keddy, 1988). Display of wealth or economic power is a standard technique humans use to maintain authority over neighbours (Kautsky, 1982), the culmination of which is perhaps seen in the arms races during and after the Second World War (Keegan, 1989). There is no need here to impress the audience with a Cray 2 supercomputer when only a pocket calculator will make the point. Similarly, there is no need to show the helicopter or float plane that it took to reach the study site.

(b) Complexity is impressive ('I know more natural history than you')
Many people know a great deal about nature. Keddy could, for example, lecture for hours (and write for pages) upon the plants that grow in wetlands or in Lanark County where he lives. If he did so in a sonorous and authoritative voice, he might even convince you that your time listening was well spent. But he would mislead if he tried to pass off his delightful knowledge of natural history under the term assembly rules. Merely knowing where organisms are found, and describing it in exquisite detail, is not the study of assembly rules. Otherwise, Tansley and Adamson (1925) would have to be credited with assembly rules for British grassland. In their study of grazing, they reported on factors controlling composition in three patches along a gradient of 'progressive increase in height and density of vegetation, with an increasing number of species'. In one paragraph they observe:

'Of the mosses, *Barbula cylindrica, B. unguiculata* and *Bryum capillare* decrease and disappear with complete closure and increasing depth of the turf; *Brachythecium purum*, one of the most ubiquitous of chalk grassland mosses, though not a 'calcicole' species, appears in (b) and increases in (c); while *B. rutabulum, Mnium undulatum* and *Thuidium tamariscinum* first appear in the damper conditions of (c)'

These sorts of descriptions are possible for all manner of ecological systems; the real problem is to discover the generalities that underlie the remarkable diversity of life (Mayr, 1982). General Montgomery (1958) says 'It is absolutely vital that a senior commander should keep himself from becoming immersed in details ... If he gets involved in details ... he will lose sight of the essentials which really matter; he will then be led off on side issues which will have little influence on the battle ...' (p. 86).

In general, simple hypotheses are preferable to complex ones (Aune, 1970). We need to remember the late Rob Peters' warning (1980a,b) that there is an important difference between natural history and ecology. It might be recalled that even an omniscient and omnipotent deity felt it necessary to give Moses only ten rules to guide all the complexities of human conduct.

(c) Sycophancy ('My friends are more important than yours')
Humans are primates, and the primate mind appears to have evolved for survival in social/tribal settings (e.g., Leakey & Lewin, 1992). It is therefore entirely natural for authority figures to be created, like the old silver back males in the Gorilla tribe, and then deferred to, whether people live in the world's most powerful democracy (Dye & Zeigler, 1987) or a Stalinist prison camp in Siberia (Shalamov, 1982).

There have been sycophants as long as there have been authority figures. Socrates is often presented as a courageous man of science who died rather than appease the Athenian rulers. But, in a compelling re-evaluation of the historical records, I.F. Stone (1989) reached quite a different conclusion. At the time, he notes, Athens was repeatedly threatened by tyrants; in both 411 and in 404, the conduct of the aristocratic dictators was 'cruel, rapacious and bloody'. Socrates was strangely silent while all this happened. Stone looks in vain for Socrates to show compassion for the poor, or opposition to tyranny, and notes that the famous *Republic*, later written by his student Plato, advocates a tyranny remarkably close to the modern Communist state, complete with internal security police ('guardians'). Socrates, concludes Stone, was a sycophant for the tyrants. He did not plead for freedom of speech because it was a principle he did not believe in, and he could not bring himself to argue before the democrats using their own principles. If Stone's interpretation is correct, using Socrates as an example for scientists may have far darker connotations than has been realized, and comes rather close to Saul's (1992) contemporary view that intellect is too often used to reinforce authority rather than challenge it.

Apparently it is natural for humans to divide themselves into tribes and attack one another (Ignatieff, 1993). It further seems natural to kick and bite (or at least ostracize) members of our tribe who do not seem to want to groom the same dominant that the others do (Browning, 1992). They may even favor their own gender (Gurevitch, 1988). The fact that this was natural behavior for primates on the African plains does not mean that it is defensible or useful behavior for contemporary scientists. Indeed, in other settings, such as religion, people frequently react with annoyance when they see blind obedience and the exercise of authority. Ape instincts such as sycophancy seem obvious when they creep out of the tribe and into our of politics, but perhaps less so when then appear in the sciences (Keddy, 1989). Appeals to authority, the selective citation of friend's work, and division into tribal units do not contribute to the advancement of science. All are all to be avoided here.

(d) Self-indulgence ('Playing with yourself is harmless fun')

It may be true that masturbation does not cause blindness, but neither is it necessarily a healthy substitute for a relationship with another human being. Similarly, self-indulgent work aimed at building up egos may not cause irreparable harm to science, but it certainly is no substitute for thoughtful goals, collegiality, care, respect and social responsibility. The motivation that is brought to our work will necessarily influence the way in which it develops. It is therefore necessary to think about where we are going and how our discoveries might be of use to society. The alternative is unpleasant for everyone. The Roman historian Plutarch describes how one Roman emperor after another 'lavished away the treasures of his people in the wildest extravagance'. One was killed by a tribune amongst his own guards, another was strangled, another killed himself, but the waste of resources caused by this self-indulgence caused immense human suffering and the diminution of the resources of the Roman empire. Are precious scientific resources being squandered in a similar manner today? Is *Ecology* just *National Geographic* without the color plates?

Hofstadter (1962) suggests that the current American environment of anti-intellectualism is responsible for isolation of scientists from their society: 'Being used to rejection, and having over the years forged a strong traditional response to society based upon the expectation that rejection would continue, many of them have come to feel that alienation is the only appropriate and honorable stance for them to take.' (p. 393) Further, he continues, it is easy to take this to the next logical step and slip into the assumption that alienation has some inherent value itself. As an example, a science reporter tried to interview one of our participants by asking 'What is the practical importance of this work on assembly rules?'. The answer 'None, I hope' perhaps illustrates Hofstadter's point.

Guarding against this self-indulgence is necessary not just because it is other humans who pay our salaries. *The attempt to explain ourselves to others and to solve real world problems actually forces us to do better science.* It may be harmful to indulge in simple natural history, but it is equally dangerous to take the other extreme and retreat entirely to a cosy world of abstraction. When an engineer builds a bad bridge, it falls down. This is a simple example of the pragmatic method, which James (1907) described as 'primarily a method of settling metaphysical disputes that might otherwise be interminable' (p. 10) In contrast, when an ecologist builds a pointless model or publishes a bad paper, it is possible for it to persist and cause unnecessary debates for decades.

(e) Lack of historical context ('My ideas are new and unprecedented')
Life is impermanent, and one way to try escape our fears is self-aggrandization. The more history is ignored, the more important it can seem to be. It has been argued that since the 1960s, ecologists have tended to inflate our self-importance by ignoring the roots of our discipline (Gorham, 1953; Jackson, 1981). Consider a recent example. In 1970 Walker tested whether wetland vegetation along lakes showed a progressive and predictable series of developmental stages using 159 observed transitions between vegetation types extracted from 20 published pollen diagrams (Fig. 7). One dominant course was observed (darker lines). Reed swamp (5) was an essential stage through which all successions must pass. Some reversals occur (17% of all transitions), 'but are usually short-lived and might frequently be due to small changes in water level, temperature or trophic status of the lake water ...' Do such observations, perhaps slightly reworded, not qualify as assembly rules? The scale in time and space, and the replication is impressive relative to many other published studies on assembly rules. So, why is work such as Walker so consistently overlooked? See if you can find it cited in the assembly rules literature.

A more recent example of historical revisionism is provided by Gotelli and Graves (1996) in their book on null models. The theme of whether plant communities are discrete communities or random assemblages can be traced back through writings by Tansley, Clements, Gleason, Ellenberg and Whittaker throughout the period from 1900 to 1970; indeed, the study of species distributions along gradients has probably been the defining feature of research in plant ecology. In the early 1970s, E.C. Pielou developed a number of null models for plant distributions along gradients, which she summarized in her 1975 book. There have been only five main papers that report the application of her null models in the field (Pielou & Routledge (1976) in salt marshes, Keddy (1983) on lake shores, Dale (1984) on marine rocky shores, Shipley and Keddy (1987) in marshes, and Hoagland and Collins (1997) in wet prairies), and in each case the null model has been rejected. Yet Gotelli and Graves do not have a chapter on gradient models, nor does their one paragraph on Pielou and Routledge (1976) explain the significance of their work – that the rejection of Pielou's null model constitutes the first demonstration that communities occur in discrete clusters rather than random (individualistic) associations. On page 1, Gotelli and Graves (two male American zoologists) even opine that the term null models for communities originated with two male American Zoologists at a Florida symposium in 1981! Chris Pielou was a woman, she was Canadian, and she was a plant ecologist: to which of these should her erasure from the ecological record be attributed?

		1	2	3	4	5	6	7	8	9	10	11	12	T
							SUCCEEDING VEGETATION							
ANTECEDENT VEGETATION	1	.	.	3	2	.	.	.	.	.	1	.	.	6
	2	.	.	.	2	2	.	.	.	.	.	.	.	4
	3	1	.	.	4	7	.	.	.	.	1	.	.	13
	4	1	.	1	.	9	.	3	1	3	5	.	.	23
	5	.	.	.	2	.	1	8	6	7	11	4	.	39
	6	.	.	.	.	.	.	1	.	.	.	.	.	1
	7	.	.	.	.	2	.	.	2	8	2	3	.	17
	8	.	.	.	.	1	.	1	.	1	2	3	.	8
	9	.	.	.	1	2	.	1	.	.	1	10	.	15
	10	1	.	.	1	1	.	.	2	.	.	10	.	15
	11	.	.	.	.	.	.	.	.	1	1	.	1	3
	12	.	.	.	1	9	.	4	.	1	.	.	.	15
	T	3	0	4	13	33	1	17	12	21	24	30	1	159

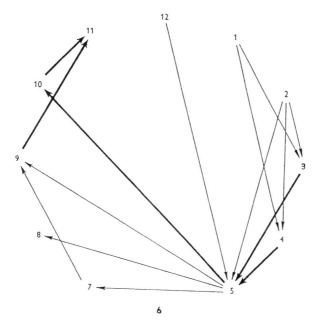

Fig. 7. Frequencies of transition between 12 vegetation stages (ranging from open water (1) through reed swamp (5) to bog (11) to mixed marsh (12)) in 20 pollen cores from a range of wetlands including small lakes, valley bottoms, and coastal lagoons in the British Isles. *Left*, tabulated frequencies, *right*, transition diagram (after Walker 1970).

In the broader sense, one could argue that our entire culture is becoming focused upon the ephemera of the present where everyone will have their 15 minutes of fame; a consequence is a loss of context or significance for properly evaluating current events (O'Brien, 1994; Saul, 1995)). Booth and Larson (this volume) suggest that most of the apparent problems with assembly rules arise out of a studied ignorance of ecological principles prevalent in the early part of this century.

Moving forward

Ending with a list of obstacles could be seen as unnecessarily defeatist or negative. The only reason to study obstacles is to avoid them. It can be hoped that, by consciously avoiding these tendencies, everyone will be able to avoid repeating patterns of behavior that limit our personal contribution to the progress of ecology, or spread discord among our colleagues. By defining warning regions, regions which are positive and valuable are equally defined (Fig. 8). It is not, then, that a single narrow goal is being advocated, nor a single path forward.

Rather than insist that there is only one goal for assembly rules, it is asked that each participant explicitly state their goal; then and only then will it be possible to understand their tactics and judge their degree of success in attaining their goal. To return to our travel analogy, some may be aiming for Paris and others for London; so long as they justify their destination, their trip can be judged only against the stated goal. Of course, if someone asserts that they intend to take us to the Louvre, but then drives us towards London instead, it may be gently suggested from the back seat that they review their travel plans, consult the map, or else choose another destination.

Nor would it be desirable to imply that there is room for only a single style of science. Within our community it can be accepted that different practitioners have different strengths and weaknesses; progress in ecology depends upon exploitation of, and respect for, these differences. So while Fig. 8 defines some regions as obstacles, it should be apparent that there is still plenty of room remaining for a diversity of approaches. Some of the relatively sterile arguments in ecology may arise out of simple lack of tolerance for different styles. It is therefore necessary to be clear in our discussions whether disagreement has arisen because of (a) different views on goals, (b) different views on tactics and style, (c) different interpretations of data, or (d) indulgence in one of the above obstacles. On one hand, it is necessary to be open minded to avoid pointless debates that really hinge on differences in personality or style. Equally however, being open minded does not mean that our brains must be allowed to fall out and outright self-indulgence be tolerated.

Having explicitly considered the need for clearly stated goals, some obstacles

Obstacles

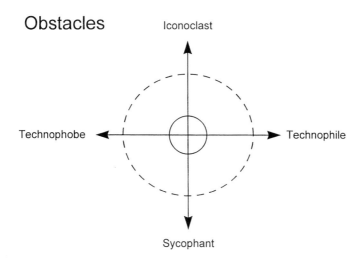

Opportunities

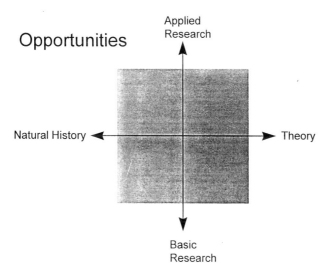

Fig. 8. A variety of psychological obstacles can hamper progress in ecology. By explicitly describing these obstacles, a region is simultaneously defined within which progress is possible; within this region, differences in style enrich collaboration.

to progress, some ways in which researchers may differ in opinion, and the merits of tolerating different approaches, let us return to the study of assembly rules. The problem remains. How do we get from the pool to the community? The introductory talk ended with Fig. 9, so this same figure

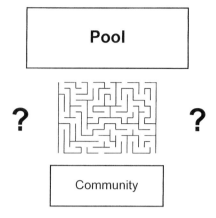

Fig. 9. Participants in the symposium were each asked to chart out their own course though the maze of possibilities that takes us from the pool to the community.

terminates this written introduction. See what the authors have to say in response. In the words of Peter Weiss (1965): 'We ask your kindly indulgence for a cast never on stage before coming to Charenton. But each inmate I can assure you will try to pull his weight.'

References

Aune, B. (1970). *Rationalism, Empiricism, and Pragmatism*. NY: Random House.
Browning, C.R. (1992). *Ordinary Men: Reserve Police Battalion 101 and the Final Solution in Poland*. NY: Harper Collins.
Dale, M.R.T. (1984). The contiguity of upslope and downslope boundaries of species in a zoned community. *Oikos* 47: 303–308.
Diamond, J.M. (1975). Assembly of species communities. In *Ecology and Evolution of Communities*, ed. M.L. Cody & J.M. Diamond, pp. 342–444. Cambridge, MA: Belknap Press of Harvard University Press.
Dye, T.R. & Zeigler, H. (1987). *The Irony of Democracy*. 7th edn. Monterey, CA: Brooks/Cole.
Gaudet, C.L. & Keddy, P.A. (1988). Predicting competitive ability from plant traits: a comparative approach. *Nature* 334: 242–243.
Gorham, E. (1953). Some early ideas concerning the nature, origin and development of peat lands. *Journal of Ecology* 41: 257–274.
Gotelli, N.J. & Graves, G.R. (1996). *Null Models in Ecology*. Washington: Smithsonian Institution Press.
Gurevitch, J. (1988). Differences in the proportion of women to men invited to give seminars: is the old boy still kicking? *Bulletin of the Ecological Society of America* 69: 155–160.
Hoagland, B.W. & Collins, S.L. (1997). Gradient models, gradient analysis and hierarchical structure in plant communities. *Oikos* 78: 21–30.
Hofstadter, R. (1962). *Anti-Intellectualism in American Life*. NY: Vintage Books.

Ignatieff, M. (1993). *Blood and Belonging*. Toronto: Penguin Books Canada Ltd.

Jackson, J.B.C. (1981). Interspecific competition and species distributions: the ghosts of theories and data past. *American Zoologist* 21: 889–901.

James, W. (1907). *Pragmatism*. Reprinted In 1990, *Great Books of the Western World*. ed. M.J. Adler, Vol. 55, pp. 1–89. Chicago, IL: Encyclopedia Britannica.

Kautsky, J.H. (1982). *The Politics of Aristocratic Empires* Chapel Hill: University of North Carolina Press.

Keddy, P.A. (1983). Shoreline vegetation in Axe Lake, Ontario: effects of exposure on zonation patterns. *Ecology* 64: 331–344.

Keddy, P.A. (1989). *Competition*. London: Chapman and Hall.

Keddy, P.A. (1991). Thoughts on a festschrift: what are we doing with our scientific lives? *Journal of Vegetation Science* 2: 419–424.

Keddy, P.A. (1992). Thoughts on a review of a critique for ecology. *Bulletin of the Ecological Society of America* 73: 231–233.

Keddy, P.A. (1993). On the distinction between ad hominid and ad hominem and its relevance to ecological research. A reply to Scheiner. *Bulletin of the Ecological Society of America* 74: 383–385.

Keegan, J. (1989). *The Second World War*. NY: Viking Penguin.

Lawton, J. (1992). Predictable plots. *Nature* 354: 444.

Lawton, J.H. & Hassell, M.P. (1981). Asymmetrical competition in insects. *Nature* 289: 793–795.

Leakey, R. & Lewin, R. (1992). *Origins Reconsidered. In Search of What Makes Us Human*. New York: Doubleday.

Lewontin, R.C. (1974). *The Genetic Basis of Evolutionary Change*. NY: Columbia University Press.

Mayr, E. (1982). *The Growth of Biological Thought. Diversity, Evolution and Inheritance*. Cambridge, MA: Belknap Press of Harvard University Press.

Montgomery, B. (1958). *The Memoirs of Field-Marshall the Viscount Montgomery of Alamein*. St James's Place, London: K.G. Collins.

O'Brien, C.C. (1994). *On the Eve of the Millennium*. Ontario: Anansi, Concord.

Paglia, C. (1992). *Sex, Art and American Culture*. NY: Random House.

Peters, R.H. (1980a). Useful concepts for predictive ecology. *Synthese* 43: 257–269.

Peters, R.H. (1980b). From natural history to ecology. *Perspectives in Biology and Medicine* 23: 191–203.

Peters, R.H. (1991). *A Critique for Ecology*. Cambridge: Cambridge University Press.

Pianka, E.R. (1992). The State of the Art in Community Ecology. Proceedings of the First World Congress of Herpetology. In *Herpetology: Current Research on the Biology of Amphibians and Reptiles*, ed. K. Adler, pp. 141–162. Oxford, OH: Society for the Study of Amphibians and Reptiles.

Pielou, E.C. (1975). *Ecological Diversity*. NY: John Wiley.

Pielou, E.C. & Routledge, R. (1976). Salt marsh vegetation: latitudinal gradients in the zonation patterns. *Oecologia* 24: 311–321.

Pirsig, R.M. (1974). *Zen and the Art of Motorcycle Maintenance*. NY: Morrow.

Saul, J.R. (1992). *Voltaire's Bastards*. Toronto, Ontario: Penguin Books.

Saul, J.R. (1995). *The Unconscious Civilization*. Ontario: Anansi, Concord.

Scheiner, S. (1993). Additional thoughts on A Critique for Ecology. *Bulletin of the Ecological Society of America* 74: 179–180.

Scheiner, S. (1994). Why ecologists should care about philosophy: a reply to Keddy's reply. *Bulletin of the Ecological Society of America* 75: 50–52.

Shalamov, V. (1982). *Kolyma Tales*. NY: W.W. Norton & Co. (trans. from Russian by J. Glad).

Shipley, B. & Keddy, P.A. (1987). The individualistic and community-unit concepts as falsifiable hypotheses. *Vegetatio* 69: 47–55.

Stone, I.F. (1989). *The Trial of Socrates*. NY: Anchor Books.

Tansley, A.G. & Adamson, R.S. (1925). Studies of the vegetation of the English chalk. III. The chalk grasslands of the Hampshire–Sussex border. *Journal of Ecology* XIII: 177–223.

van der Valk, A.G. (1981). Succession in wetlands: a Gleasonian approach. *Ecology* 62: 688–696.

Walker, D. (1970). Direction and rate in some British post-glacial hydroseres. In ed. D. Walker, & R.G. West, *Studies in the Vegetational History of the British Isles*. pp. 117–139. Cambridge: Cambridge University Press.

Weiher, E. & Keddy, P.A. (1995). Assembly rules, null models and trait dispersion: new questions from old patterns. *Oikos* 74: 159–164.

Weiss, P. (1965). *The Persecution and Assassination of Marat as Performed by the Inmates of the Asylum of Charenton under the Direction of the Marquis de Sade*. (translated by G. Skelton; verse adaptation by A. Mitchell) London: Calder & Boyards.

Part I:

The search for meaningful patterns in species assemblages

1

The genesis and development of guild assembly rules

Barry J. Fox

Introduction

Diamond's assembly rule

The idea that there were rules to govern how communities might be assembled was first explored by Jared Diamond (1975) in his treatise on 'assembly of species communities'. Although there were inklings of what was to come in earlier papers (Diamond 1973, 1974) the 1975 paper built on his mountain of observational data on the distribution of bird species on the many islands surrounding New Guinea to produce 'incidence functions'. These incidence functions were then used to deduce or infer the 'role' or 'strategy' of that species, such as the supertramp strategy already described by Diamond (1974). Diamond's ideas were further developed (covering over 100 pages) culminating in his assembly rules predicting which species were able to co-exist on islands in the New Guinea archipelago, in terms of allowed and forbidden combinations. An abbreviated version of his reasoning is shown in Fig. 1.1, which matches the resource utilization curves of four species of birds (dashed lines) to the availability of resources (resource production curves – solid lines) on islands with different levels of resources. By subtracting individual resource utilization curves from resource production curves, it is possible to obtain estimates of the distribution of the remaining resources allowing one to see which additional species could survive and which species requirements would exceed the resource levels available. In this way, Diamond was able to predict allowed and forbidden combinations of species. These examples have been used here to illustrate which combinations of the four species from one guild might be assembled on islands of different sizes (one, two or four units of resources, see Fig. 1.1). Diamond concluded that one would expect to find only species 3 on small (one unit of resource) islands; species 2 and 4 on medium (two units of resource) islands;

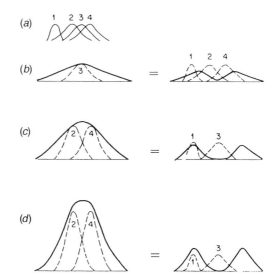

Fig. 1.1. Resource utilization functions are shown for four species (1, 2, 3 and 4) from the same guild (*a*), together with the resource production curves (as a thick line) for the resources used by this guild, for a set of islands of increasing size which produce one unit (*b*), two units (*c*), or four units (*d*) of resources. The right-hand side of each part of the figure shows the curve for the resources remaining when the utilization curves for species shown on the left-hand side are subtracted from the resource production curve. (Figure adapted from Diamond, 1975: 425.)

and species 1, 2 and 4 on large (four units of resource) islands (for the complete explanation see Diamond 1975: 425). This seminal paper appeared in the volume edited by Martin Cody and Jared Diamond (1975) that was dedicated to Robert MacArthur to commemorate the bountiful legacy he left in so many aspects of ecology, particularly in legitimizing the fledgling discipline of 'community ecology'. The theory of island biogeography that MacArthur developed with E.O. Wilson (1967) was clearly the most important contribution to Diamond's ideas, but the whole concept of niche theory and the pre-eminent place afforded to interspecific competition were also central. The whole volume (Cody & Diamond, 1975) had an electric effect on the development of the discipline and galvanized the careers of many ecologists, including myself, and was obligatory reading because of the amount of outstanding and interesting research it contained. However, it was the assembly rule paper that was the focus of controversy over the next two decades.

M'Closkey's demonstration of the mechanism

Jared Diamond's (1975) approach was empirical, the rules were deduced from observational data and the analysis of the distributional patterns formed. The most important part of these rules was the existence of 'forbidden' and 'allowed' states as they were described by Diamond. The next step in the development came not with the the continuing study of birds, but with a study of Sonoran desert seed-eating rodents. By accumulating extensive information on their habitat niche and food niche, Bob M'Closkey (1978) most elegantly demonstrated how his 'observed' assemblages were those with minimum measured niche separation, which also maximized resource utilization. All the other assemblages with greater niche separation he called 'imaginary' assemblages, as he did not observe them in any of his field sites. This can be seen in Fig. 1.2 which illustrates the mean niche separation of assemblages of

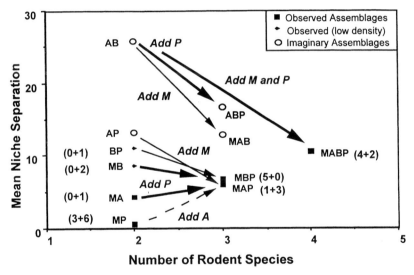

Number of Rodent Species

Fig. 1.2. Mean niche separation as a function of species richness for assemblages of desert rodents. Numbers in brackets for each point (shown as A + B) represent the number of sites from which this assemblage was observed from (A) Saguaro National Monument, Arizona, and (+ B) Organ Pipe Cactus National Monument. Letters with each point represent the combination of species in the assemblage. Imaginary sites that were not observed are shown as open circles, observed sites shown with filled squares, and observed sites with low density are shown with filled diamonds (these sites had one species at no more than 1 ha^{-1}, whereas at all other sites densities were up to 10 or 11 ha^{-1}). The species are *Dipodomys merriami* (M), *Perognathus penicillatus* (P) {now *Chaetodipus penicillatus*}, *P. amplus* (A) and *P. baileyi* (B) {now *Chaetodipus baileyi*}. (The data were adapted from M'Closkey, 1978.)

Sonoran desert seed-eating rodents comprising two, three or four species in Saguaro National Monument, Arizona and includes data from a second area in Organ Pipe Cactus National Monument. Two main points arise from M'Closkey's work: observed assemblages all have low mean niche separation, and the low diversity assemblages are precursors of the higher diversity assemblages, in other words the sites are 'nested' (see Patterson, 1990; Patterson & Brown, 1991). In all cases the greater niche separation of these 'imaginary' assemblages made them subject to invasion, and the addition of another species converted them into a new higher diversity 'observed' state.

For example, in a total of 13 sites in Saguaro National Monument there was only one observed combination of two species (*Dipodomys merriami* (M) and *Perognathus penicillatus* (P) – the latter is now *Chaetodipus penicillatus*.). This pair of species had the smallest niche separation of any pair from this species pool (see Fig. 1.2 first number = number of sites in Saguaro National Monument) and was observed at three sites. No other two-species sites were observed in Saguaro National Monument, although there were some in Organ Pipe Cactus National Monument (see below). The three-species and four-species sites were obtained from the addition of appropriate species, which resulted in a marked drop in the mean niche separation. A niche separation of over 25 units identified the only 'imaginary' pairs of species (A–B) (*P. amplus* (A) and *P. baileyi* (B) {the latter is now *Chaetodipus baileyi*}) that could not be converted to one of the two observed triplets (M–A–P and M–B–P) by the addition of a single species. Instead, the addition of either of the remaining two species to the A–B pair produced one of the two imaginary triplets (A–B–P or M–A–B), both with niche separations over 10 units, whereas the addition of both produced the four species assemblage (M–A–B–P) which was observed (see arrows Fig. 1.2).

There were 15 sites studied in Organ Pipe Cactus National Monument (Fig. 1.2, second number near each point). Two sites with four-species assemblages (M–A–B–P), three sites with triplets (M–A–P) and seven sites with pairs (six with M–P and one with M–A). Two other sites had the pair M–B and one the pair B–P, but in both cases the density of one of the species was 1 ha^{-1} or less, while other species' densities were up to 10 or 11 ha^{-1}.

Close analysis of these data clearly reinforces the concepts of both assembly and invasion, as these communities are assembled by the invasion of species to fill the large niche separations found in imaginary assemblages. M'Closkey's contribution, which was confirmed by his later paper (M'Closkey, 1985), was an important advance as it provided a mechanism to explain how the rule proposed by Diamond might work.

Connor and Simberloff's null models

Controversy arose with the publication of a critique by Ed Connor and Dan Simberloff (1979) that challenged the central importance of interspecific competition and asserted that the patterns observed could equally well be attributed to chance events. This controversy continued for over a decade, and although at different times the focus has been on methodological differences and statistical arguments, ultimately it revolves around whether the patterns observed are the result of deterministic or stochastic processes. These arguments will not be canvassed in detail here as one of the down sides of the controversy has been the amounts of time, energy, resources and paper that have been consumed. Instead, the positive aspects of the controversy will be concentrated on, those that have led to increased knowledge and understanding. The first of these that must be recognized was the introduction of null models (Connor & Simberloff, 1979). This was an important step, and led to the general acceptance of the need to demonstrate conclusively, using appropriate statistical tests, that observed patterns were significantly different from what might be expected by chance. Randomization tests and Monte-Carlo simulations have since became increasingly important and necessary components of many ecological studies spreading to much wider fields than assembly rules.

This approach was used to investigate the relationship between niche parameters and species richness (Fox, 1981). For the small-mammal community in patches of heathland habitat from one site on the eastern coast of Australia, Monte-Carlo computer simulations were employed to demonstrate that mean niche overlap decreased with increasing species richness, significantly more than expected by chance. Mean niche separation was also demonstrated to increase at a greater rate than expected by chance in habitat patches with increasing species richness. However, tests with appropriate null hypotheses excluding interaction between species clearly demonstrated that similar decreases in mean niche breadth with increasing species richness as had been observed in the field. These results were important as such decreases in niche breadth had been assumed to result from interspecific competition. Outcomes of the same hierarchical set of null hypotheses and field observations of niche separations demonstrate that increasing levels of interspecific interactions lead to increased niche separations (Table 1.1).

Colwell and Winkler's null models for null models

While most would agree that some of the methodological arguments were unedifying (see the extended exchanges between Jared Diamond and Dan

Table 1.1. *Niche separation measured as the slope (± one standard error) of mean spatial niche separation as a function of the number of species present in each habitat patch (n = 13, each with from 2 to 7 species)*

Null model	Description and interpretation of model	Niche separation
Equal model	All species equally abundant (excludes relative abundances, habitat selection and species interactions)	0.018 ± 0.028 (a)
Total model	Relative species abundances as observed; random distribution between habitats (exclude habitat selection and species interactions)	0.062 ± 0.027 (a, b)
Patch model	Relative abundances and distribution between habitats as observed; random within habitat patches (exclude species interactions)	0.102 ± 0.020 (b, c)
Field values	Observed captures of all individuals at all trap stations in all habitat patches (include all species interactions)	0.160 ± 0.011 (c)

Data adapted from Fox (1981).
Values with the same letter (in parentheses) do not differ significantly at $P = 0.05$.

Simberloff in the volume by Strong *et al.*, 1984), there were many useful advances stemming from these and others that have led to much more focused testing and in some cases also more focused thinking (Strong *et al.* 1984; Diamond & Case, 1986). One paper had an important bearing on the design of appropriate null models. Robert Colwell and David Winkler (1984) used computer simulation programs to establish artificial hierarchical arrangements of evolving species that specifically include or exclude interspecific competitive interactions. Species subsets from these were drawn to represent communities on imaginary islands in imaginary archipelagos, each with explicitly designed treatments. These communities were then tested against null hypotheses which identified three effects that can confound studies of assembly rules: (a) the Narcissus effect (sampling from the post-competition pool underestimates the role of competition, since its effect is already reflected in the pool); (b) the Icarus effect (correlations between vagility and morphology can obscure the effects of competition in morphological comparisons of mainland and island biotas); (c) the J.P. Morgan effect (the weaker the taxonomic constraints on sampling, the harder it becomes to detect competition) (Colwell & Winkler, 1984). The authors also emphasized the need to consider both type I and type II errors when selecting the appropriate null hypotheses. Grant and Abbott (1980) had already brought attention to the dangers of these problems but it was this study by Colwell and Winkler (1984) that conclusively demonstrated the impact these effects can have.

There was also a marked increase in the use of experimental ecology that seemed to have also been influenced by this continuing controversy, which made clear to all researchers that it was necessary to have strong experimental and statistical support for the inferences they drew from their results. One piece of experimental work had a very marked impact. Michael Gilpin, Patricia Carpenter and Mark Pomerantz (1986) were able to demonstrate in laboratory experiments that competitive interactions between species of *Drosophila* played a most important role in determining which species were able to form viable communities. They used 28 species and found, from the 378 possible pairwise comparisons, that there were only 46 (12%) that were able to coexist. They then excluded the strongest and weakest competitors and used ten species from the intermediate competitors in a further set of 30 trials, each with these ten species introduced simultaneously, but with different initial frequencies. After 35 weeks they found that the ten-species systems had relaxed to smaller systems: three species in 7 trials, two species in 21 trials and one species in 2 trials, so that never more than three species were found to coexist. There are 45 possible pairwise combinations for ten species, but only three of these were found to coexist in the 21 trials, 18 of the trials ending with the the same pair of species. Of the 120 possible trios from ten species only three were observed. There was a strong concordance between the results from the ten-species trials and the outcomes of the pairwise comparisons. The outcomes from these laboratory experiments provided strong support for the view that assembly rules similar to those proposed for birds by Diamond (1975) were also operating to form these *Drosophila* communities.

An assembly rule for guilds

The genesis of the guild assembly rule

The author's doctoral thesis (Fox, 1980) examined small-mammal communities in Myall Lakes National Park, NSW, Australia. It was influenced by the work of Diamond (1975) and M'Closkey (1978) and it considered the assembly of the small mammals that coexisted in the patches of habitat examined in the analysis of niche parameters and species richness (Fox, 1981). M'Closkey's paper was influential because it provided an elegant demonstration of the mechanism that structured his desert rodents, determining how they were assembled with minimum niche separation. However, the amount of information that had to be obtained for each species made this a daunting task and led to the proposal of a much simpler rule in which species from the same genus were considered, from taxonomically related groups, or from guilds

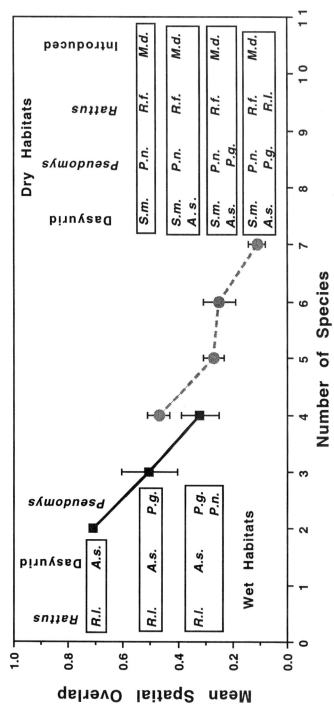

Fig. 1.3. Mean spatial niche overlap as a function of species richness for seven habitat patches (three wet habitats and four dry habitats). The structure and order of species packing is shown for species from the groups: *Rattus*, Dasyurid and *Pseudomys*, with the introduced *Mus* shown in all four dry habitats. Species abbreviations are R.f. = *Rattus fuscipes*, R.l. = *R. lutreolus*, A.s. = *Antechinus stuartii*, S.m. = *Sminthopsis murina*, P.g. = *Pseudomys gracilicaudatus*, P.n. = *P. novaehollandiae*, M.d. = *Mus domesticus*. (Figure adapted from Fox, 1989.)

and used the term 'functional group' to encompass all of these classifications. Although these data had been observed during my doctoral studies (1974–1980), the generality of these findings was not confirmed. The author was particularly influenced by M'Closkey's notions that species invade 'imaginary' assemblages because of the large niche separation, and his suggestion that assemblages with unused resources were vulnerable. These views were instrumental in developing the rule.

Figure 1.3 (adapted from Fox, 1980) examines how the species were assembled from taxonomic groups that had entered Australia at very different times: dasyurid marsupials (more than 38 mybp), Conylurine rodents represented by the genus *Pseudomys* (more than 4.5 mybp); and native rats, *Rattus* (1 mybp). When the rule was first presented (Fox, 1985a), the different evolutionary histories of the three groups influenced the thinking on how species might be structured in communities considering their different diets which were, respectively; insectivore, omnivore/granivore (depending on species), and herbivore/omnivore (depending on species).

Statement of the guild assembly rule

The rule was conceived as a resource-based rule and made the simplifying assumption that: the most usual distribution of resources would have roughly similar amounts of resources available to each of the three different trophic groups considered: insectivore, omnivore, herbivore. However, the effect of unequal resource availability was considered (see below and Fox, 1987). Based on the assumption of equal resources, a rule was postulated that specified the functional group from which species should come, rather than identifying the individual species in an assemblage. For these Australian mammal assemblages, the rule stated: 'There is a much higher probability that each species entering a community will be drawn from a different genus (or other taxonomically related group of species with similar diets) until each group is represented, before the rule repeats' (Fox, 1987: 201). The only input required was an *a priori* knowledge of how the species in the pool are divided into functional or taxonomic groups (Fox, 1989). Assemblages for which the rule was followed were termed 'favored', those for which the rule was not followed were termed 'unfavored'. Stated mathematically: (a) 'favored' states are those for which differences between the number of species from each functional group are never more than one; or (b) 'unfavored' states are those with a difference of more than one between the number of species from each functional group.

The assembly of species reflecting this rule do so because any assemblage lacking a species from any one of these groups, but with more than one from another group ('unfavored'), would be subject to invasion by a species able to capitalize on the unused resource type, which would then convert the assemblage to a 'favored' state.

An example of the guild assembly rule

Further analyses of heathland (Fox, 1981, 1982) and forest succession (Fox & McKay 1981) strengthened this view, as did an analysis of patterns in small-mammal diversity in eastern Australian heathlands (Fox, 1983). A review of small mammal communities in Australian temperate heathlands and forests (Fox, 1985b) finally provided me with two separate sets of data to test the generality of this empirical rule (Fox, 1985a; Fox, 1987, 1989).

The detailed analysis of 80 forest sites (Fox, 1987) includes a much more complete explanation of how the simulations were carried out, together with speculation on a possible operational mechanism for the assembly rule (discussed below). Another important part of that paper was an illustration of how 'favored' and 'unfavored' states were determined, including a case with a skewed resource availability curve (Fox, 1987: Fig. 1.1). Bastow Wilson (1989) developed an idea of guild proportionality with a method for testing whether the proportions of species from different guilds were relatively constant across sites, which he applied to a New Zealand temperate rainforest.

The effect of skewed resource availability

To investigate the effect of skewed resources, and emphasize the way in which the guild assembly rule relates to resource availability (see also the outcome from Morris and Knight 1996, described below), an example has been built on from the Nevada data set analyzed by Fox and Brown (1993). In this example the resource types represented are: arthropods (insectivores), seeds (granivores) and plant material (herbivores), but the resources are not uniformly distributed among resource types. The distribution of available resources is substantially skewed toward seeds. An example has been used with the abundance of seeds three times larger than that for plant material available to herbivores or the arthropod resources available to insectivores (Fig. 1.4). This is a reflection of the diversity of seed eating rodents found in this Nevada desert and other southwestern deserts. These communities have been recognized to contain three different functional groups, each with a different method of foraging for seeds (Brown, 1987).

Four simplistic examples of assembled communities have been shown, two are 'favored' states, the first has five units of available resources with one insectivore species (I_1), one herbivore species (H_1) and three granivore species (G_1, G_2, G_3), similar to the distribution of resources (Fig. 1.4(*a*)). For ease of presentation these examples are shown as rectangular profiles, but the same outcomes would be expected with more realistic bell-shaped curves (as shown much more elegantly for Diamond's example in Fig. 1.1, where the curves shown are mathematical representations of the actual resource production curves and the outcomes of subtracting the utilization curves). In these examples the amount of resource use is proportional to the area under the curve for each species, and the resource availability is represented by the heavy line enclosing all of the species in each example. The distribution of available resources is represented by the length of the heavy lines below the *x*-axis with appropriate labelling. The values on the *y*-axis representing the total resource available. The second example shows double the available resources (ten units), sufficient for two complete cycles assembling two insectivore species, two herbivore species and six granivore species (Fig. 1.4(*b*)), each species' use of resources has the same area, twice as high but half as wide as in Fig. 1.4(*a*), with an implication of greater specialization.

Two examples of 'unfavored' states are also shown. The first 'unfavored' state also has ten units of resources sufficient for a total of ten species (Fig. 1.4(*c*)), distributed as two units of arthropods, two units of plant material and six units of seeds (the same as in Fig. 1.4(*b*)), but in this case there are two insectivore species, four herbivore species and four granivore species, shown by the different fill and labelling. As there would be sufficient seed resource for six granivores (as in Fig. 1.4(*b*)), the community will be vulnerable to invasion by granivorous species. In addition, there are four herbivore species present, but with sufficient resources only for two, which would lead to an intensification of interspecific competition between herbivores.

The second 'unfavored' state has seven and a half units of resource available, nominally sufficient for seven species, but distributed as one and a half units each of arthropods and plants, with four units of seeds (Fig. 1.4(*d*)). However, the species occupying the site are five granivores, two insectivores, but no herbivores (Fig. 1.4(*d*)). Hence, the plant resource would be underutilized, the community would be subject to invasion by herbivore species, while both insectivores and granivores exceed their available resource, again intensifying interspecific competition.

The existence of communities such as these described in Fig. 1.4, and the behavior demonstrated with increasing species richness, (Fig. 1.3), helped inspire a graphical model for mammalian assembly and evolution (Fox, 1987), that can be illustrated with the following examples. The original model was

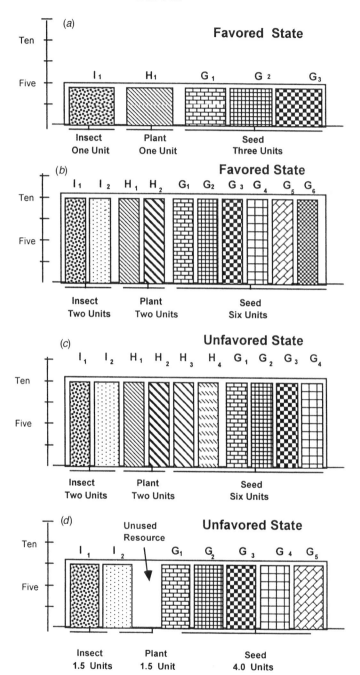

conceived from observations of Australian communities from heathland including dasyurid marsupials, insectivores that can select either wet or dry habitats for this example, with respectively brown antechinus (*Antechinus stuartii*) and common dunnart (*Sminthopsis murina*) from Myall Lakes, NSW (Fox, 1982, 1983). This can be thought of as a sorting process on an ecological timescale, and for the total community the number of species that coexist is dependent on the adequacy and abundance of resources.

However, in lowland heath at Kentbrook Heath in Victoria, two species of antechinus are found in wet heath habitats, the swamp antechinus (*A. minimus*), that would be considered to have evolved as a wet habitat specialist, in addition to the brown antechinus (Braithwaite *et al.*, 1978; Fox, 1983). At an evolutionary timescale, this should be considered a parallel process to that at the ecological timescale (Fox, 1987) and one that can occur only when sufficient resources are available.

An adaptation of this model is shown in Fig. 1.5, referring specifically to the case of granivores from the desert rodent communities of the Nevada Test Site (see Fox & Brown, 1993). Here, granivorous species (heteromyids and cricetids) should have some advantages in terms of digestive physiology, gut morphology and tooth morphology that allow them to efficiently use seed in their diet, in the same way that they will be constrained from being able to obtain sufficient benefit from eating plant material, without the ability to obtain energy from digesting cellulose. The different foraging behaviors exhibited by these guilds results in effective habitat partitioning, as they select shrubby and open macrohabitats differentially, although these habitats are intimately mixed in the southwestern deserts.

An example of within guild evolution in this case is illustrated with the Great Basin kangaroo rat (*Dipodomys microps*) that has evolved adaptations for leaf eating, with incisors adapted for peeling salt-laden epidermis cells from saltbush before eating mesophyll cells (Kenagy, 1973). While this is an extreme example, that shifts *Dipodomys microps* from the granivore guild to

Fig. 1.4. An illustration of the possible favored and unfavored states and the effects of skewed resources. (*a*) Five units of available resource (one arthropod, one plant, three seed), with five species: one insectivore (I_1), one herbivore (H_1) and three granivores (G_1, G_2, G_3); *favored*. (*b*) Ten units of available resource (two arthropod, two plant, six seed), with ten species: two insectivores (I_1, I_2), two herbivores (H_1, H_2) and six granivores (G_1, G_2, G_3, G_4, G_5, G_6); *favored*. (*c*) Ten units of available resource (two arthropod, two plant, six seed), with ten species: two insectivores (I_1, I_2), four herbivores (H_1, H_2, H_3, H_4) and four granivores (G_1, G_2, G_3, G_4); *unfavored*. (*d*) Seven units of available resource (one and a half arthropod, one and a half plant, four seed), with seven species: two insectivores (I_1, I_2), no herbivore and five granivores (G_1, G_2, G_3, G_4, G_5); *unfavored*.

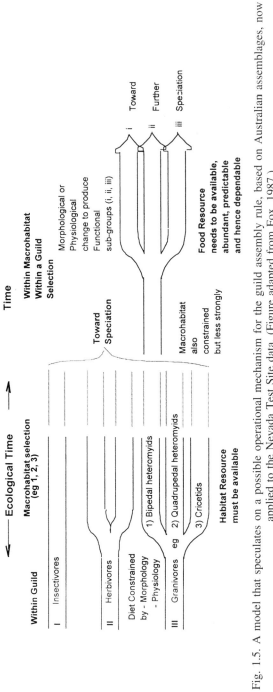

Fig. 1.5. A model that speculates on a possible operational mechanism for the guild assembly rule, based on Australian assemblages, now applied to the Nevada Test Site data. (Figure adapted from Fox, 1987.)

the folivore guild, it does demonstrate how on an evolutionary timescale it is possible to produce morphological or physiological change to allow further partitioning of the food resource, which might even in some cases produce further speciation (Fig. 1.5). Clearly, for such specialization to occur the food resource needs to be abundant, predictable and hence dependable. In this way the amount of resource available will directly affect the number of groups, subgroups and species forming the pool of species available and the number and type of species able to coexist in any one place, in the way described for 'favored' as opposed to 'unfavored' states.

General applicability of the guild assembly rule

In order to consider the generality of the guild assembly rule, the brief findings from a number of tests of the rule will be summarized, noting the biogeographical area and the taxonomic groups involved. All of the studies in this list have been published and the wide range of biogeographical regions in which this rule is applicable greatly strengthens my confidence in the operation of the rule.

Australia: rodents and marsupials in forest and heath

Two studies have been carried out for southeastern Australia that included marsupials (insectivores from the family Dasyuridae); native mice (granivores and omnivores from the old endemics, the conylurine rodents); and the more recently arrived native rats (herbivores and omnivores from the genus *Rattus*). The analyses were based on these three functional groups or guilds. One study demonstrated highly significant departure from random assembly ($P < 0.001$) for 80 small mammal assemblages sampled from eucalypt forest sites extending over a latitudinal range from 27° S to 43° S along the east coast of Australia (Fox, 1987). The other study, over a similar latitudinal range, also showed that the mammalian assemblages assessed from 52 coastal heathland sites had significantly more favored states and significantly fewer unfavored states than expected from Monte-Carlo simulations ($P < 0.01$), thus rejecting the null hypothesis that they had been assembled randomly (Fox, 1989).

North America: shrews in NE forests, rodents in SW deserts and boreal forests

On the North American continent three sets of data have been analyzed. One study demonstrated significant departure from random assembly ($P < 0.001$)

for the soricid communities from 43 temperate forest sites in the New England area of northeastern USA (Fox & Kirkland, 1992). In this study the shrews were divided into three guilds based on body size.

The second study was from the southwestern deserts of USA, and comprised three parts (Fox & Brown, 1993). First, data were collated from over 202 sites from the Chihuahuan, Great Basin, Mojave and Sonoran deserts with a pool of 28 species of granivorous desert rodents in three different foraging guilds, bipedal and quadrupedal heteromyids and cricetids (Brown & Kurzius, 1987). There were significantly more favored states observed than the number expected from 10000 Monte-Carlo random simulations, and the null hypothesis of random assembly was rejected ($P < 0.01$). Another data set from the nearby Nevada Test site (Jorgensen & Hayward, 1965) was also analyzed, this time at two scales: (a) three granivore guilds and (b) five rodent guilds. Ten species from the three granivore guilds formed the pool from which the assemblages that occupied 115 transects were drawn. Significantly more favored states were observed than expected from random computer simulations, thus rejecting ($P < 0.001$) the null hypothesis (Fox & Brown, 1993). When herbivorous and insectivorous species were added to form five guilds, the results were similar, 74 favored states were observed while random simulations produced 41.6 ± 4.2, again the null hypothesis was rejected ($P < 0.001$).

The third study was conducted over an area of 100 km^2 in boreal forest in northern Ontario, Canada (Morris & Knight, 1996). Three herbivorous voles, two diurnal chipmunks, two jumping mice and one cricetid made up the species pool of four feeding guilds from which communities of rodents were drawn. Favored states were observed for 67% of the assemblies when the expected value was 28.4%, so that random assembly was rejected ($P = 0.0008$, cumulative binomial test).

South America: rodents at the Valdivian rainforest–Patagonian steppe interface

Data were collected from 31 communities in southern Chile (46° S) along the boundary between Valdivian temperate rainforest and Patagonian steppe, and 12 species of native sigmodontine species of rodents formed the species pool for the analyzes (Kelt *et al.*, 1995). Functional groups from four trophic categories were recognized with two omnivores, six herbivores, two fungivore–insectivores, and two granivores. Their analyzes enabled them to reject the null hypothesis for random assembly ($P = 0.001$) and also demonstrate that there was a significant level of interspecific competition with a statistical power of 93–97%. Further details of these analyzes will be provided in the section on assessment of interspecific competition, and the same techniques are applied by Kelt and Brown (this volume).

Madagascar: lemurs in evergreen rainforests and dry forests

Jurg Ganzhorn (1997) has recently demonstrated that communities of arboreal lemurs from evergreen rainforest habitats in Madagascar obeyed the guild assembly rule when compared to 10 000 Monte-Carlo simulations for neutral models ($P = 0.001$). From a pool of 20 species (6 folivores, 8 frugivores and 6 omnivores) communities were assembled with from 3 to 13 species, indicating that the rule had operated through up to four, and in some cases five cycles of species additions. The high degree of concordance with the rule was quite impressive, as examples from other biogeographical areas and with other taxonomic groups have generally only proceeded through two or in some cases three cycles at most. The fit for dry forests was less good, although still significant ($P = 0.029$), and this was attributed to recent natural and anthropogenic changes over the 2000 years of human occupation that have been very apparent in dry forests, but not in the undisturbed evergreen rainforests. Ganzhorn (1997: 534) also reported that '[this guild assembly rule] is also consistent with the composition of European small mammal communities' and cites Schröpfer (1990) as his authority.

Summary of the applicability of the guild assembly rule

This guild assembly rule has been statistically tested against Monte-Carlo computer simulations for seven data sets of small mammal assemblages from different biogeographical regions on four continents: Australia, North America, South America and Madagascar. In all cases the observed assemblages have been shown to differ significantly from the random simulations, and to be consistent with the composition of small mammal communities on a fifth continent, Europe. These tests have examined a wide range of mammalian taxa: dasyurid marsupials, lemurs, shrews, voles, heteromyid, cricetid and sigmodontine rodents. The guilds analyzed have been based on: taxonomic relatedeness, body size, diet, and foraging behavior. The analyzes have been between diet guilds (insectivore, granivore and herbivore); within a diet guild (insectivores, granivores); and finally for a combination of between and within guild analysis (the 5-guild Nevada data set in Fox & Brown, 1993). Spatial scales have ranged from 3500 km^2 Nevada Test Site, 640 000 km^2 in four southwestern deserts to the largest covering 16° of latitude along the eastern

Australian coast. The tests have also included a wide range of biomes from deserts to boreal forests in North America, moist evergreen forests in north-eastern USA, eucalypt forest and heathlands in Australia, Valdivean rain-forest and Patagonian steppe in South America, dry forests and evergreen rainforests in Madagascar. This great breadth of biogeographical regions, taxonomic groups, vegetation types and spatial scales most surely attests to the generality of this guild assembly rule.

In rejecting all these null hypotheses of random assembly one must accept an alternative hypothesis, and the guild assembly rule, based on resource availabilty, resource partitioning, and interspecific competition offers the best alternative hypothesis proposed. This view is substantially strengthened by the recent derivation of the rule from Tilman's consumer-resource model (Morris & Knight, 1996) described below.

Methodological considerations

The two decades since Diamond (1975) published his species assembly rule paper have been marked by critiques and rebuttals of the methods used for analyses of assembly rules, as I mentioned in the introduction. Many of these have focused on the appropriateness of null hypotheses, selection of appro-priate species pools, marginal totals constraints, initial assumptions and the simulation techniques used in the analyses. The guild assembly rule has been the target of similar critiques, and will no doubt continue to be subjected to them. While these arguments are often targeted at statistical methods, they seem to basically arise from differences of opinion on what should represent the initial conditions or assumptions for the null hypotheses used in these analyses, and this has been most eloquently demonstrated by Colwell and Winkler (1984).

The major debate on the guild assembly rule comes down to whether one should maintain the distribution of species richness across sites, which reflects the availability of resources at those sites; or whether one should retain the distribution of each species' frequency of occurrence across sites, which reflects the interactions determining which sites each species can occupy and which species coexist at each site. The most sensible course would seem to be to agree to differ on these points and get on with research activities that advance our knowledge more. This is a better alternative than spending time discussing what would seem to be relatively trivial points on which are the most appropriate statistical techniques to use, as they are unlikely to be resolved by these debates when the real difference is a philosophical one about what assumptions are made in constructing null hypotheses.

Wilson (1995a), Fox and Brown (1995) and Wilson (1995b)

Bastow Wilson (1995a) challenged the methods used by Fox and Brown (1993) to analyze the Nevada data set claiming they would give significant results with randomized data. A reply to Wilson's (1995a) critique was written (Fox & Brown, 1995) which reaffirmed the validity of the guild assembly rule and the methods used in the analyzes and will not be further canvassed here. However, Wilson (1995b) wrote a second critique claiming that the random data sets used by Fox and Brown (1995) were not random. As the editor of *Oikos* would not allow a reply to this second critique, a brief comment should be made here.

Bastow Wilson (1995b) is correct in indicating that the method used by Fox and Brown (1995) to generate the random data set was flawed. Random data sets had been generated by randomly allocating 115 sites to the 36 cells in the matrix (0, 0, 0) to (2, 2, 3) (see Fox & Brown, 1995). As Wilson points out this effectively considered all 36 cells equiprobable, which was incorrect and hence needed to be corrected.

To rectify this, 200 new random data sets were generated, in each case considering separately each of the 115 sites to form a matrix for the nine species encountered, with each species identified and in the same order. For each site, the number of species observed was retained but the presence or absence of each species was randomly allocated across all nine species, with no consideration of which guilds were represented, until the observed richness for the site was achieved. In this way the row constraints of the matrix were preserved but not the the column constraints, whereas Wilson maintained the column constraints but not the row constraints.

For each random matrix generated, the observed guild structure (3, 2, 4 species in each of the three guilds) was reimposed across the nine species to calculate the status of each simulated site as either favored or unfavored. Each random data set was then analyzed against a null hypothesis with the species pool of 3, 3, 5 species in each guild for the Nevada Test Site as used by Fox and Brown (1993, 1995), so as to include the two additional species caught across the whole Nevada test area, but not on any of the set of 115 transects used in the analysis. The matrix of possible outcomes has also been constrained to the maximum observed number of species from each guild (2, 2, 3). In these analyses rather than use the Monte-Carlo simulation previously used, an analytical approach developed by R.J. Luo (R.J. Luo & B.J. Fox, unpublished data), has been used and the exact probability from the Binomial test calculated. The result for three examples, all with the same matrix constraint (2, 2, 3), but different species pools are shown in Figs. 1.6(*a*)–(*c*).

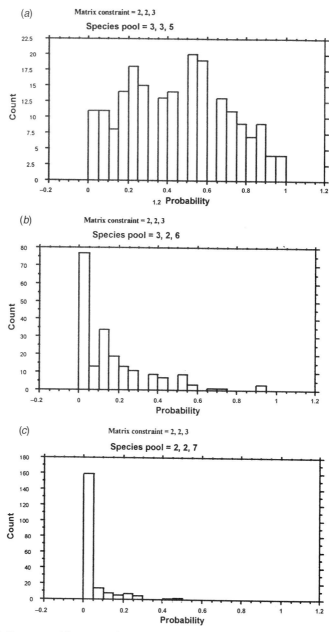

Fig. 1.6. Frequency histograms for the probability distribution that the null hypothesis would be accepted when 200 randomly generated data sets for 115 sites were analyzed. Each random data set was generated (see text) to correct the flaw in the data sets used by Fox and Brown (1995). (*a*) to (*c*) are all constrained to have maxima of 2, 2, 3 species in each guild and match the figures presented by Fox and Brown (1995)

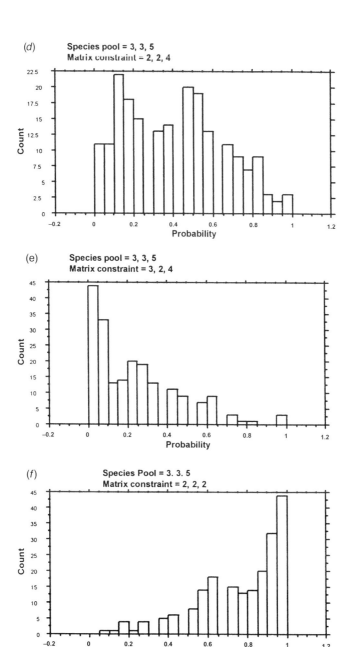

Fig. 1.6. (continued)
with species pools of *a* 3, 3, 5; *b* 3, 2, 6; *c* 2, 2, 7. (*d*) to (*f*) all have the same species pool (3, 3, 5) but show the effects of less restrictive matrix constraints, to have a maximum number of species in each guild of: (*d*) 2, 2, 4; and (*e*) 3, 2, 4; but also the effects of a more restrictive matrix constraint (*f*) 2, 2, 2.

Figure 1.6(*a*) illustrates a relatively even probability distribution as one might expect with the observed pool (3, 3, 5) for such random data. Probability distributions for two more extremely skewed species pools for 11 species are shown in Fig. 1.6(*b*) (3, 2, 6) and Fig. 1.6(*c*) (2, 2, 7), to match those illustrated by Fox and Brown (1995: Fig. 1). The conclusion reached is qualitatively little different to that reported by Fox and Brown (1995). So, while Wilson (1995b) correctly identified a flaw in the construction of random data sets used by Fox and Brown (1995), after correct generation, when they were reanalyzed, they still clearly demonstrate that random data do produce relatively even probability distributions, using the same null hypothesis as Fox and Brown (1995), and that these may become biased with extreme species pools.

To further illustrate the conservative nature of the null hypotheses used by Fox and Brown (1993, 1995) another two examples have been considered (Fig. 1.6(*d*), (*e*)), with the observed species pool (3, 3, 5), but the matrix size constraint has been relaxed to accommodate maxima in the simulation: of 2, 2, 4, allowing up to eight species to occupy 45 cells (Fig. 1.6(*d*)); and of 3, 2, 4, allowing up to 9 species to occupy 60 cells (Fig. 1.6(*e*)), although in fact none of the 115 sites had maxima of more than 2, 2, 3, allowing up to seven species to be present in 36 cells. Relaxing these constraints means that more cells (9 and 24, respectively) are added to the analysis, even though they were never observed. However, they will have to be given a random expectation from the null hypothesis and as more of these cells will represent unfavored states there will be an increase in the probability of rejecting the null hypothesis. This is the reason why matrix size constraints have been used in the past to exclude these unrealistic cells, to make the test more conservative. As predicted, in each case the results are shifted to the left, thus confirming that the use of these constraints on the maximum size of the matrix does lead to more conservative tests. To emphasize this, an illustration of the effect of constraining the matrix to maxima of 2, 2, 2, has also been included allowing up to six species in 27 cells (Fig. 1.6(*f*)). There is a marked shift to the right in the probability distribution, making it virtually impossible to reject the null hypothesis for these random data. It should be pointed out that 113 of the sites could meet this constraint (see Fox & Brown, 1993).

One should note here that, for the observed matrix of 115 sites and nine species, using column constraints to preserve the species frequencies means that a great deal more of the actual structure will be smuggled into the null hypothesis, and this is the basis of the Narcissus effect described by Colwell and Winkler (1984). On the other hand, preserving the distribution of species richness values for each site (row constraints) smuggles much less information into the null hypothesis to be used, and is directly related to the resources available at each site.

Stone, Dayan and Simberloff (1996)

Lewi Stone, Tamar Dayan and Dan Simberloff (1996:1013) reanalyzed the data sets examined by Fox and Brown (1993), and they concluded:

Our analysis failed to find evidence that interspecific competition or deterministic assembly rules shaped local community composition ... ... the only possible structure noted in our study was that attributed to three or four widespread species ... ... a study of spatial distribution alone provides little evidence for competition between these species ... ... Because ecological field and experimental studies (for review see Brown, 1987; Kotler & Brown, 1988) strongly imply the role of interspecific competition among granivorous desert rodents, it seems conceivable that such a process might be better detected with more discriminating statistical techniques.

This is not the appropriate place for a detailed rebuttal of these comments but it should be pointed out that the methodology used by Stone *et al*. (1996 and this volume) incorporates the Narcissus effect, as the authors recognize in part. The fact that some species are widespread while others have more restricted distributions cannot be decoupled from the relative competitive abilities of these species. In turn, the species competitive abilities may or may not have influenced their distributions, so that use of this distributional information may still have the potential to introduce competitive effects into their null model.

It should also be pointed out that, while the last part of the above quote recognizes reality, interspecific competition plays an important role in these communities, the authors have ignored the existence of two important studies which address the point they make (Kelt *et al*, 1995; Morris & Knight, 1996). Both of these are dealt with in some detail in the next section and the reader is directed to the chapter in this volume which specifically applies Kelt's methodology to questions raised by Stone *et al*, (1996). Doug Kelt and Jim Brown (this volume) have explicitly dealt with the geographic ranges question by calculating separate species pools for each trapping site used in the Fox and Brown (1993) analyses. Detailed discussion of these analyses will be left to their chapter, but it should be pointed out that they very convincingly reject the null hypothesis of no interaction ($P < 0.0005$), accepting the alternative hypothesis of a maximum likelihood estimate for the mean strength of the observed negative association of 0.33, with a power of 94% (probability of being correct).

Further development of the guild assembly rule

Morris's derivation from the MacArthur–Tilman consumer-resource model

Doug Morris demonstrated that the guild assembly rule is a probabilistic consequence of adding guild structure to models of consumer-resource competition (Morris & Knight, 1996). The graphical models of consumer-resource dynamics were developed initially by Robert MacArthur (1972) and then much more fully by David Tilman (1982). The author's guild assembly rule had been developed in an empirical manner, as shown above, from observations of the structure of small mammal communities occupying different patches of habitat (Fox, 1981; 1987). A potential theoretical mechanism had been set out, with speculation on how the rule operated (Fox, 1987). However, it was not able to provide any theoretical derivation (although the mechanism was linked to resource availability, with a strong implication that competition for these resources played a major role). It was also suggested that the mechanism was an extension of the niche compression hypothesis that MacArthur and Wilson (1967) applied to the optimal use of patchy habitats.

To emphasize the main points made by Morris and Knight (1996), an attempt has been made to devise a simplified, if not simplistic, illustration of Tilman's model that retains the most essential components. On a graph of the amounts of two resources (L and N), three species each from two guilds that specialize on each of these resources have been represented. (Fig. 1.7*a*). The ratio of resource N to resource L can be shown as a vector that represents the combined resource consumption rate for each species. The direction of these consumption vectors will more closely parallel the direction of the resource axis as the species become more specialized. Species from guild L will therefore use more of resource L. For multi-guild systems one would expect that evolutionary pressure will result in consumption vectors for species in a guild tending to be further from the 45° line of equal resource use and closer to the axis for the resource used by that guild. This situation has been shown for the two-guild system (Fig. 1.7(*a*) with species consumption vectors labeled with letters for guild L and labeled with numbers for guild N.

For simplicity the zero net growth isocline (ZNGI) for each species has been omitted, that would be displayed as a pair of rectangular axes representing the minimum amounts of each essential resource required for each species' survival (see Tilman, 1982; Morris & Knight, 1996). To concentrate on the slope of each consumption vector the supply points (the levels of resources available in the absence of consumption) from which each con-

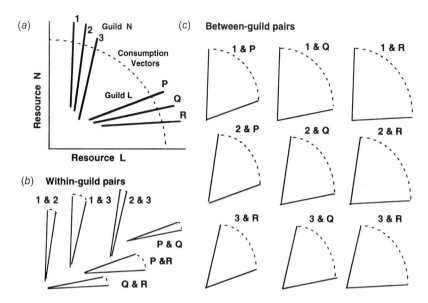

Fig. 1.7. A simplified version of how Morris and Knight (1996) applied Tilman's (1982) models of consumer resource competition. (*a*) Species 1, 2, and 3 from guild N and species P, Q, and R from guild L are represented by their consumption vectors (lines of equal resource ratios for the two resources N and L). The area within which supply points are available is bounded by a dashed arc. For simplicity, the zero net growth isocline (ZNGI) has been omitted for each species, that would be displayed as a pair of rectangular axes representing the minimum amounts of each essential resource required for each species survival (for complete details, see Tilman, 1982, Morris & Knight 1996). (*b*) Regions of stable coexistence for within-guild pairs of species are defined by the angle between the consumption vectors, out to the dotted arc. The area of each region shown can then be equated to the probability of finding appropriate supply points allowing the coexistence of the two species indicated. (*c*) Regions of stable coexistence for between-guild pairs of species are defined by the angle between the consumption vectors, out to the dashed arc. The area of each region shown can then be equated to the probability of finding appropriate supply points allowing the coexistence of the two species indicated. This leads to the conclusion that the probability of finding pairs of species from the same guild coexisting is much smaller than for pairs of species from different guilds.

sumption vector would emanate. However, a dashed arc has been used to enclose a possible range of available supply points. Regions of stable coexistence for pairs of species would be defined by the angle between their consumption vectors, taking the area as extending from the dotted arc to the point of intersection of the vectors. The point of intersection of the vectors would be determined by the relative positions and intersection of the appropriate ZNGIs. For simplicity, those points have been omitted in Fig. 1.7.(*a*), but their actual positions do not alter the conclusions that follow here (see Morris & Knight, 1996). The area of each region of stable coexistence can then be equated to the probability of finding appropriate supply points allowing the coexistence of the two species indicated.

Representations of the relative areas of the regions that would allow coexistence for each of the 15 possible species pairs are shown for the six within-guild pairs (Fig. 1.7(b)) and for the nine between-guild pairs (Fig. 1.7(c)). In all cases the areas of the regions of coexistence are markedly smaller for intra-guild pairs than for inter-guild pairs. This leads to the conclusion that the probability of finding pairs of species from the same guild coexisting is much smaller than for pairs of species from different guilds. This is the main message that Morris and Knight (1996) convey, although they make a number of other important points that will not be followed up here.

Interpretation of the Nevada data in terms of a consumer-resource model

A reanalysis of the data set from the Nevada test site that was used by Fox and Brown (1993) provides an interesting illustration of, and further support for, Morris and Knight (1996) linking the guild assembly rule to the resource competition model. A regression technique, was developed separately by Schoener (1974) and Pimm (Crowell & Pimm, 1976). The technique used the density of one species as the dependent variable and focused on the use of multiple regression to remove all of the variance associated with independent habitat variables before allowing the densities of other potentially competing species to enter the multiple regression equations as independent variables. The regression coefficients for species variables were then considered as the coefficients of competition between species. Rosenzweig *et al.* (1985) questioned the technique and demonstrated some inconsistencies in analyses, and the technique fell into disuse after that. This regression technique has recently been revived by Fox and Luo (1996). They confirmed the artefact identified by Rosenzweig *et al.* (1985), but demonstrated how to overcome the problem by making use of standardized data, thus providing a useful method to extract competition coefficients from census data and habitat measurements.

Rodent density data were available from 95 trapping plots in four different habitats: *Larrea-Franseria* (39 plots), *Coleogyne* (27), *Grayia-Lycium* (15) and *Salsola* (14). Ten species were found on enough plots to be included in these regression analyses. Two quadrupedal heteromyids (QH) *Perognathus formosus* and *Pg. longimembris*; two bipedal heteromyids (BH) *Dipodomys merriami* and *D. ordii*; three quadrupedal non-heteromyids (QN) *Peromyscus maniculatus, Pe. crinitis* and *Ammospermophilus leucurus*; two folivorous herbivores (F) *Neotoma lepida* and *Dipodomys microps*; and one insectivore *Onychomys torridus*. The interaction coefficients resulting from this multiple regression technique are shown in Table 1.2. There were six significant inter-action coefficients shown for within-guild species pairs (from a total of 12 possible), all significant values were negative with a mean value of -1.820 ± 0.0295 (1 s.e.). For between-guild pairs, in the 10 equations there were 18 significant interaction coefficients (from the 78 possible), all significant val-ues were positive (7 of these with $P < 0.005$) with a mean value of $+2.494$ ± 0.0215 (1s.e.). This is many more significant results than might have been expected by chance. Even more important is the distribution of the results. The strong negative outcome within-guild and the linked positive outcome for between-guild pairs clearly demonstrates the importance of interspecific competition. This pattern confirms the expectation from the operation of a guild assembly rule derived from Tilman's consumer-resource model (Morris & Knight, 1996), as shown in Fig. 1.7.

Kelt's inclusion of habitat component assessment of interspecific competition

Doug Kelt, Mark Taper and Peter Meserve (1995), produced an important development for guild assembly rules when they introduced specific alterna-tive hypotheses to the null hypothesis that enabled them to make an estimate of the power of their test. In addition to the total species pool that is usually used, they introduced a geographic species pool that took account of differ-ent geographic distributions and a habitat species pool that took account of different habitats that might be encountered by species in any one area. While some account had already been taken of how each species geographic distri-bution influenced the overall or total species pool (Fox, 1987, 1989; Fox & Brown, 1993), there had been no successful attempt to incorporate geographic effects or to introduce habitat effects at individual sites. The inclusion of a habitat component improved resolution for determining species pools because habitat selection by species will influence which species may be present at any site. In other words, trapping sites in the same area, but in different habi-tats may have different species pools because of habitat selection. In addi-tion to including geographic and habitat components these simulations were run a number of times, each with a different level of interspecific competi-tion (Θ) incorporated, ranging from $\Theta = -1.0$ to $\Theta = +1.0$. The values of Θ tested then represent a range of alternative hypotheses. They were able to reject the null hypothesis for the total species pool analysis ($P = 0.001$) and demonstrate that there was a significant level of interspecific competition

Table 1.2. *Interaction coefficients among species in local community using multiple regression technique originally developed by Crowell and Pimm (1976) and modified by Fox and Luo (1996)*

Dependent variable		Independent variables										Habitat variables	Overall equation	
Guild	Species	Pg. for	Pg. lon	D. mer	D. ord	Pe. man	Pe. cri	A. leu	N. lep	D. mic	O. tor		R^2	P value
QH	Pg. formosus		**−0.100**	ns	ns	ns	0.464***	ns	ns	ns	ns	Significant	0.495	<0.0001
	Pg. longimembris	**−0.145**		0.182+	ns	ns	ns	0.199*	ns	ns	0.292**	Significant	0.257	<0.0001
BH	D. merriami	ns	0.189+		ns	ns	ns	ns	ns	ns	ns	Significant	0.060	0.058
	D. ordii	ns	ns	ns		ns	ns	ns	ns	ns	ns	Significant	0.093	0.011
QN	Pe. maniculatus	ns	ns	ns	ns		−0.168+	ns	ns	ns	ns	Significant	0.137	0.0012
	Pe. crinitis	0.411***	ns	ns	ns	**−0.137**			0.273**	0.178*	ns	ns	0.343	<.0001
	A. leucurus	ns	0.228*	ns	ns	ns	ns			0.172+	ns	Significant	0.169	0.0007
Flvr	N. lepida	ns	ns	ns	ns	ns	0.352***	ns		**−0.271****	0.174*	Significant	0.348	<.0001
	D. microps	ns	ns	ns	ns	ns	0.188+	0.156+	**−0.271****		0.255***	Significant	0.477	<.0001
Inst	O. torridus	ns	0.346***	ns	ns	ns	ns	ns	0.171+	0.260**		ns	0.415	0.0001

Habitat heterogeneity was accounted for using habitat variables. Species across the row are independent variables for species shown as dependent variables down the left-hand column. Interaction coefficients are marked according to their significance ($+P < 0.1$, $*P < 0.05$, $**P < 0.01$, $***P < 0.0001$). Independent variables for species from the same guild were forced into equations when $0.2 > P > 0.1$ and the interaction coefficients are shown in bold typeface with the same significance symbols.

(with the best estimate $\Theta = 0.60$). The statistical power of this rejection was 93–97%. For the geographic species pool analysis, they were able to reject the null hypothesis ($P = 0.023$) and demonstrate that there was a significant level of interspecific competition ($\Theta = 0.45$) and that the statistical power of the rejection was 56–71%.

Doug Kelt's further development of the assembly rule confirms that inter-specific competition is a major factor in the operation of the guild assembly rule as described (Fox, 1987). Field removal experiments have also corroborated the empirical pattern analyses (Fox & Pople, 1984; Fox & Gullick, 1989; Higgs & Fox, 1993; Thompson & Fox, 1993). Doug Kelt's analysis also provides support for Doug Morris's demonstration that the guild assembly rule is a probabilistic consequence of adding guild structure to models of consumer-resource competition (Morris & Knight, 1996). Doug Kelt and Jim Brown have explicitly applied Kelt's methods (see chapter in this volume) by calculating separate species pools for each trapping site used in the Fox and Brown (1993) analyses. With this technique they use spatial distributions to provide convincing evidence for competition between the species in these communities, for which there is abundant experimental evidence (e.g., see Munger & Brown, 1981; Brown & Munger, 1985; Heske *et al.*, 1994; Valone & Brown, 1995; and references in the Kelt and Brown chapter, this volume).

What future for guild assembly rules?

One point that should be emphasized here, is that the guild assembly rule has both deterministic and probabilistic components, as was emphasized by Fox and Brown (1993). Replacing the terms 'favored' states and 'unfavored' states with terms like 'high probability' states and 'low probability' states would reinforce this point. The message in this is that all states are possible, but some have a higher probability of occurrence than others (see Fig. 1.7). This seems to better represent the reality of the processes at work, on both temporal and spatial scales. In the extreme case a site might be switched from 'high probability' to a 'low probability' state by the chance addition of a single individual, which may only persist for a short period before it dies or moves on. On the other hand, it might be joined by other individuals of the same species and also from another additional species, either way it could again switch the site to a 'high probability' site. Hence, the site passes through states of differing probability for different lengths of time that reflect the probability of finding those states. This is the stochastic component overlain on an otherwise deterministic process.

Another thing to be emphasized in the guild assembly rule is the importance of process. This empirically based rule grew out of an attempt to understand the mechanisms that operate in assembling communities. The guild assembly rule has provided a powerful tool to aid this understanding, particularly with relation to consumer–resource competition. One of the rule's strengths has been its simplicity, all that is required is knowledge of which species belong to which functional groups or guilds, the distribution of species richness across sites and the appropriate species pool for each guild. One of the drawbacks is that individual species in the communities are not identified, but this is the cost of not being required to have a mass of detailed information on each species. One point that should be emphasized: it is the principle of resource allocation that is involved. This is the generalization of the rule, not a requirement that there be a similar range of guilds in all situations. This point becomes clear when considering the way in which the question of allocation of resources among species and guilds has been addressed by Morris and Knight (1996), as illustrated in Fig. 1.7.

The further development of the rule by Doug Kelt provides a most interesting way to proceed. By incorporating specific alternative hypotheses we are provided with a powerful means of testing mechanisms that might be involved. This should be limited only by the ingenuity of researchers to devise appropriate null models and appropriate means to test them in the field. One interesting possibility here is to incorporate the detailed approach taken by Bob M'Closkey (1978) with that used by Doug Kelt (Kelt *et al.*, 1995; Kelt and Brown, this volume). If such an approach were fruitful it would certainly provide us with an excellent opportunity to further advance our understanding of the ways in which communities are structured.

The other area that might prove fruitful is the incorporation of the regression technique with standardized data. As has been briefly illustrated here (Table 1.2), this technique has the ability to easily provide information on the competition between species if additional habitat information is available. This technique provides another interesting means of testing the degree of interaction in assemblages to offer an independent method to that used in the Kelt technique.

Conclusions

Twenty-five years after Robert MacArthur's death, we are still working on the wealth of ideas that he left as his legacy to ecology, and geographical ecology in particular. The theory of island biogeography provided a framework for the study of species richness on islands which led to Jared Diamond's exposition of his species assembly rules first published in the 1975 memorial volume to Robert MacArthur. Diamond's rule arose from his study of birds from the archipelagoes near New Guinea, but these were soon followed by an application to small mammals by Bob M'Closkey in 1978 that elegantly demonstrated a mechanism for how the rule might operate. Despite a decade of sometimes acrimonious debate the assembly rule concept has survived and prospered to move on to a further stage.

In 1985 an idea for an assembly rule was presented that was based on functional groups rather than individual species, these functional groups were equivalent to guilds. The main advantage to this rule was its simplicity as it required only a knowledge of which guilds were present and the pool of species available for each guild, from which the communities were assembled. The rule was tested against rigorously devised conservative null hypotheses for a wide range of taxonomic groups, in many different habitats over a range of spatial scales on four continents.

Doug Kelt provided a very interesting further development for the guild assembly rule, by introducing specific alternative hypotheses to the null hypothesis to make an estimate of the power of the test. This was a very elegant test of the degree to which interspecific competition was involved as a mechanism in the operation of the rule. Incorporating habitat and creating a separate species pool for every site is a powerful extension of these analyses. Doug Morris demonstrated that the author's guild assembly rule is a probabilistic consequence of adding guild structure to models of consumer-resource competition, providing the necessary theoretical underpinning that had been missing from his rule.

The most recent development has involved an application of the Schoener–Pimm regression technique that has been recently revived by Barry Fox and Roger Luo, using standardized data. When applied to the Nevada communities, that already demonstrated adherence to the author's guild assembly rule, all six significant interaction coefficients found for within-guild species pairs (from 12 possible) were negative. For between-guild pairs there were 18 significant interaction coefficients (from 78 possible), all were positive. This interesting result supports interspecific interaction as one part of the mechanism for the guild assembly rule, and provided important field confirmation of the theoretical derivation made by Doug Morris.

The guild assembly rule should be seen as a combination of stochastic and deterministic components. Use of the terms 'high probability' states and 'low probability' states in place of 'favored' and 'unfavored' should reinforce this. The way forward offers at least three paths: (a) use of specific alternative null hypotheses as active tests of mechanisms that might structure communities; (b) the use of habitat data and species densities with the standardized regression technique to demonstrate interactive effects; (c) incorporation of the detailed species specific information as used by Bob M'Closkey. These all offer exciting opportunities to further develop the guild assembly rule in the future.

Acknowledgments

I have already acknowledged the intellectual debt I owe to the people whose ideas most influenced me in developing the guild assembly rule. However, I must mention that Bob M'Closkey's contribution was extremely influential, not only in converting me from a solid-state physicist to an ecologist but also in focusing my interest on why communities are structured as they are.

I would also like to offer my thanks to Rodger Luo for many useful and interesting discussions and ideas about assembly rules, and for completing the additional analyses on the Nevada data for me, particularly with his development of the analytical solutions and construction of random data sets. I would also like to thank Marilyn Fox, Vaughan Monamy and Doug Morris for discussions about ideas and suggestions on reading the manuscript. I thank Paul Keddy and Evan Weiher for the opportunity to write this chapter, and I also appreciated the constructive critical comment from Evan Weiher.

References

Braithwaite, R.W., Cockburn, A., & Lee, A.K. (1978). Resource partitioning by small mammals in lowland heath communities of southeastern Australia. *Australian Journal of Ecology* 3: 423–445.
Brown, J.H. (1987). Variation in desert rodent guilds: patterns, processes and scales. In *Organization of Communities: Past and Present*, ed. J.H.R. Gee & P.S. Giller, pp. 185–203. Oxford: Blackwell.
Brown, J.H. & Munger, J.C. (1985). Experimental manipulation of a desert rodent community: food addition and species removal. *Ecology* 66: 1545–1563.
Brown, J.H. & Kurzius, M. (1987). Composition of desert rodent faunas: composition of co-existing species. *Annals Zool. Fenn* 24: 227–237.
Cody, M.L. & Diamond, J.M. (eds.) (1975). *Ecology and Evolution of Communities*. Cambridge, MA: Harvard University Press.
Colwell, R.K. & Winkler, D.W. (1984). A null model for null models in biogeography. In *Ecological Communities: Conceptual Issues and the Evidence*, ed. D.R Strong, D Simberloff, L.G. Abele & A.B. Thistle, pp. 344–359. Princeton, NJ: Princeton University Press.
Connor E.H. & Simberloff, D. (1979). The assembly of species communities: chance or competition? *Ecology* 60: 1132–1140.
Crowell, K.L. & Pimm, S.L. (1976). Competition and niche shifts of mice introduced onto small islands. *Oikos* 27: 251–258.
Diamond, J.M. (1973). Distributional ecology of New Guinea birds. *Science* 179: 759–769.

Diamond, J.M. (1974). Colonization of exploded volcanic islands by birds: the supertramp strategy. *Science* 184: 803–806.

Diamond, J.M. (1975). Assembly of species communities. In *Ecology and Evolution of Communities*, ed. M.L. Cody & J.M. Diamond, pp. 342–444. MA: Harvard University Press.

Diamond, J.M. & Case, T.J. (1986). *Community Ecology*. New York: Harper & Row.

Fox, B.J. (1980). The ecology of a small mammal community: secondary succession, niche dynamics, habitat partitioning, community structure and species diversity. PhD Thesis, Macquarie University, Sydney.

Fox, B.J. (1981). Niche parameters and species richness. *Ecology* 62: 1415–1425.

Fox, B.J. (1982). Fire and mammalian secondary succession in an Australian coastal heath. *Ecology* 63: 1332–1341.

Fox, B.J. (1983). Mammal species diversity in Australian heathlands: the importance of pyric succession and habitat diversity. In *Mediterranean-Type Ecosystems: The Role of Nutrients*, ed. F.J. Kruger, D.T. Mitchell, & J.U.M. Jarvis, pp. 473–489. Berlin: Springer-Verlag.

Fox, B.J. (1985a). Small-mammal community pattern in Australian heathland. In *Fourth International Theriological Congress Abstracts*, ed. W.A. Fuller, N.T. Nietfeld, & M.A. Harris, Edmonton: ITC4.

Fox, B.J. (1985b). Small mammal communities in Australian temperate heathlands and forests. *Australian Mammalogy* 8: 153–158.

Fox, B.J. (1987). Species assembly and the evolution of community structure. *Evolutionary Ecology* 1: 201–213.

Fox, B.J. (1989). Small mammal community pattern in Australian heathland: a taxonomic rule for species assembly. In *Patterns in the Structure of Mammalian Communities*, ed. D.W. Morris, Z. Abramsky, B.J. Fox & Willig, M. pp. 91–103. Lubbock, TX: Texas Tech Museum Special Publication Series.

Fox, B.J. & McKay, G.M. (1981). Small mammal response to pyric successional changes in eucalypt forest. *Australian Journal of Ecology* 6: 29–42.

Fox, B.J. & Pople, A.R. (1984). Experimental confirmation of interspecific competition between native and introduced mice. *Australian Journal of Ecology* 9: 323–334.

Fox, B.J. & Gullick, G. (1989). Interspecific competition between mice: a reciprocal field manipulation experiment. *Australian Journal of Ecology* 14: 357–366.

Fox, B.J. & Kirkland, G.L., Jr. (1992). North American soricid communities follow Australian small mammal assembly rule. *Journal of Mammalogy* 73: 491–503.

Fox, B.J. & Brown J.H. (1993). Assembly rules for functional groups in North American desert rodent communities. *Oikos* 67: 358–370.

Fox, B.J. & Brown J.H. (1995). Reaffirming the validity of the assembly rule for functional groups or guilds. *Oikos* 73: 125–32.

Fox, B.J. & Luo, J. (1996). Estimating competition coefficients in the field: a validity test of the regression method. *Oikos* 77: 291–300.

Ganzhorn, J.U. (1997). Test of Fox's assembly rule for functional groups in lemur communities in Madagascar. *Journal Zoology London* 241: 533–542.

Gilpin, M.E., Carpenter, P. & Pomerantz, M.J. (1986). The assembly of a laboratory community: multispecies competition in *Drosophila*. In *Community Ecology*, ed. J.M. Diamond & T.J. Case, pp. 23–40. New York: Harper & Row.

Grant, P.R. & Abbott, I. (1980). Interspecific competition, island biogeography and null hypotheses. *Evolution* 34: 332–341.

Heske, E.J., Brown, J.H., & Mistry, S. (1994). Long-term experimental study of a Chihuahuan desert rodent community: 13 years of competition. *Ecology* 75: 438–445.

Higgs, P. & Fox, B.J. (1993). Interspecific competition: a mechanism for rodent succession after fire in wet heathland. *Australian Journal Ecology* 18: 193–201.

Jorgensen, C.D. & Hayward, C.L. (1965). Mammals of the Nevada test site. *Brigham Young University Science Bulletin Biology Series* 6(3): 1–81.

Kelt, J.A., Taper, M.L., & Meserve, P.L. (1995). Assessing the impact of competition on the assembly of communities: a case study using small mammals. *Ecology* 76: 1283–1296.

Kenagy, G.J. (1973). Adaptations for leaf eating in the Great Basin Kangaroo Rat *Dipodomys microps*. *Oecologia* 12: 383–412.

Kotler, B.P. & Brown, J.S. (1988). Environmental heterogeneity and the coexistence of desert rodents. *Annual Reviews of Ecology Systems* 19: 281–307.

MacArthur, R.H. (1972). *Geographical Ecology*. New York: Harper & Row.

MacArthur, R.H. & Wilson, E.O. (1967). *The Theory of Island Biogeography*. Princeton, NJ: Princeton University Press.

M'Closkey, R.T. (1978). Niche separation and assembly in four species of Sonoran Desert rodents. *American Naturalist* 112: 683–694.

M'Closkey, R.T. (1985). Species pools and combinations of heteromyid rodents. *Journal of Mammalogy*. 66: 132–134.

Morris, D.W., Abramsky, Z., Fox, B.J. & Willig, M. (ed.) (1989). *Patterns in the Structure of Mammalian Communities*. Lubbock, TX: Texas Tech Museum Special Publ. Series.

Morris, D.W. & Knight, T.W. (1996). Can consumer-resource dynamics explain favored versus unfavored states of species assembly. *American Naturalist* 147: 558–575.

Munger, J.C. & Brown, J.H. (1981). Competition in desert rodents: an experiment with semipermeable exclosures. *Science* 211: 510–512.

Patterson, B.D. (1990). On the temporal development of nested subset patterns of species composition. *Oikos* 59: 330–342.

Patterson, B.D. & Brown, J.H. (1991). Regionally nested patterns of species composition in granivorous rodent assemblages. *Journal of Biogeography* 18: 395–402.

Rosenzweig, M.L., Abramsky, Z., Kotler, B. & Mitchell, W. (1985). Can interaction coefficients be determined from census data? *Oecologia* 66: 194–198.

Schoener, T.W. (1974). Competition and the form of the habitat shift. *Theoretical Population Biology* 6: 265–307.

Schröpfer, R. (1990). The structure of European small mammal communities. *Zoologische Jahrbüch Systematik* 117: 355–367.

Stone, L., Dayan, T. & Simberloff, D. (1996). Community-wide patterns unmasked: the importance of species' differing geographical ranges. *American Naturalist* 148: 997–1015.

Strong, D.R., Simberloff, D., Abele, L.G. & Thistle, A.B. (1984). *Ecological Communities: Conceptual issues and the Evidence*. Princeton, NJ: Princeton University Press.

Thompson, P.T. & Fox, B.J. (1993). Asymmetric competition in Australian heathland rodents: a reciprocal removal experiment demonstrating the influence of size-class structure. *Oikos* 67: 264–278.

Tilman, D. (1982). *Resource Competition and Community Structure*. Princeton, NJ, Princeton University Press.

Valone, T.J. & Brown J.H. (1995). Effects of competition, colonization, and extinction on rodent species diversity. *Science* 267: 880–883.

Wilson, J.B. (1989). A null model of guild proportionality, applied to stratification of a New Zealand temperate rainforest. *Oecologia* 80: 263–267.

Wilson, J.B. (1995a). Null models for assembly rules: the Jack Horner effect is more insidious than the Narcissus effect. *Oikos* 74: 543–544.

Wilson, J.B. (1995b). Fox and Brown's 'random data sets' are not random. *Oikos* 72: 139–143.

2

Ruling out a community assembly rule: the method of favored states

Daniel Simberloff, Lewi Stone, and Tamar Dayan

Introduction

Arguments have raged for two decades over the relative importance of interspecific competition and individual species' responses to the physical environment in determining community composition and about the nature of evidence on this matter (references in Stone, 1996). This debate is one of the most noteworthy features of modern community ecology. Several adherents of the view that competition plays a key role have sought support in patterns detectable in local communities, patterns they feel reflect a governing role for competition in the assembly of communities. The focus here is on one of the most recent such assembly rules, that of Fox and coauthors (1985, 1987, 1989; Fox & Kirkland, 1992; Fox & Brown, 1993), who have modified the assembly rule of Diamond (1975). Applying a null hypothesis approach (*cf.* Connor & Simberloff, 1979; Gilpin & Diamond, 1982), Fox proposed that a competitive assembly rule, described below, determines how sequential addition constructs communities. His analyses of several data sets seem to confirm this assembly rule and to imply for these communities that competition largely governs their composition.

Fox and Brown (1993) applied a variant of this rule to local communities of North American granivorous desert rodents, finding strong confirmation of Fox's assembly rule (1987). Many people have studied these rodent communities (references in Brown, 1987; Kotler & Brown, 1988), and it is evident that interspecific competition is occurring. Thus, if co-occurrence patterns in any regional communities manifest competition, this should be one such community. However, the fact that competition occurs between these species need not mean that competition governs local community composition. So it would be particularly interesting to find an assembly rule that showed that it does. Stone *et al.* (1996) reexamined Fox's assembly rule using the same data studied by Fox and Brown (1993). Their results are summarized and extended here. A similar study by Fox and Kirkland (1992) of a community of six shrew species coexisting over a wide region of the northeastern United States and eastern Canada is also reexamined.

Functional groups and favored states

Fox (1987, 1989) suggested that, if competition determines which species can enter a developing local community, one can foresee the outcome of community assembly in terms of 'functional groups': sets of ecologically similar species, like guilds. Fox's strategy suggests that local community composition can be predicted in terms of numbers of species in different functional groups without extensive knowledge of every species, so long as how to assign species to functional groups is known. Species have often been assigned to functional groups and guilds based solely on systematics, although in principle both categories are supposed to be determined by the types of food eaten, and guilds by the way the food is gathered as well (Simberloff & Dayan, 1991). So, dividing a biota into functional groups is no trivial matter, though the functional group assignments for the rodents studied by Fox and Brown (1993) appear to have an ecological basis (see below), even though they follow taxonomic lines.

Fox's assembly rule is simple: 'There is a much higher probability that each species entering a [local] community will be drawn from a different functional group until each group is represented, before the cycle repeats' (Fox 1987, p. 201). Functional groups should be equally represented in local communities derived from a larger regional pool. The rule is based on interspecific competition, primarily for food (Fox & Brown, 1995): if some functional group becomes disproportionately represented in a local community, competition lowers the probability that the next species to colonize will belong to that group and raises the probability that it will belong to one of the other groups.

A local community is in a 'favored state' whenever all pairs of functional groups have the same number of species or differ by at most one (Fox & Brown, 1993). For example, consider the numbers of species in three different functional groups at a hypothetical site. Favored states include (2,2,2), (2,3,2), and (1,1,0): functional groups are evenly represented. By contrast, a local community is in an 'unfavored state' if the number of species in any pair of functional groups differs by more than one. The configurations (1,0,2), (2,1,4), and (4,3,1) are unfavored states.

If local communities are assembled according to this rule, more of them should be in favored states than if the species already present do not affect which other species will be added next. Fox (1987; Fox & Brown 1993) thus proposed a test of this assembly rule that uses the null hypothesis that species enter local communities independently of their functional groups. He simulated the assembly of each local community by randomly drawing species from the regional pool, without replacement, and irrespective of which other species have already been drawn. The simulated local community is complete when the same number of species are present as are in its real analog.

Fox and his coworkers have used several algorithms to assign probabilities that the next species drawn in a random local community comes from a particular functional group (references in Stone *et al.*, 1996). Fox and Kirkland (1992) made the initial probability for each functional group proportional to the number of species in that group, but sampling was with replacement. The focus here is on the most recent algorithm, that of Fox and Brown (1993), who made the initial probability for each functional group proportional to the number of species in the group and sampled without replacement. Kelt *et al.* (1995), examining assembly of South American small mammal communities, used a similar method. From the null distribution of favored states among the simulated communities, one can test whether the observed data set differs from expected, as would be expected if Fox's rule is at work.

The data

Fox and Brown (1993) examined data for two regional North American mammal communities:

(a) Data of Jorgensen and Hayward (1965) on mammals of the Nevada Test Site, from surveys in Nevada for the 'evaluation of the effects nuclear weapons testing, peaceful use of nuclear weapons, and nuclear warfare may have on mammal populations' (Foreword, p. 1). They trapped small mammals for six years at 115 sites in an area of 3500 km². If Fox's assembly rule works, this finding would be remarkable, as it would mean that the assembly rule overrides the effects of a series of nuclear explosions near these sites four years before the survey (Stone *et al.*, 1996)!

(b) Data of Brown and Kurzius (1987) on 28 rodent species at 202 sites across 640 000 km² of the arid Southwest.

Granivorous rodents have been variously partitioned into guilds (Simberloff & Dayan, 1991). Generally, heteromyids are separated from cricetids because the former have large external cheek pouches and different foraging behavior. The quadrupedal heteromyids (pocket mice) are usually separated from the bipedal ones (kangaroo rats and mice) both because they move differently and because the former tend to forage under shrubs, while the latter forage mostly in open areas between widely spaced plants. Stone *et al.* (1996) assigned the granivorous species of each data set to functional groups exactly as Fox and Brown (1993) did. For the Nevada site, the nine granivorous rodents fell into three functional groups: (a) bipedal heteromyids = BH (two species); (b) quadrupedal heteromyids = QH (three species); (c) quadrupedal non-heteromyids = QN (four species). Fox and Brown (1993, 1995) included in the species pool two species found at none of the sites. Stone *et al.* (1996) did not; the results of Stone *et al.* (1996) are unlikely to be changed by whether or not these species are seen as part of the pool. For a new analysis detailed below, they are included, so the numbers of species in the three functional groups are 3,3, and 5, respectively. The 28 southwestern granivorous species were likewise partitioned into three groups: (a) bipedal heteromyids (7 species); (b) quadrupedal heteromyids (11 species); (c) cricetids (10 species).

Fox and Kirkland (1992) used data from Rickens (1974) plus Kirkland's unpublished surveys of six shrew species at 43 sites in Nova Scotia, New Brunswick, New York, Pennsylvania, and West Virginia. They assigned these species to three functional groups of two species each, based on body mass: large, medium, and small. We have adopted these assignments.

The original results

Of the 115 local communities at the Nevada Test Site, Fox and Brown (1993) found 92 to be in favored states, while Stone *et al.* (1996) found 93. Fox and Brown (1993), from their random-draw simulations, found an expected number of favored states of 62.5, and the probability of a number as high as the observed value to be $P < 0.001$.

For the 202 local communities in the southwestern data set, Fox and Brown (1993) found 128 favored states, while Stone *et al.* (1996) found 126. Fox and Brown (1993) calculated an expectation of 111.3 favored states and, using their random-draw simulation, the probability of a value as high as that observed to be $P < 0.01$.

Of the 43 local shrew communities, Fox and Kirkland (1992) reported that 12 were in unfavored states and 31 in favored states. Their own data (Fox & Kirkland, 1992, Table 2.1) show 11 unfavored and 32 favored ones. Their random draw simulations yielded an expectation of 20.8 unfavored states and 22.2 favored states, and they found the observations to differ from expectations at $P = 0.02$ by Fisher's exact test. This result would be even more extreme for the correct number of favored states.

Table 2.1. *Observed numbers of favored states and those expected for species randomly assigned to functional groups for desert rodents of the Nevada Test Site and American Southwest (after Stone* et al. *1996) and shrews of the northeastern US and eastern Canada*

Data set	Observed # favored states	Expected # favored states	Standard deviation	% tail of observation
Nevada	93	64.9	22.6	11.8
Southwest	126	105.6	18.0	13.8
Shrews	32	29.0	7.3	46.7

The matrix randomization method

Beginning with Connor and Simberloff (1979), many people have used presence–absence matrices to deal with data of this sort (references in Gotelli & Graves, 1996). Columns represent sites and rows represent species. A '1' in the (i,j)th entry means species i is present at site j, while a '0' means it is absent. The number of species on the jth site is the jth column sum, and the number of sites occupied by the ith species is the ith row sum.

The simulated communites are a sample from the set of all matrices having the dimensions of the observed matrix. However, to preserve some biological realism, we follow Connor and Simberloff (1979) and require the species and sites of each simulated matrix to satisfy constraints: (a) A simulated site contains as many species as its analog does in nature. Thus, some simulated sites support more species than others do, as in the real data. These differences reflect such ecological patterns as the species–area effect. (b) A simulated species occupies as many sites as its analog does in nature. Some species are thus more widespread than others. In nature, this variation is caused by species differences such as dispersal abilities and physical factor tolerances; it could also reflect competitive ability. An important component of these simulations is that species' geographic ranges are not preserved. Real ranges are usually continuous, while in simulated matrices species ranges tend to be very discontinuous and can include areas far outside the real ranges (Stone *et al.*, 1996).

Matrices drawn equiprobably from the set of all matrices with the same row and column sums thus represent 'random' communities. Because we have incorporated these constraints to preserve realism, the 'null model' may also no longer be null with respect to some hypotheses. For example, because interspecific competition can affect the species richness of a site or the distribution of some species, fixed row and column sums may embed the effects of competition in each 'random' community. So this test, as used by Stone *et al.* (1996), is a narrow one; it is not about whether interspecific competition occurs, but about whether, for whatever reasons, the entry of species into local communities accords with Fox's functional group model.

The random communities of Fox and Brown (1993) are not assembled by filling in such a binary matrix, but, if they were, they would have column constraints only – each site has the number of species observed on its analog in nature. However, because species were not randomly drawn (only functional groups), species are not restricted to particular sites or to a particular number of sites. Thus, to the extent that species' ranges and site occupancies in nature reflect biological traits as noted above, the assembly algorithm is unrealistic. Implicitly, all species in a functional group are the same.

It is often unrealistic to simulate a regional community as if all species can occupy all sites. For example, in the southwestern data set, *Dipodomys merriami* and *Peromyscus maniculatus* are very widespread, occupying about half the 202 possible sites. But almost half of the species occupy fewer than ten sites. Each functional group has at least one widespread species and several that are much more narrowly distributed. It seems unrealistic to assume that all species in a functional group are interchangeable, because the observed data show otherwise. Possible reasons why different species occupy different numbers of sites include differing dispersal abilities, differing tolerances of physical factors, differing histories, and differing competitive abilities. The test of Fox and Brown (1993) does not address this issue. An unrealistic model need not be useless for testing a particular hypothesis; this would depend on the particular nature of the lack of realism and the hypothesis. Stone *et al.* (1996) found that, in this instance, the lack of realism could have been crucial, as we discussed below.

Reanalyzing the Nevada test site and southwestern data sets by randomizing matrices, Stone *et al.* (1996) fixed row and column sums. Although a constrained row sum means that each simulated species occupies the same number of sites as in nature, it does not limit species' ranges, as noted above, because each species can still occur on any site. For each data set, Stone *et al.* (1996) produced 10000 random matrices and computed the number of favored states, as well as the standard deviation and percent tail of the observation. The number of favored states in the Nevada matrix is in the 28.3% tail, while that in the Southwest is in the 17.8% tail. So, by this method, the null hypothesis of independent colonization would not be rejected whereas Fox and Brown (1993, 1995) find significantly more favored states than expected in both data sets.

Clearly, if these results are to be taken as a guide, a null hypothesis that species colonize sites independently of one another cannot be rejected. By contrast, Fox's method shows that the local communities of both data sets include significantly more favored states than expected, just as predicted by his assembly rule (Fox & Brown, 1993). Wilson (1995) independently published a very similar study to ours, using matrix randomization, and came to the same conclusion: the Fox null model will nearly always find an observed data set significantly different from expectation. He attributes this finding to the fact that species frequencies are rarely equal in nature. He argues that randomly constructed data sets would show an excess of favored states, so long as species frequencies are not all the same. He called this the 'Jack Horner effect' because the obvious is demonstrated: in this case, that species frequencies generally differ.

Is the Fox and Brown (1995) test biased?

Rejecting Wilson's criticism of Fox and Brown (1993), Fox and Brown (1995) sought to show that, contrary to Wilson's assertions, randomly constructed data sets would not show a surplus of favored states relative to expectation. However, they chose highly non-random data sets to do this.

Their method is exemplified by their treatment of the Nevada test site data. Observing that the maximum numbers of species in the three functional groups BH, QH, and QN in any local community are 2,2, and 3, respectively, they constructed 36 different combinations to represent the various possible numbers of species in the functional groups in the 115 local communities. That is, in their null assignment of sets of species to the local communities, each community could have 0,1, or 2 species in the first functional group, 0,1, or 2 species in the second group, and 0,1,2 or 3 species in the third group. Thus, irrespective of species' identities, there are $3 \times 3 \times 4 = 36$ possible communities, and each of the 115 local communities was uniform randomly assigned to one of these 36 combinations. Each such random assignment of the 115 local communities constituted a random data set. They then asked how likely it is that the Monte-Carlo method of Fox and Brown (1993) described above would show more favored states than expected for such a random assignment. In fact, they found the opposite – usually the null hypothesis was accepted.

By uniform randomly assigning the communities to the 36 possible combinations, they were using Bose–Einstein statistics, which are appropriate only if the species are indistinguishable (Feller, 1968), and they are not. Even if species within a functional group are ecologically equivalent, as assumed by Fox and Brown (1993), they are still distinguishable, in that one can be distinguished from the other (for example, by morphology). Thus, a random assignment of local communities would make the probability that a community falls in each of the 36 cells proportional to the number of possible ways that cell can arise (Maxwell–Boltzmann statistics; Feller, 1968). Thus, for example, the cell (0,0,0) can arise only one way, the cell (0,0,1) five ways (because there are five species in group QN), and the cell (2,2,3) 90 ways (because there are three species each in the other two groups, $_2C_3 = 3$, $_3C_5 =$ 10, and $3 \times 3 \times 10 = 90$). In fact, there are 1274 distinguishable arrangements for the 36 cells. Favored states are represented by 16 cells, and these 16 cells contain 681 distinct arrangements. Thus, 681/1274 = 53.5% of all possible arrangements are favored states, but, on average, Fox and Brown (1995) would find only 16/36 = 44.4% of the random communities to be favored states in any of their randomized data sets.

It is not known exactly how this fact would affect the expected number of favored states when the Monte-Carlo algorithm of Fox and Brown (1993) is applied using these randomized data sets as the observed data, but one might expect it to bias the test towards finding that the expected number of favored states is at least as large as the observed. Consider: if one happened to assign the 115 random local communities to the 36 cells such that none fell in favored states, one would subsequently calculate the expected number of favored states, based on these particular observed ones, to be much greater. This is because, in the Monte-Carlo procedure, one still randomly draws species from groups that, in the species pool, have sizes of 3,3, and 5 species, respectively. A truly random, Maxwell–Boltzmann assignment of the 115 local communities to the 36 cells might not show fewer random states in a randomized set of local communities than would be expected by the Monte-Carlo procedure, but the Bose–Einstein assignment method of Fox and Brown (1995) is certainly biased rather than conservative.

In fact, the maximum numbers of species in the three functional groups BH, QH, and QN are *not* (2,2,3), as in Fox and Brown (1995). Rather, they are either (3,2,3) or (3,2,2), depending on whether one counts one QN species collected too infrequently for Jorgensen and Hayward (1965) to calculate densities. Apparently Fox and Brown (1995) included this species, as this inclusion is necessary to find 5 QN species in the regional pool. With maximum numbers of species in the local communities as (3,2,3), there are $4 \times 3 \times 4$ = 48 cells, of which 18 representing favored states contain 741 of the 1456 arrangements and 30 representing unfavored states contain 715 arrangements. So 741/1456 = 50.9% of the possible arrangements are favored states, but only 18/48 = 37.5% of the random communities will be favored states in the randomized data sets drawn by the Bose–Einstein method. Again, when Fox and Brown (1995) subsequently perform their Monte-Carlo procedure on each randomized data set, there will be a bias towards accepting the null hypothesis that the observed (random) data set has no more favored states than expected.

Species ranges and co-occurrence patterns

The discrepancy between the results of Stone *et al.* (1996) and those of Fox and Brown (1993) led Stone *et al.* (1996) to ask whether the apparent structure Fox and Brown detected may arise partly because their model does not incorporate species' differing numbers of occurrences (as hypothesized by Wilson, 1995) and/or different ranges. If species have very different geographic ranges, the random matrix approach might lead to an erroneous conclusion because it would permit co-occurrences that are, in fact, very unlikely. This would not appear to be a problem for the Nevada test site data, as the study area is small and the ranges of all species encompass it. On the other hand, for the southwestern data set, the fact that the study area is huge and includes four different deserts suggests that ranges of various pairs of species overlap only slightly or not at all. Of 378 species pairs in the set, Stone *et al.* (1996) found only nine sharing more than 20 sites. Either *D. merriami* or *P. maniculatus*, the two most widespread species, appears in each of the nine pairs. If these two species are considered jointly with *Perognathus parvus*, it is clear how the random matrix method can produce a very unrealistic result (Stone *et al.*, 1996).

   Pg. parvus occupies 35 sites, of which *Pm. maniculatus* occupies 34. One might have expected, just from the numbers of occurrences, that *Pg. parvus* and *D. merriami* would also co-occur on many sites, but they do not. They co-occur on only five sites. The reason for the different co-occurrences between *Pg. parvus* on the one hand, and either *Pm. maniculatus* or *D. merriami* on the other, has to do with the species' geographic ranges. *Pm. maniculatus* has a very large range and is distributed throughout the entire southwestern study area. However, although the range of *D. merriami* includes many sites, it is much smaller than that of *Pm. maniculatus*, and it barely intersects that of *Pg. parvus*. There were only about five sites censused by Brown and Kurzius (1987) in the region of overlap between *D. merriami* and *Pg. parvus*, and these are the five sites at which they co-occur. But, in the random matrices, where both species can be placed on any site, they share an average of 18.1 sites (Stone *et al.*, 1996).

   Now, one could argue that the reason these two species share so few sites is competition. Given the arrangement of the sampling sites, it would be just as reasonable to say that their geographic ranges are quite different, so that the reason they share so few sites is whatever causes their ranges to differ. This could be past or present competition, but it could also be differences of dispersal ability, physical factor tolerances, and various historical factors. One would need evidence independent of the ranges to judge the reasons the ranges barely overlap.

Because of greatly differing species' ranges, Fox and Brown's functional group assembly rule would also be unrealistic in exactly the same cases in which the matrix randomization approach is unrealistic. Their null probability that a particular functional group is drawn to populate a simulated local community is proportional to the number of undrawn species in that functional group. This is equivalent to making all species equally likely to be drawn, no matter how widely or narrowly distributed they are. Again, in the Nevada test site data this problem should not greatly affect results by either method, as the area is small and all species are part of the species pool for all sites. For the southwestern data set, however, both methods would produce unrealistic null distributions for at least this reason.

Brown has wagered that, if the random draws to populate local communities in the Fox–Brown method are restricted for each local community to just those species whose ranges include the site for the southwestern data set, their result (Fox & Brown, 1993) of an excess of observed favored states will stand (J. Brown, pers. comm.). This will be an interesting study.

Although the Nevada test site locations were all within the geographic ranges of all the species, it is quite possible that habitat differences between the species would make certain random draws, by either Fox's method or matrix randomization, extremely unrealistic, and such lack of realism could warp the expected number of favored states. For example, *Pg. parvus* appeared in the test site study to be found in low densities in many plant communities, but in high density only in two types, of which one is pinyon–juniper. *Pg. longimembris*, on the other hand, was collected from all plant communities except pinyon–juniper (Jorgensen & Hayward, 1965). Of the 115 local rodent communities tabulated by Jorgensen and Hayward (1965), *Pg. parvus* was found in only two, while *Pg. longimembris* occurred in 89; they shared only one. The description of the plant communities of these 115 sites shows that, unsurprisingly, they had few or no pinyon–juniper communities among them. Other sites not included in this tabulation included much pinyon–juniper (Allred *et al.*, 1963). These are two of only three species in the QH functional group in the species pool. The particular selection of sites for the co-occurrence study apparently predisposed strongly against any local community containing three QH species. This predisposition probably increased the number of favored states relative to that in a random draw from the species pool according to Fox's method, irrespective of habitat. As with the case of exclusive ranges, it may be that habitat exclusivity reflects the 'ghost of competition past' (Connell, 1980), but, if this were the case, it would have nothing to do with the operation of Fox's assembly rule in the present.

Randomizing functional group assignments

The assembly rule of Fox and Brown (1993) rests crucially on how species are assigned to functional groups. In fact, the only input required to test for the operation of the rule, other than which species are where, is how the species in the pool are divided into functional groups (Fox & Brown, 1993). Whether an observed local community is in a favored state depends on which functional groups its species belong to. Thus, if finding a high number of favored states among observed local communities is evidence of Fox's assembly rule, the number of favored states should be much lower if species are randomly assigned to functional groups. Stone *et al.* (1996) randomly assigned species to functional groups for the two rodent data sets. They did this by simply randomizing the functional group designations, maintaining the observed numbers of species in each functional group. Then, for each set of random functional groups, they looked at the observed local communities to see if they were in favored states. Since every species stays on its observed sites, the problem of unrealistic ranges and co-occurrence frequencies in the null matrices discussed above is eliminated.

For both rodent data sets, Stone *et al.* (1996) randomized the functional groups 10 000 times and recorded the distribution of favored states (Table 2.1). Although the observed number of favored states exceeds expectation for both data sets, the differences between observed and expected are not nearly significant at the 0.05 level. In other words, if one randomly concocts functional groups from the same species pool, completely independently of their biology, the expected number of favored states among the observed local communities would not differ from the number found when the real functional groups are used.

A similar result arises for the shrew data set of Fox and Kirkland (1992). Here, because there are only six species in the pool, two in each functional group, there are only $6!/2!2!2! = 90$ possible assignments of species into functional groups, and we calculated the number of favored states explicitly for each of these, rather than simulating. Taking these assignments as equiprobable, we found the observed 32 favored states to be approximately that expected (29.0) if the species had been randomly assigned to functional groups (Table 2.1). Of the randomized functional group assignments, 46.67% produced at least as many favored states as were observed.

In sum, for all three data sets, there is no indication that the particular division of the species pool into functional groups produces more favored states among the observed local communities than a randomly chosen division of the species pool would have. Because Fox's assembly rule rests on partitioning the species pool into functional groups of species that are especially likely to compete with one another (Fox & Brown, 1993), the conclusion of Fox and his coauthors for these three communities – that competition causes an unusually high number of favored states – is unconvincing.

Effects of widespread species

As noted above, Stone *et al.* (1996) found that variation among species' biogeographic ranges can greatly affect co-occurrence patterns. They tested the effect of the fact that a few species have very large ranges and most are greatly restricted on the distribution of favored states. With the matrix randomization method, they fixed the widespread species on their actual sites and randomized the other species. With the randomized functional group method, they assigned the widespread species to their correct functional groups and randomly assigned the others. Then, with either method, they determined the expected number of favored states. For both rodent data sets, Stone *et al.* (1996) found that fixing either the locations or the functional groups of the few widespread species caused the randomizations to produce approximately as many favored states as the observed number. For both methods, it is as if 'unfixing' the locations or functional groups of the widespread species is largely responsible for the fact that the expected number of favored states is frequently lower than the observed number in our previous simulations. The other species simply add background 'noise' to this pattern.

The shared-site null hypothesis

Like Fox and Brown (1993), Wright and Biehl (1982) argued that one should focus on particular pairs or trios of species suspected of competing. In their view, the most efficient way to do this is to fix the numbers of occurrences of each putative competitor, then uniform-randomly distribute these occurrences among all the sites. In this way, they argue, they have mimicked different dispersal abilities among species. On the other hand, they allow the column sums (site richnesses) to vary while maintaining the row sums (species occurrences). That is, the numbers of species in each site are not fixed as species are randomly arranged on the sites. They do not view the resulting violation of species–area relationships as a major problem. As noted by Connor and Simberloff (1979, 1983, 1984), if one identifies in advance which groups of species are hypothesized to compete, rather than simply scanning a matrix for exclusively distributed pairs, it would be a fair test of the hypothesis to look only at those groups. It is not so clear whether a procedure that does not constrain numbers of species at each site is appropriate. However, in the Nevada test site data, because all sites have so few species of any functional group anyway, perhaps it is instructive to proceed as suggested by Wright and Biehl (1982).

Each of the functional groups is identified by Fox and Brown (1993) as a likely locus of competition, and in the 115 sites there are three species in each of the groups BH and QH, and 4 in the group QN. (Jorgensen and Hayward, 1965 did not tabulate densities of one species in BH, which was thus not used by Stone *et al.*, 1996; it is included in the present analysis.) The explicit calculation of the probability that X sites will be occupied by two or more species is sketched out by Wright and Biehl (1982). However, these probabilities and the resultant tail probabilities are more easily calculated by computer simulation. For each of the three groups, the species were randomly distributed 10 000 times among the 115 sites, filling as many sites for each species as it occupies in nature.

For group BH, 46 sites are occupied by two or more species (four sites have three species each). For randomly distributed species, one would have expected 46 or fewer pair plus trio co-occurrences 26.9% of the time (and one would have expected four or fewer trio co-occurrences 69.7% of the time). For group QH, 40 sites contain two species, and none contain three. One would have expected 40 or fewer sites with two or more species 36.8% of the time. For group QN, 21 sites contain two species (of which two contain trios). One would have expected 21 or fewer sites with two or more species 53.5% of the time (and two or fewer sites with three or more species 76.1% of the time). In short, for all three functional groups, species co-occur about as frequently as expected if they colonize sites independently of one another.

An analogous analysis was performed with the shrew data. For the two species in the functional group comprising small species, which occupy 22 and 7 of the 43 sites, respectively, six sites are occupied by both species. That is, almost all the occurrences of the least common of the two species are on sites occupied by the other species. It is thus not surprising that, rather than the species appearing to exclude one another, only 0.5% of the time would they co-occur this often if they were independently distributed at random. Similarly, the medium-sized species (which occur on 33 and 10 sites, respectively) co-occur on seven sites. For independent random colonization, one would have expected seven or fewer sites to have been occupied by both species 41.9% of the time. The large species occupy 40 and 12 sites, respectively, and they co-occur on 11 sites. Had they colonized independently and randomly, they would have co-occurred on 11 or fewer sites 97.8% of the time. In short, there is again no indication that species within a functional group are excluding one another.

Discussion

No statistical test of a single presence–absence data set can conclusively demonstrate the presence or absence of interspecific competition. There are two main reasons (Stone *et al.*, 1996). First, even if the data have a pattern consistent with the workings of competition, other plausible hypotheses can usually be found that would predict the same pattern. Second, the statistical test of the biological hypothesis might be affected by competition (the 'Narcissus effect' of Colwell and Winkler, 1984). An example is the possibility, discussed above, that row and/or column sums fixed in the matrix randomization method used here were themselves partially determined by interspecific competition. Thus, it is not being claimed that the rodents or shrews in these communities do not or did not compete. For the rodents, at least, it is known that they do from prior research cited above. This fact makes it all the more important that these data do not implicate Fox's assembly rule. It is also emphasized that, in principle, one could find patterns that would be consistent with the operation of Fox's assembly rule. That is, one could conceivably find a set of local communities that, when properly tested, were found to have more favored states than expected if species colonized independently of which other species are present or absent. All that is being claimed is that the three communities discussed here do not show such patterns.

The abiding search for simple rules that determine community structure, particularly rules that rest on single, relatively easily measured traits, like morphological size or functional group, has so far been fruitless. Fox's favored state approach is one such type of rule. In addition to ignoring habitat, range, and historical differences that might be as crucial as competition for food in determining which species now coexist in which local communities, Fox's assembly rule does not see order of species entry as important. Some models of community assembly by sequential addition of species do (e.g., Drake, 1991), building on priority effects in some empirical studies (e.g., Alford & Wilbur, 1985; Robinson & Dickerson, 1987). Even the framework of Fox's model – sequential colonization of empty sites – may not be appropriate for some data sets. For example, in regions as large as that encompassed by the southwestern data set, it is possible that geological events could cause whole groups of species to be joined almost simultaneously. If competition then acted to determine local community composition, how would this competition affect the distribution of favored states? No doubt an analog of the Fox model could be produced for sequential deletion as for those generated by sequential addition, but the statistical tests used by Fox and Brown (1993) would be as problematic for communities produced by deletion as by addition.

If a simple rule turned out to implicate an assembly mechanism strongly, our skepticism about the search would be alleviated. But Fox's rule has not passed this test.

The failure of simple, general assembly rules to date does not mean that no rules govern community assembly. With sufficient knowledge of the basic biology of each species in a local species pool, including far more information than is usually brought to bear on the precise habitat preferences and requirements, it would probably be possible, in a narrowly defined region, to predict which species will be found at which site and which species will not coexist. The necessary research, including not only matters commonly viewed as ecological but also much that is ethological, generally falls under the rubric of 'natural history'. Of course, natural history is not very fashionable nowadays, to a large extent because it is thought to be superseded by such generalizations as assembly rules. Perhaps the failure of general assembly rules to pass close scrutiny indicates that natural history should not be so quickly discarded. In any event, the subtle responses of species to one another and to minute changes in the physical environment, as well as substantial phenotypic and genetic differences among populations of some species, suggest that assembly rules, if they exist, will be quite local. Not only will assembly rules not be simple, they will not be very general.

References

Alford, R.A & Wilbur, H.M. (1985). Priority effects in experimental pond communities: competition between *Bufo* and *Rana*. *Ecology* 66: 1097–1105.

Allred, D.M., Beck, D.E. & Jorgensen C.D., (1963). Nevada Test Site study areas and specimen depositories. *Brigham Young University Science Bulletin, Biology Series* 2(4):15 pp.

Brown, J.H. (1987). Variation in desert rodent guilds: patterns, processes, and scales. In *Organization of Communities: Past and Present*. ed. J.H.R. Gee & P.S. Giller, pp. 185–203. Oxford: Blackwell Scientific.

Brown, J.H. & Kurzius, M. (1987). Composition of desert rodent faunas: combinations of co-existing species. *Annales Zool. Fenn.* 24: 227–237.

Colwell, R.K. & Winkler, D.W. (1984). A null model for null models in biogeography. In *Ecological Communities. Conceptual Issues and the Evidence*. ed. D.R. Strong, Jr., D. Simberloff, L.G. Abele, & A.B. Thistle, pp. 344–359. Princeton, NJ: Princeton University Press.

Connell, J.H. (1980). Diversity and the coevolution of competition, or the ghost of competition past. *Oikos* 35: 131–138.

Connor, E.F. & Simberloff, D. (1979). The assembly of species communities: chance or competition? *Ecology* 60: 1132–1140.

Connor, E.F. & Simberloff, D. (1983). Interspecific competition and species co-occurrence patterns on islands: null models and the evaluation of evidence. *Oikos* 41: 455–465.

Connor, E.F. & Simberloff, D. (1984). Neutral models of species' co-occurrence patterns. In *Ecological Communities. Conceptual Issues and the Evidence*, ed D.R. Strong, Jr., D. Simberloff, L.G. Abele, & A.B. Thistles, pp. 316–331. Princeton, NJ: Princeton University Press.

Diamond, J.M. (1975). Assembly of species communities. In *Ecology and Evolution of Communities*, ed. M.L. Cody & J.M. Diamond, pp. 342–444. Cambridge, MA: Harvard University Press.

Drake, J.A. (1991). Community-assembly mechanics and the structure of an experimental species ensemble. *American Naturalist* 137: 1–26.

Feller, W. (1968). *An Introduction to Probability Theory and Its Applications*, 3rd edn., Vol. 1. New York: Wiley.

Fox, B.J. (1985). Small mammal communities in Australian temperate heathlands and forests. *Australian Mammalogy* 8: 153–158.

Fox, B.J. (1987). Species assembly and evolution of community structure. *Evolution and Ecology* 1: 201–213

Fox, B.J. (1989). Small-mammal community pattern in Australian heathland: a taxonomically based rule for species assembly. In *Patterns in the Structure of Mammalian Communities*, ed. D.W. Morris, Z. Abramsky, B.J. Fox, & M.R. Willig, Special Publication 28, pp. 91–103. Lubbock, TX: Texas Tech University.

Fox, B.J. & Brown, J.H. (1993). Assembly rules for functional groups in North American desert rodent communities. *Oikos* 67: 358–370.

Fox, B.J. & Brown, J.H. (1995). Reaffirming the validity of the assembly rule for functional groups or guilds: a reply to Wilson. *Oikos* 73: 125–132.

Fox, B.J. & Kirkland, Jr., G.L. (1992). An assembly rule for functional groups applied to North American soricid communities. *Journal of Mammalogy* 73: 125–132.

Gilpin, M.E. & Diamond, J.M. (1982). Factors contributing to non-randomness in species co-occurrences on islands. *Oecologia* 52: 75–82.

Gotelli, N. & Graves, G. (1996). *Null Models in Ecology*. Washington, DC: Smithsonian Institution Press.

Jorgensen, C.D. & Hayward, C.L. (1965). Mammals of the Nevada test site. *Brigham Young University Science Bulletin, Biology Series* 6(3): 1–81.

Kelt, D.A., Taper, M.A. & Meserve, P.L. (1995). Assessing the impact of competition on the assembly of communities: a case study using small mammals. *Ecology* 76: 1283–1296.

Kotler, B.P. & Brown, J.S. (1988). Environmental heterogeneity and the coexistence of desert rodents. *Annual Review of Ecology and Systematics* 19: 281–307.

Rickens, V.B. (1974). Numbers and habitat affinities of small mammals in northwestern Maine. *Canadian Field Naturalist* 88: 191–196.

Robinson, J.V. & Dickerson, Jr., (1987). Does invasion sequence affect community structure? *Ecology* 68: 587–595.

Simberloff, D. & Dayan, T. (1991). The guild concept and the structure of ecological communities. *Annual Review of Ecology and Systematics* 22: 115–143.

Stone, L., Dayan, T. & Simberloff, D. (1996). Community-wide assembly patterns unmasked: the importance of species' differing geographic ranges. *American Naturalist* 148: 997–1015.

Wilson, J.B. (1995). Null models for assembly rules: the Jack Horner effect is more insidious than the Narcissus effect. *Oikos* 72: 139–144.

Wright, S.J. & Biehl, C.C. (1982). Island biogeographic distributions: testing for random, regular, and aggregated patterns of species occurrence. *American Naturalist* 119: 345–357.

3

Community structure and assembly rules: confronting conceptual and statistical issues with data on desert rodents

Douglas A. Kelt and James H. Brown

Introduction

Two main themes have dominated empirical research in community ecology for the last several decades. On the one hand, the enormously successful experiments of British plant ecologists and of intertidal ecologists such as Connell (1961a, b) and Paine (1966) stimulated many ecologists to perform controlled, replicated, manipulative field studies to investigate direct and indirect interactions. On the other hand, the insightful observations of Hutchinson (1957) and MacArthur (1958) stimulated many other ecologists to explore the relationships between patterns of community organization and the mechanistic processes that produce them. These two approaches have tended to diverge in the kinds of organisms and the spatial and temporal scales studied. The experimentalists tended to work with terrestrial plants, freshwater fishes and amphibians, and intertidal marine organisms, to manipulate on small spatial scales, and to study ongoing, proximate ecological processes. In contrast, those studying patterns of community organization tended to study terrestrial reptiles, birds, and mammals, to do comparative studies over geographic spatial scales, and to be concerned with evolutionary, as well as ecological, processes and timescales. Most studies of community-level patterns and processes have focused on two phenomena: the organization of food webs (e.g., Cohen *et al.*, 1993; Polis & Strong, 1996; papers in Special Feature section of *Ecology* 69, pp. 1647–1676) and the structure of 'guilds' (sets of functionally similar and closely related species; e.g., Cody & Diamond, 1975; Diamond & Case, 1986; Simberloff & Dayan, 1991). The latter studies have often invoked 'assembly rules' to describe the apparently non-random patterns of local coexistence of species, and they have often hypothesized that the rules reflect the ecological and evolutionary consequences of interspecific competition.

Much of the early, influential work on assembly rules focused on three 'model systems': Caribbean *Anolis* lizards (e.g., Schoener, 1969, 1970; Roughgarden, 1995), island birds (e.g., Lack, 1947; Diamond, 1975; Grant, 1986), and desert rodents (e.g., Rosenzweig & Winakur, 1969; Brown, 1975; Reichman & Brown, 1983). There was much in common among these studies. They relied heavily on non-experimental, comparative geographic studies in which some variables, such as the influences of phylogenetic and paleoenvironmental history on the composition of the regional species pool, were kept as constant as possible, while other variables, such as the number, identity, morphology, diet, and habitat affinity of coexisting species, were measured, analyzed, and interpreted. They focused on patterns of non-random assembly, i.e., the ways in which local assemblages of species differed from a random sample of the regional species pool. And, to explain such community structure, they often invoked interspecific competition as the most likely process to have influenced both ecological coexistence and evolutionary divergence of species within a guild.

In the present chapter the extensive information on the seed-eating desert rodents of southwestern North America is used to evaluate the assembly rules that have been proposed for these organisms. We first review the empirical patterns of community structure found in desert rodents. In addition to summarizing the existing literature, several new analyzes are presented which provide more statistically robust conclusions. From an ecological perspective, North American desert rodents are one of the most thoroughly studied groups of organisms, and the extensive data on composition of local communities are complemented by information on many other characteristics, including systematics, phylogenetic and biogeographic history, morphology, physiology, behavior, genetics, life history, and population dynamics (e.g., Reichman & Brown, 1983; Genoways & Brown, 1993). Work in North America stimulated studies of ecology of desert rodents elsewhere in the world. These investigations revealed many similarities, but also many differences. Environmental differences among the deserts and influences of phylogenetic and biogeographic history on the rodents have affected the ways that communities are organized (e.g., Kelt *et al.*, 1996). For the sake of simplicity therefore, our attention will be confined to North American assemblages.

Second, the data and analyses from studies of these desert rodents are used as a basis to discuss some important conceptual and statistical issues that are fundamental to many studies of community assembly and community structure. The conceptual issues concern the nature of assembly rules, and the relationship between empirical patterns and underlying mechanistic processes. Statistical and methodological issues include: the formulation of null hypotheses that are, on the one hand, interesting and informative, and, on the other hand, statistically rigorous and testable; the nature of species pools and the problems of defining them; the spatial and temporal scales at which the rules can be observed and the mechanisms operate; and the problems that arise because the variables are often not independent and the hypothesized mechanisms are not mutually exclusive.

Finally, the information on assembly rules for the desert rodents is used as a basis to comment on two issues: on the one hand, on the contribution of comparative geographic studies of community structure to our current understanding of the interactions within these guilds, and on the other hand, on the more general contributions of the assembly rule approach to our current understanding of the phenomena of coexistence, convergence and divergence, and diversity of species.

Before proceeding, it is necessary to be clear what we are talking about. For the purposes of this chapter, assembly rules will be defined simply as empirical patterns of community organization, synonymous with what will also be called community structure or community organization. Thus, the existence of an assembly rule implies a non-random pattern of community structure. To claim the existence of a rule implies that the rule has been tested against the null hypothesis of random assembly and rejected, usually by comparing the observed community organization with that expected on the basis of random assembly from some appropriate species pool. It must be emphasized that assembly rules are descriptive, empirical characterizations of community composition. While the existence of a non-random pattern implies the operation of some deterministic mechanism, an assembly rule is not taken as necessarily implying the operation of any particular process, such as interspecific competition, nor as implying the operation of any particular temporal sequence of species assembly or successional process. In desert rodents, as in other organisms, investigators have claimed to have demonstrated several different kinds of assembly rules characterized by different variables, measured in a variety of different ways. Examples of assembly rules are presented that describe three kinds of empirical patterns: in species richness, in morphological and functional composition, and in species identity.

The rules: patterns of community structure

Patterns of species richness

Perhaps the simplest measure of community structure is species richness, a tally of the number of species present. Such a tally avoids the complications associated with indices of diversity which include measures of relative abundance of the species and often creates problems with statistical analysis and inter-pretation (see Pielou, 1969; Magurran, 1988). Both diversity indices and simple species richness measured for a local community will depend on the sampling effort and the number of individuals tallied. Such sampling problems can largely be avoided, however, by restricting comparisons to communities that have been sampled with equal effort and using some standardized con-vention to tally only the common species. When this has been done for desert rodent assemblages, three kinds of patterns have been documented.

Local species richness increases with increasing complexity of habitat structure, and other characteristics of the soil and vegetation

Some variables, such as composition of the species pool and climate, can be held approximately constant and features of the habitat allowed to vary, by restricting comparisons to a local region. When this is done, rodent species diversity generally varies along gradients of soil structure, from relatively few species on shallow, rocky soils, to larger numbers on deeper, sandy soils (e.g., Rosenzweig & Winakur, 1969; Brown, 1975; Brown & Harney, 1993). Species richness also varies with vegetation structure. In general, there is a positive relationship between rodent species diversity and foliage height diversity or some other measure of structural complexity of vegetation (e.g., Rosenzweig & Winakur, 1969; Rosenzweig, 1973; Rosenzweig *et al.*, 1975; Price, 1978; Thompson, 1982a).

The relationship between rodent species diversity and characteristics of habitat heterogeneity apparently reflect the differential abilities of species to exist and coexist in distinctive habitats and microhabitats (Price, 1978; Thompson, 1982b). This is apparent at different spatial scales. Among habi-tats within regions, certain species and larger functional and taxonomic groups vary in habitat specificity. For example, the rock pocket mouse (*Chaetodipus intermedius*) is typically confined to boulder-strewn hillsides, while bipedal kangaroo rats (*Dipodomys* spp.) and kangaroo mice (*Microdipodops* spp.) are absent from rocky soils, and the deer mouse (*Peromyscus maniculatus*) occurs in an enormous range of habitats. Within these kinds of macrohabitats, rodents forage differentially in distinctive microhabitats. For example, the bipedal species tend to use open patches of bare ground, while pocket mice forage under vegetative cover. As a consequence of both macro- and microhabitat selection, highest rodent species diversity typically occurs on stabilized sand dunes and other deep, sandy soils, where the vegetation exhibits both vertical and horizontal heterogeneity: a spatial mosaic composed of patches of bare ground and plants of sizes ranging from small herbs to large shrubs or small trees.

Local species richness increases with increasing productivity

In arid regions productivity is closely correlated with actual evapotranspiration and with precipitation (Rosenzweig, 1968; Brown *et al.*, 1979). Productivity can be varied, and other variables held relatively constant by making comparisons along gradients of increasing precipitation within a major desert region with a similar species pool and among structurally similar habitat types. Brown (1973, 1975; see also Hafner, 1977) made such comparisons along productivity gradients, comparing sand dune habitats in the Mojave and Great Basin Deserts and rocky hillside and sandy flatland habitats in the Sonoran Desert. He found that about 50% of the variance in the number of common species was positively correlated with mean annual precipitation. This positive relationship holds so long as the comparisons are restricted to desert habitats characterized by a spatial mosaic of bare ground and isolated perennial plants (predominantly woody shrubs and succulents). But, as precipitation and productivity continue to increase, desert shrubland habitat is replaced by semi-arid grassland and rodent species richness declines abruptly (Brown, 1975; Owen, 1988).

Local species richness is lower in isolated areas where the size of the species pool is reduced

In geographically isolated desert basins, habitats with otherwise similar soils and vegetation possess fewer species than expected on the basis of productivity (Brown, 1973, 1975). These areas have been isolated since the Pleistocene by barriers of inhospitable habitats that have prevented the colonization of several desert rodent species. The influence of this reduced species pool is reflected in reduced species richness in local communities.

Patterns of morphological, phylogenetic, and ecological similarity

Since Darwin's commentary (1859) and Lack's (1948) classic study on character displacement in Darwin's finches (see also Hutchinson, 1959), many observers have noted that those species that coexist in local communities appear to be more different in structure and function than closely related populations that occur allopatrically. This pattern is readily apparent in desert rodents, and can be characterized and quantified in several ways.

Local communities contain species that are more different in body size and other morphological characteristics than expected by chance

Differences in size among coexisting desert rodent species have been noted by many investigators (e.g., Grinnell & Orr, 1934; Rosenzweig & Sterner,

1970; Brown & Lieberman, 1973; Brown, 1975; M'Closkey, 1976, 1978; Price, 1978, 1983). Bowers and Brown (1982) compiled data on 95 local communities of granivorous desert rodents, and showed that species of similar body size co-occurred significantly less frequently than expected on the basis of chance. Bowers and Brown (1982) also showed, however, that species of similar body size had less overlap in their geographic ranges than would be expected on the basis of chance. This is especially obvious in the case of species in both the largest and smallest body size category, which have almost perfectly non-overlapping geographic distributions. This is an important point, because it means that the regional pool of species from which local communities are assembled is already structured with respect to body size (and other attributes as shown below).

Since Bowers and Brown's study, accumulation of larger data sets and ease of performing computerized randomization tests of null hypotheses have permitted more rigorous analyses. In Table 3.1 a similar analysis is applied to two data sets: rodents coexisting in 115 samples of local habitat on the Nevada Test Site (Jorgensen & Hayward, 1965), and in 201 local sites from throughout the arid southwestern United States (Brown & Kurzius, 1987, as emended in Morton *et al.*, 1994). In the analysis of the southwestern US data set the species pool for each sample site has also been restricted to just those species whose geographic ranges overlap that locality, thus removing the influence of the non-overlapping geographic distributions on the composition of the regional pool. (The Nevada Test Site is such a restricted geographic area – approximately 3500 km^2 – that all of the sample sites can be assumed to be potentially accessible to all species in the pool.) The results of these new analyses mirror those of Bowers and Brown (1982): species of similar body size (ratio of masses < 1.5) coexist much less frequently than expected on the basis of random associations of the species for both the regional (Nevada Test Site) and larger geographical (southwestern North America) assemblages.

Dayan and Simberloff (1994) have shown that two regional heteromyid faunas, from southeastern Arizona and northwestern Nevada, exhibit highly non-random ratios of morphological traits. The most regular pattern is in the width of the incisors, but this trait is correlated with several others, including body size and cheek pouch volume. This result is largely complementary to the patterns in body size (above) and other characteristics (below), but it raises two interesting issues. First, unlike ratios of body size, which show the clearest patterns within local communities, the ratios of incisor widths are more even among the species in the entire regional pools than among the actual assemblages. Second, since Dayan and Simberloff also show that the sizes of all morphological characteristics of these rodents tend to be posi-

Table 3.1. *Test of the null hypothesis that local coexistence and geographic overlap or granivorous rodents are independent of body size*

		Frequency of pairwise co-occurrence of species of	
		similar size (body mass ratio < 1.5)	different size (body mass ratio > 1.5)
Nevada test site 115 sites, 10 species	Observed	9	228
	Expected (Potential)	57.93 (1265)	179.07 (3710)
		$\chi^2 = 54.70\ P < 0.001$	
Southwestern US 201 sites, 28 species	Observed	217	679
	Expected (Potential)	270 (4703)	626 (10 904)
		$\chi^2 = 14.89\ P < 0.001$	

Reanalyzed from Jorgensen and Hayward, (1965) and Brown and Kurzius, (1987). Expected values were calculated as expected$_i$ = ((potential$_i$ / Σ potential) × Σ observed), where potential$_i$ is the number of potential pairwise co-occurrences of similar or of different-sized species, based on the pool of species with access to each site. For the Nevada test range, all sites shared a common species pool. For the Southwestern US data set, species pools included those granivorous rodent species whose geographic ranges (from Hall, 1981) included the site. Because the number of potential co-occurrences (in parentheses) is vastly greater than the observed, expected values were proportionalized to sum to the same number as to the observed.

tively correlated, the observation that different traits appear to show clearer patterns at different spatial scales raises intriguing questions about the underlying mechanisms producing the observed community structure.

Closely related species coexist less frequently than expected by chance

Because phylogenetically related species often share many ecological and morphological characteristics, they might be expected to co-occur less frequently than they would with more distantly related species. Using the Nevada test site and the southwestern US data sets, the frequency with which closely congeneric and more distantly related species co-occurred in local communities was compared (Table 3.2). For both data sets, it is clear that congeners co-occur significantly less frequently than expected by chance. As in the previous analysis, the species pool for each sample community in the southwestern US analysis included only those species whose geographic ranges overlapped the site.

Table 3.2. *Test of the null hypothesis that local coexistence and geographic overlap or granivorous rodents are independent of taxonomic relatedness*

		Frequency of pairwise co-occurrence of species in	
		Same genus	Different genus
Nevada test site 115 sites, 10 species	Observed	144	372
	Expected (Potential)	100.96 (2070)	415.04 (8510)
		$\chi^2 = 22.82 \; P < 0.001$	
Southwestern US 201 sites, 28 species	Observed	222	1686
	Expected (Potential)	312.34 (5019)	1595.67 (25 641)
		$\chi^2 = 31.24 \; P < 0.001$	

Reanalyzed from Jorgensen and Hayward (1965) and Brown and kurzius (1987). Expected values were calculated as expected$_i$ = (potential$_i$ / Σ potential) $\times$ Σ observed), where potential$_i$ is the number of potential pairwise co-occurrences, based on the pool of species with access to each site. For the Nevada test range, all sites shared a common species pool. For the Southwestern US data set, species pools included those granivorous rodent species whose geographic ranges (from Hall, 1981) included the site. Because the number of potential co-occurrences (in parentheses) is vastly greater than the observe, expected values were proportionalized to sum to the same number as to the observed.

Species in the same functional group coexist less frequently than expected by chance

Fox (1987; see also Fox & Brown, 1993; Kelt *et al.*, 1995; Fox, this volume) developed an assembly rule for Australian small mammals that specified that species tended to be drawn sequentially from different taxonomic or functional groups. Functional groups are composed of species which use their environment in a similar manner (see below). Fox's rule divides all possible combinations of functional groups into two categories: favored states, in which the number of species in the different functional groups differs by no more than one; and unfavored states, in which the number of species in the different groups differs by two or more. Monte-Carlo techniques are used to draw species at random from the pool to evaluate the null hypothesis that neither favored nor unfavored states are observed more frequently than expected by chance.

Fox and Brown (1993; see also Wilson, 1995; Fox & Brown, 1995) applied this rule to North American desert rodents (both the Nevada Test Site and southwestern US data sets), classifying the granivorous species *a priori* into one of three functional/morphological groups: (a) bipedal with cheek pouches (kangaroo rats and kangaroo mice); (b) quadrupedal with cheek pouches (pocket mice); or (c) quadrupedal without cheek pouches (cricetine rodents). For both data sets, favored states were observed significantly more frequently and unfavored states significantly less frequently than expected by chance. Local communities tended to be comprised of functionally dissimilar species. But, in their analysis of the southwestern US data set, the species pool for all sites included all of the 28 species that inhabited any of the 200 sample sites throughout the large geographic area.

Stone *et al.* (1996, see also Simberloff *et al.* this volume) challenged the assembly rule presented by Fox and Brown (1993), claiming that their results were artifactual and that their analysis '... failed to find evidence that inter-specific competition or a deterministic assembly rule shaped local community composition.' They also claimed that geographical structuring (the Narcissus effect; Colwell & Winkler, 1984) had a much greater influence on local community structure than did local interspecific interactions. While implying that a Narcissus effect could account for the degree to which local assemblages follow the assembly rule (presumably obviating the analysis of Fox and Brown, 1993), Stone *et al.* made no attempt to provide a more refined analysis, claiming that '... detailed information on species' ranges cannot always be determined from the literature.' This simply is not true. Although compiling such information is somewhat tedious and time consuming, it is readily accomplished. To address concerns raised by these authors, the southwestern US data has been reanalyzed by defining species pools for each site as consisting only of those species whose geographic ranges encompass the site (this is the analysis that Simberloff *et al.* (this volume) claimed 'will be an interesting study' and upon which he and Brown bet a beer (see Simberloff, this volume)). This analysis, based on 2000 simulations, clearly demonstrated that these communities were significantly non-random assemblages (Fig. 3.1a). The observed number of favored states was significantly greater than would be expected if species were drawn at random according to the number of species in the regional pool in each functional group. The methodology of Kelt *et al.* (1995) was also applied and simulations conducted that assumed various levels of negative interspecific association among the species within functional groups (Fig. 3.1a–d; see Appendix for details), to obtain a maximum likelihood estimate for the mean strength of the observed associations. Since this estimate was an average value over all 201 sites and 28 species, it is most useful in allowing the statistical power of our analysis to be determined,

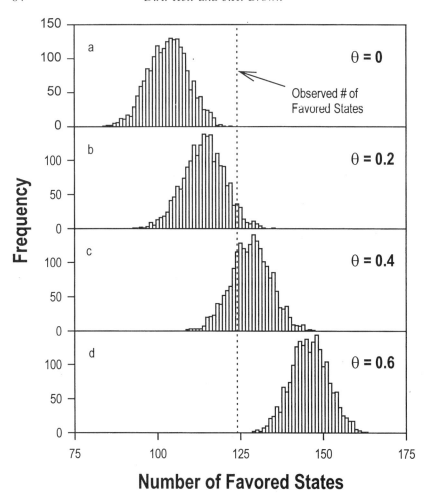

Number of Favored States

Fig. 3.1. Frequency distributions of the number of communities expected to be in favored states, using four levels of interspecific interaction. Each histogram represents 2000 iterations of the null model, in which 201 artificial communities were simulated using the species pools derived from species' geographic ranges. The vertical dotted line indicates the observed number of favored states. Panel a demonstrates that the null hypothesis of no interspecific interactions ($\theta = 0$) is rejected. Increasing levels of interspecific interaction are represented by increasing values of θ.

i.e, the ability to detect existence of favored states. The maximum likelihood strength of negative association was estimated as $\hat{\theta} = 0.33$ (Fig. 3.2). From this, it was determined that the statistical power of our estimate was extremely high, above 94% (Fig. 3.3). This gave us confidence that these communities are highly structured non-random assemblages. Indeed, a relatively simple

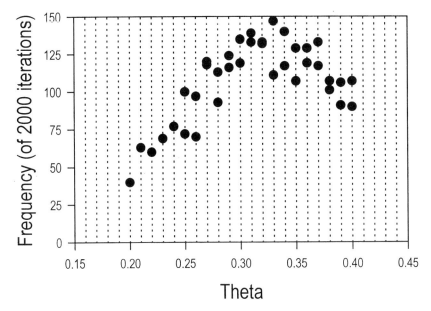

Fig. 3.2. Relation between the number of iterations of 201 sites that yielded 123 favored states and the value of theta. The peak of this distribution represents the value of theta that best describes the observed set of sites. This peak occurs at approximately $\theta = 0.33$, and provides an maximum likelihood estimate of the real strength of interaction in these communities.

model of community assembly, based on a moderate level of negative associations among functionally similar species, described the structure of these communities surprisingly well.

This methodology has also been applied to the Nevada test site data (Brown *et al.* ms). For this data set we also developed two sets of species pools. One species pool was based on the geographic distribution of species and was conceptually identical to the species pools used for the southwestern US analysis. The second set of species pools was based on the habitat requirements of these species, and consisted of all species found within the Nevada Test Site who were known to occur in the type of habitat present at a given site. Because the spatial distribution of two species (*Dipodomys deserti* and *Reithrodontomys megalotis*) were not delineated by Jorgensen and Hayward (1965) we have not included them in the present analyses; however, similar analyses have been conducted with these species conservatively allocated to sites with appropriate habitat, and the results were qualitatively identical to those presented here. The observed number of favored states in every analysis was significantly greater than expected at random (Table 3.3), suggesting

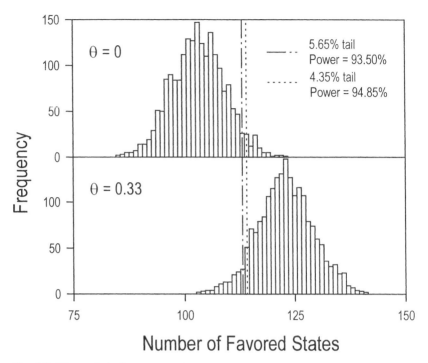

Number of Favored States

Fig. 3.3. The statistical power can be estimated by determining the number of θ = 0.33 simulations that lie within the critical (5%) region of θ = 0 simulations. Using the southwestern US data set, it can be seen that the null hypothesis of no interaction is rejected with a power of about 94%. In other words, there is a 94% probability of being correct when the null hypothesis of no interaction among species during community assembly is rejected.

that even at this reduced geographical scale these communities are highly structured and non-random assemblages.

Finally, we note that the idea that a Narcissus effect occurs for these rodents is hardly a novel observation. Bowers and Brown (1982) pointed out that 'species of similar (body) size ... coexist less frequently in local communities and overlap less in their geographic distributions than expected on the basis of chance, suggesting that their co-occurrence is precluded by interspecific competition.' Additionally, as has been summarized above, there is a tendency for similar species to occur less frequently than expected by chance, whether similarity is measured in terms of body size, taxonomic affinity, or functional group membership, and these patterns apply at the spatial scales of both local communities and regional faunas (Brown & Bowers 1984; Brown, 1987; Brown & Harney, 1991, this chapter). What few authors appear

Table 3.3. *Nested subset structure of small mammal communities of 13 sand dunes in the Mojave and Great Basin Deserts*

Species	Dune (sample site) number													Total dunes
	1	5	6	7	2	3	4	8	9	10	11	12	13	
Dipodomys deserti	X	X	X	X	X	X	X	X	X	X	X	X	X	13
Dipodomys merriami	X	X	X	X	X	X	X	X	X					9
Perognathus longimembris	X	X	X	X	X	X	X			X				8
Peromyscus maniculatus	X		X	X	X	X	X							6
Dipodomys ordii	X	X	X	X	X									5
Reithrodontomys megalotis	X	X	X		X		X							5
Microdipodops pallidus		X	X	X		X								4
Dipodomys microps				X	X									2
Perognathus parvus	X													1
Chaetodipus penicillatus									X					1
Peromyscus crinitus		X												1
Microdipodops megacephalus	X													1
Total species	8	7	7	7	6	5	5	3	3	2	1	1	1	

Reanalyzed from Brown, 1973.
As with the sites discussed in the text, these communities show highly nested structure (system temperature = 7.2°, $P(T < 7.2°) = 1.69 \times 10^{-6}$, based on 1000 iterations of the program (Patterson & Atmar, 1995).

to consider is that patterns occurring at local and geographical scales are linked by both 'top-down' and 'bottom-up' processes. The former include such factors as a Narcissus effect. However, the latter include such mechanisms as interspecific interactions (including competition) which may feed upwards to influence the structure and composition of regional biotas (e.g., Bowers & Brown 1982), thereby *producing* a Narcissus effect.

Patterns in the identities of species

So far, our concern has been with characteristics of communities, total species richness, or morphological, phylogenetic, or functional similarity among species. But, there are also patterns in the frequency with which particular species occur in local communities. As implied in our discussion of species richness and habitat use, species differ in the sizes of their geographic ranges and in the proportion of sample sites within their ranges where they occur. Many of the assembly rules described in early studies of community structure were concerned with the associations of Latin binomials, i.e., with the patterns of occurrence and co-occurrence of particular species and combinations of species (e.g., Diamond, 1975, 1984; Connor & Simberloff, 1979, 1983, 1984; Gilpin & Diamond, 1981, 1982, 1984; Diamond & Gilpin, 1982). Two patterns are particularly apparent in desert small mammals.

The communities exhibit a nested subset structure

One pattern that appears pervasive across many communities is a nested subset or core-satellite structure (e.g., Table 3.3; see Hanski, 1982; Patterson & Atmar, 1986; Patterson, 1987, 1990; Patterson & Brown, 1991). Some species are widely distributed and occur in many local communities, whereas other species have more restricted distributions and occur only in a subset of the local samples. Patterson and Brown (1991) demonstrated significant nestedness for the North American desert rodent fauna at the scale of major desert regions and the entire southwestern US. Here, the 201-sample southwestern US data set has been reanalyzed using an improved method for calculating deviation from nestedness (Atmar & Patterson, 1995; see Atmar & Patterson, 1993) and results are similar to those presented by Patterson and Brown (1991), except that even more of the assemblages are significantly more nested than expected from random simulations (compare Tables 3.4 and 3.5 to Tables 1 and 2 in Patterson & Brown, 1991).

Certain pairs and other combinations of species coexist either less or more frequently than expected by chance

Another species specific pattern of assembly is that certain combinations of species are strongly positively or negatively associated with respect to frequency of co-occurrence in local communities. In general, data on desert rodents have not been thoroughly analyzed in this way (but see Bowers & Brown, 1982; Brown & Kurzius, 1987; Patterson & Brown, 1991; Morton *et al.*, 1994), but this pattern is implicit in many of the other patterns described above. Thus, if, in general, congeners and species of similar body size and functional group tend to coexist less frequently than expected by chance, it follows that particular pairs of species that belong to the same genus, are of similar size, and/or are members of the same functional group, will often show negative associations. Conversely, local communities will tend to be comprised of species that often have positive associations and are in different genera, functional groups, and body size categories.

Table 3.4. *Representation of 201 sites and 29 species of granivorous rodents among deserts and habitats, in the revised data set*

	All habitats	Desert shrub	Desert grassland	Sand dunes	Shrubsteppe
All deserts	201,29	136,27	29,15	20,14	16,6
Great Basin	56,14	21,13	5,3	14,13	16,6
Mojave	52,14	49,14		3,5	
Sonoran	48,14	45,14		3,1	
Chihuahuan	45,17	21,14	24,14		

Data from Brown and Kurzius (1987), as modified in Morton *et al.* (1994).

The processes: evaluating alternative mechanisms

Statistical, phylogenetic, and functional relationships

The previous section makes clear something that should be obvious by now: the various patterns documented above are not independent of each other. Species in the same genus should be closely related phylogenetically, and they would be expected to be more similar in body size and other morphological characteristics and in ecological attributes than more distantly related species. A general feature of desert rodent community assembly therefore, is that species that are similar by some measures tend also to be similar in other ways, and similar species tend to overlap in geographic ranges and to coexist in local communities less frequently than more different species – and less frequently than expected by chance. The fact that assemblages tend to be formed of well-differentiated species, presumably with correspondingly distinctive requirements for resources (including macro- and microhabitats), contributes to the patterns of nested subset structure and species diversity.

It is trendy to try to pull out the component of similarity in attributes of species that can be attributed to phylogenetic relationships and to suggest that the remaining similarities and differences reflect 'ecology'. We have reservations about these approaches in general, and in particular about their utility for disentangling the correlations among those traits that are important in community assembly. Removing effects of phylogenetic relationships may simultaneously remove effects of ecological similarity. While it is possible to identify correlations between phylogenetic affinity and similarity in many traits, it is much more difficult to attribute causality. (Do the powerful kidneys shared by most heteromyid rodents reflect the constraint of common ancestry or the selection of a common arid environment?) There are, however, some cases where phylogenetic analysis could contribute importantly to understanding community assembly. This topic will be returned to below.

Table 3.5. *Patterns of community structure in 201 sites in arid North America*

	All habitats	Desert shrub	Desert grassland	Sand dunes	Shrub-steppe
All deserts	15.37°*; −13.97*	14.96°; −11.66	16.06°; − 3.33	21.25°; − 4.02	12.27°; −3.57
Great Basin	12.52°; − 9.27	14.21°; − 5.14	6.88°; − 0.83	25.08°; − 3.45	12.27°; − 3.51
Mojave	21.26°; − 6.16	20.47°; − 6.02		42.45°; + 1.1	
Sonoran	15.14°; − 5.68	13.96°; − 6.02		0°; 0	
Chihuahuan	16.87°; − 5.97	22.92°; − 3.61	13.42°; − 3.91		0

*Because the nestedness temperature calculator (Patterson & Atmar, 1995) is limited to arrays no larger than 200×200 cells, the analyses for all habitats required the removal of one site for analysis. The results presented for this analysis are mean values from three simulations in which a different random site was removed from the data set.

Presented are system temperatures, followed by the number standard deviations between observed system temperatures and the mean of 1000 simulated systems. All analyses were significantly more structured than simulated systems with the exception of Mojave sand dune rodents ($P = 0.0864$) and Sonoran sand dune rodents, for which the sample size precluded any statistical power.

It is also important to recognize that similar patterns of community structure characterize desert rodent assemblages across a range of spatial scales. The same differences among species that are apparent at the level of communities of species coexisting within a few hectares of relatively homogeneous habitat also tend to characterize the desert rodent faunas that inhabit much larger geographic regions, areas of millions of square kilometers. From an analytical perspective, this means that, while local communities represent non-random assembly of species from the regional species pool, the regional pool is itself a non-random assemblage of species. This is presumably because local ecological processes influence the distributions of species on larger spatial scales and thereby affect the composition of regional and geographic species pools, and vice versa. More about this later.

Causal mechanisms: interspecific competition and/or allopatric speciation

So far, assembly rules have been discussed simply as empirical patterns. The patterns have been tested against appropriate null hypotheses and shown to require some kind of non-random assembly. As stated above, however, none of the rules assumes the operation of interspecific competition or any other mechanistic process. This seems the most logical and rigorous way to proceed. Having demonstrated the patterns, the possible mechanisms can now be evaluated as alternative, but not necessarily mutually exclusive, hypotheses.

The most obvious mechanism is interspecific competition. If competition can be assumed to be most severe between species that are closely related and similar in morphology, physiology, and behavior, and therefore in ecology, then it provides a single, parsimonious explanation for nearly all of the assembly rules described above. This includes patterns of nestedness. If competition is greatest among similar species then species-poor communities will be comprised largely of dissimilar species. As local diversity increases, however, more similar species are able to coexist, and a nested pattern emerges. On the other hand, if competition causes a checkerboard or similar pattern of species co-occurrences, then a nested structure is not likely. In the data set on North American desert rodents, the former scenario is true, and competition apparently contributes to a nested community structure.

Further, the competition hypothesis does not require attempts to disentangle the correlations among different kinds of 'similarity', because similarity in any or all of the above characteristics of species (e.g., body size, phylogenetic relationship, functional group) can be expected to increase competition and decrease the probability of facilitated coexistence.

There is abundant experimental evidence that desert rodent species do compete (e.g., Rosenzweig, 1973; Schroder & Rosenzweig, 1975; Munger & Brown, 1981; Freeman & Lemen, 1983; Frye, 1983; Brown & Munger, 1985; J.S. Brown, 1989; Heske _et al._, 1994; Valone & Brown, 1996), and even that competition influences community assembly by affecting probabilities of local colonization and extinction (Valone & Brown, 1995). Thus, more recent evidence obtained by manipulating local communities provides independent verification of the mechanism hypothesized by earlier investigators to explain comparative geographic patterns of community structure.

Competition is not the only mechanism that might be invoked to explain the above patterns of faunal assembly. Another process that could explain most of the above assembly rules is allopatric speciation. If the assumption is made that species form in geographic isolation, that they shift their geographic distributions slowly, and diverge over time in their morphological, physiological, and behavioral characteristics, then it follows that the most closely related and similar species will tend not to overlap in their geographic ranges and therefore not to coexist in local communities. Such a legacy of allopatric speciation could account for the considerable structure observed at a geographic scale, in the composition of regional species pools. It is hard to see, however, how it could account for the non-random assembly of local communities from these regional pools.

Just because it is not sufficient to account for all observed community structure, allopatric speciation should not be dismissed as an important contributing mechanism. If it does play a significant role, we would predict that the most closely related species are the least likely to occur together. While the negative associations among congeneric species pairs supports this prediction, a proper phylogenetic analysis would provide much more resolution. In particular, while it would be generally expected that closest relatives would be most similar in the morphological and functional characteristics that are the basis of the above assembly rules, this need not always be true. Indeed, current reconstructions of phylogenetic relationships among the heteromyid rodents suggest that some species of similar body size and in the same functional group are not particularly closely related. For example, the large kangaroo rats (_Dipodomys_ spp. with body mass > 100 g) which have nearly perfectly contiguous and allopatric geographic ranges (Bowers & Brown, 1982), appear to be more closely related to smaller kangaroo rats, with which they occur sympatrically and coexist in local communities, than they are to each other (Patton & Rogers, 1993). Bowers and Brown (1982) found that sister species accounted for only a modest proportion of the pairs of species of similar body size that showed negative associations in both overlap of geographic ranges and frequency of local coexistence.

It is important to note that the allopatric speciation hypothesis is not mutually exclusive of, and may even be complementary to, the competition hypothesis. Everything else being equal, the most closely related species are likely to be most similar in form and function, and hence likely to be strong competitors. Thus, it is likely that competitive exclusion among close relatives plays a major role, not only in restricting membership in local communities, but also in limiting overlap in geographic ranges. This is one way that local ecological interactions can affect the composition of regional species pools.

Causal mechanisms: ecological sorting and/or evolutionary character displacement

A related mechanistic issue concerns the degree to which the above assembly rules reflect some evolutionary character displacement as opposed to ecological sorting. To what extent does the tendency of coexisting species to be more different than expected by chance reflect adaptive divergence resulting from, and perhaps serving to promote, coexistence? And, to what extent does it simply reflect the ecological compatibility – the differential ability to colonize and persist in local communities – of those species whose characteristics were shaped primarily by other selective pressures. This question warrants further attention. Brown (1975) suggested that intraspecific geographic variation in body sizes of widely distributed species may reflect character displacement due to local interactions with other species. However, analysis of geographic variation in the widely distributed kangaroo rat, *Dipodomys merriami*, suggested that body size was correlated with environmental productivity but not related to geographic overlap or local coexistence with other kangaroo rat species (J.H. Brown, unpublished data).

The issue of character displacement has been revived by Dayan and Simberloff's (1994a) demonstration of extremely uniform ratios of certain morphological traits, most notably width of incisors in rodents, within regional species pools. Since only a subset of the species in these pools coexist in any local community, and the exact combination of species varies depending on habitat (see above), what mechanism can account for the uniform ratios? Since Dayan and Simberloff analyzed several morphological traits in only two species assemblages, it is possible that the near perfectly uniform ratios in one trait, incisor width, are due to chance. If not, then the most likely explanation seems to be some kind of 'diffuse character displacement' as they suggest. If incisor width plays a major role in resource use and strongly affects interspecific competition, then having a substantial and uniform ratio of incisor widths among all the species in the regional pool would insure that all combinations of locally coexisting species assembled from that pool would differ by at least that critical amount. This is certainly a plausible mechanism, and it suggests another way in which local interactions can 'feed up' to affect the composition of the regional species pool. While both evolutionary character displacement and ecological sorting could play a role in structuring the differences among the species in the regional pool, it is hard to imagine ecological sorting alone producing such precisely uniform ratios.

Dayan and Simberloff's findings also raise interesting questions about the statistical correlations and functional relationships among different characteristics of co-occurring species. Since the dimensions of most morphological traits are correlated with each other and with overall body size, we need to determine which ones most strongly influence different kinds of ecological interactions and thereby affect coexistence, and which ones are most subject to selection for character displacement or some other evolutionary trend? There is abundant observational and experimental evidence that in desert rodents both resource use, which is important in exploitative competition, and aggressive dominance, which is critical in interference competition, are correlated with body size (e.g., Rosenzweig & Sterner, 1970; Brown & Lieberman, 1973; Brown, 1975; M'Closkey, 1976; Frye, 1983; Price, 1983). Thus, it is perhaps not surprising that body size appears to play such a large role in the assembly of local communities. This does not mean, however, that other traits, such as incisor width and cheek pouch volume, that are correlated with body size, do not also affect the outcome of ecological interactions and may be subject to selection. And certainly, traits such as mode of locomotion (e.g., bipedal vs. quadrupedal, see functional groups above), that are not closely correlated with body size, also affect microhabitat and resource use and thereby strongly influence interspecific interactions and coexistence.

Discussion and synthesis

Statistical and methodological issues: distinguishing pattern from randomness and characterizing assembly rules

The long debate over the existence of community structure and the validity of various assembly rules (e.g., Strong *et al*., 1984; Diamond & Case, 1986; Gotelli & Graves, 1995) testifies to the critical importance of rigorous statistical methods. If an assembly rule describes a nonrandom pattern of community organization, then it is important to test and reject the null hypothesis that the apparent organization is not simply a random assemblage of species. The human senses and brain are so attuned to perceive pattern, that they may perceive it when it is not there. While it might seem a simple matter to erect and test null hypotheses, there are many statistical and conceptual issues that must be addressed. These are considered in detail in the excellent recent book by Gotelli and Graves (1995). Some of these issues will be touched upon here only insofar as they are particularly well illustrated by studies of assembly rules in desert rodents.

There are many possible 'null' or alternative hypotheses

As any statistician will be quick to point out, the question of whether some empirical data differ from a random distribution must be qualified by how one defines random. Most tests of null hypotheses in community ecology involve comparison of an observed distribution of values either with a particular analytical random distribution or with a randomized distribution of data compiled by computer simulation. An example of the former is Dayan and Simberloff's (1994) test for non-random morphological ratios by comparison with the Barton and David statistic. An example of the latter is Fox and Brown's (1993) test for Fox's assembly rule by randomly drawing species from the pool to estimate the expected probability of favored and unfavored states. Both approaches are equally valid, but the applications of the former are limited to cases when a particular analytical form of the random hypothesis can be assumed.

The issue of 'random with respect to what?' also means that an observed empirical distribution can potentially be compared to multiple random distributions. This is a problem in analytical statistical tests, because different 'random' distributions can be expected (e.g., normal, lognormal, negative binomial, and broken stick, to name only a few) and different test statistics can be applied (e.g., different parametric and nonparametric tests, each of which of have unique assumptions). To reiterate using the above example, Dayan and Simberloff (1994) assume that the null distribution of morphological traits is equiprobable on a logarithmic scale, and they use the Barton–David test to compare observed and expected null distributions. This issue is also a problem with randomization tests, because different sets of data can potentially be randomized (e.g., different species pools can be assumed) and different kinds of random draws can be taken from the pool (e.g., species or their characteristics can be drawn with or without replacement, and with equal frequencies or with frequencies intentionally biased to account for the internal structure of the data, e.g., proportional to the number of sites inhabited by each species or to the total number of individuals in the sample).

Related issues concern the particular question to be addressed and the design of the study, and include: the spatial (and temporal) scale of sampling; the level of structure that may be present at larger (or smaller) scales; and the kinds of random processes that are assumed. In randomization tests a critical step is determining the appropriate species pool. There is no single 'right' answer to this question, but the null hypothesis can often be accepted or rejected depending on what choice is made. For example, local communities of desert rodents are more likely to appear to be nonrandom assemblages if they are compared to some global rather than a more regional species pool, because, as we have shown, the regional pools are themselves already 'structured' with respect to body sizes, congeners, and functional groups. Stone *et al.* (this volume) would apply very restrictive criteria in testing for structure of local communities, taking into account both the geographic ranges and the relative frequency of local occurrence of species. There is nothing wrong with this, but it deliberately ignores a high degree of structure that is already present in the assemblages. This is not necessarily the only or most appropriate way to evaluate the hypothesis that the communities are structured by interspecific competition. It has been shown above how local competitive interactions can 'feed up' to affect the composition of species pools at larger scales, and this may strongly bias such restrictive tests.

Tests differ in statistical power

The power of tests of the null hypothesis depend on the nature of the test itself (see above) and the sample sizes. Applying different tests to the same data sets can give different apparent 'answers'. For example, Simberloff and Boecklen (1981; see also Dayan & Simberloff, 1994) applied the Barton–David test to the body size distributions in several of the rodent communities reported in Bowers and Brown (1982) and found that only a few had more uniform size ratios than expected by chance. The Barton–Davis test must be applied to communities one at a time, and its power to reject the null hypothesis is low, especially if the number of coexisting species is relatively small. Rejection when using the Barton–David test, however, usually implies a highly non-random distribution. A different approach is to test for a repeated non-random pattern of body size distributions across a large sample of different communities (e.g., Bowers & Brown, 1982; Hopf *et al.*, 1993; this chapter). This method has a much greater power to detect distributions that are less conspicuously non-random. Again, neither approach is 'right' or 'wrong': they test for structure at different levels of organization (within single communities vs. across multiple communities) with correspondingly different power. However, in the aftermath of all the debate over assembly rules, it is important to remember that the power of statistical tests vary, so that failure to reject one null hypothesis does not necessarily mean that an observed community is a random assemblage.

Many of the measures of community structure may not be independent

This is a problem at several levels. As pointed out above, there are statistical correlations and functional relationships among many of the characteristics used to assess community structure, e.g., body size, incisor width, functional group, and phylogenetic relationship. This means that it is difficult to determine which trait or combination of traits provide the mechanistic basis for the ecological sorting and evolutionary divergence that has produced the observed structure. Probably the way to address this problem is to develop alternative, mutually exclusive mechanistic hypotheses based on the operation of a specific mechanism (not an easy task) and then to subject these hypotheses to rigorous independent tests, using experimental methods if possible.

A second problem is that the composition of communities and the characteristics of the environments in which the assemblages occur are typically autocorrelated over both space and time. How far apart in space or distant in time must samples be taken to insure that they are statistically independent? A related and intertwined problem is that the same single species and combinations of multiple species typically occur repeatedly, even in the most widely separated samples. Thus, for example, the kangaroo rat, *D. merriami*, and the deer mouse, *P. maniculatus*, are habitat generalists with very large geographic ranges that occur together in many communities throughout the southwestern desert region. How should we deal with such redundancy? The literature of statistics and community ecology offers little guidance for handling these problems. On the one hand, such repeated occurrences probably should not be regarded as statistically independent events, but on the other hand, the multiple co-occurrences of certain pairs or other combinations of species almost certainly conveys important information about their interactions and compatibility, and hence about community assembly. These difficult issues warrant attention from both statisticians and community ecologists. For the moment, we suggest that multivariate and computer simulation techniques offers some promise for disentangling the complex patterns of covariation.

Conceptual issues: pursuing the relationship between pattern and process

We reiterate that assembly rules are here defined to be empirical patterns of community structure. Their real value is that they provide evidence of interactions among species and between species and their environment that determine the composition of assemblages. We have gone beyond the stage of asking whether North American desert rodent communities exhibit non-random structure. With so many papers demonstrating one kind of assembly rule or another, to the extent that there is still some controversy, it tends to center around details of methodology (above) and problems of inferring process from pattern.

So, what are the interactions and how do they operate to structure local communities of desert rodents? The vast majority of the above assembly rules indicate that the species that coexist are more different than expected by chance; species in local communities are nonrandom samples from the regional species pool, and the species in the regional pools are nonrandom samples from the larger pool of species inhabiting the entire southwestern desert region. The first pattern seems uniquely consistent with a process of interspecific competition, whereas the second is consistent with mechanisms of both competition and allopatric speciation. And the experimental evidence for interspecific competition is overwhelming (e.g., Brown & Harney, 1993 and references therein; see also Heske *et al.*, 1994: Valone & Brown, 1995, 1996). Recently, phylogenetic and biogeographic analyzes have provided evidence of long-lasting legacies of allopatric speciation, which often occurred when desert basins were isolated by geological events and/or climatic changes (e.g., Riddle, 1995, 1996). Requiring further investigation, however, is the relationship between phylogenetic, biogeographic, and ecological relationships. It is possible, even likely, that interspecific competition plays a major role in preventing the products of recent speciation events from expanding their distributions to overlap in their geographic ranges and coexist in local communities.

The assembly rules raise several additional questions about the process of interspecific competition in desert rodents. For one thing, many of the assembly rules describe probabilistic, rather than absolute, patterns of co-occurrence, i.e., species in the same body size category, genus, and functional group occur together less frequently than expected by chance, but they sometimes do coexist not only in the same geographic region but also in the same local community. For example, in a clear exception to several of the above assembly rules, the closely related and morphologically and functionally similar kangaroo rats, *D. merriami* and *D. ordii*, have occurred together nearly continuously at Brown's experimental study site for the last 19 years (see Brown & Heske, 1990). This does not mean that the rules are invalid or that the two kangaroo rats do not compete. Indeed, it is easy to imagine that, even though such species might usually be the strongest competitors, in certain environments their relative competitive abilities would be almost evenly matched. In such cases they could coexist almost indefinitely, especially if the presence of each species were reinforced, either by immigration from adjacent habitats where each is the superior competitor (e.g., Schroder, 1987) or by temporal environmental shifts which alternately favor each species (e.g., Chesson, 1986).

A more telling question concerns the generality of assembly rules. Do the rules derived for North American desert rodents also apply to rodents in other deserts or to other kinds of organisms? While more evidence would be helpful, that which is available suggests that the generality is very limited. In particular, what is now known about the rodent communities inhabiting other deserts throughout the world suggests that they are structured quite differently (Kelt *et al.*, 1996). There seem to be several reasons for this. First, while differences among species may reduce interspecific competition and facilitate coexistence, similarities among species may facilitate their occurrence in the same local environments. Thus, for example, the extremely arid open deserts of central Asia support diverse rodent communities that often contain species of similar body size and bipedal locomotion (e.g., Shenbrot *et al.*, 1994). Second, different kinds of characteristics may play a role in the existence and coexistence of rodents in different deserts. Thus, the majority of North American desert rodents are granivores, and coexisting species tend to differ in body size and locomotor mode. In other deserts (Asia, Australia, South Africa) there are few granivorous rodents, and coexistence of species appears to depend much more on differences in diet and less on differences in body size and locomotion (Kelt *et al.*, 1996). Finally, the other deserts have quite different physical settings and geological and climatic histories. These abiotic and historical factors have influenced the patterns of evolutionary differentiation and ecological segregation of rodent assemblages. They are reflected not only in the trophic differences just mentioned, but also in different patterns of morphological and physiological specialization. For example, while only two genera of North American desert rodents (*Dipodomys* and *Microdipodops*) are bipedal, the more ancient and more extensive central Asian deserts contain 11 bipedal genera (*Alactodipus, Allactaga, Cardiocranius, Dipus, Eremodipus, Euchoreutes, Jaculus, Paradipus, Pygeretmus, Salpingotus, Stylodipus*), and most of them show a much higher degree of morphological specialization for bipedal, saltatory locomotion on both hard and sandy substrates. Moreover, most of the Old World bipedal taxa are not granivorous (Kelt *et al.*, 1996). Clearly, different functional groups of rodents occur in the different deserts (e.g., there are no strict quadrupedal or bipedal granivores in Australia), and when similar functional groups do occur, they tend to be represented by quite different numbers of species and genera (e.g., bipedal granivores: 19–20 species of *Dipodomys* and two of *Microdipodops* in North America (Patton, 1993; Williams *et al.*, 1993), compared to just one species of *Cardiocranius*, four species of *Jaculus* (whose diet is both granivorous and folivorous (G. Shenbrot and K. Rogovin, personal communication)), and six species of *Salpingotus* in the Middle Eastern and North African deserts (Holden, 1993; Mares, 1993)). Thus, while Fox's (1987, 1989, Fox & Brown 1993; this chapter) approach of analyzing representation of different functional groups might be applied in other deserts, different functional groups defined on the basis of different criteria would have to be specified for Fox's assembly rule to hold.

It has been more than two decades since Diamond's (1975) seminal paper on community assembly rules and Rosenzweig and Winakur's (1969) pioneering study of community structure in desert rodents. Despite a good deal of work on desert rodent community ecology within the last decade, however, most of the studies have been experimental and few have been the kinds of non-manipulative geographic comparisons that led to the formulation of assembly rules. This emphasis seems to reflect the severe limitations of the assembly rule approach to community ecology: difficulties in inferring process from pattern, beyond the straightforward evidence for some form of interspecific competition, and lack of any broad generality in the nature and applicability of the rules, despite the widespread occurrence of the predominant underlying mechanism, interspecific competition. If there are truly general patterns and processes that characterize the organization of communities, these do not appear to be reflected in any very direct way in the well-documented assembly rules that describe the structure of North American desert rodent communities.

Acknowledgments

We thank Evan Weiher and Paul Keddy for inviting us to participate in the symposium. Comments by Mark Lomolino and an anonymous reviewer helped to improve the presentation. This work was partially supported by National Science Foundation grant DEB-9553623 to J. H. Brown.

References

Atmar, W. & Patterson, B.D. (1993). The measure of order and disorder in the distribution of species in fragmented habitat. *Oecologia* 96: 373–382.
Atmar, W. & Patterson, B.D. (1995). The nestedness temperature calculator: a visual basic program, including 294 presence–absence matrices. AICS Research, Inc., University Park, NM, and The Field Museum, Chicago, IL.
Bowers, M.A. & Brown, J.H. (1982). Body size and coexistence in desert rodents: chance or community structure. *Ecology* 63: 391–400.

Brown, J.H. (1973). Species diversity of seed-eating desert rodents in sand dune habitats. *Ecology* 54: 775–787.

Brown, J.H. (1975). Geographical ecology of desert rodents. In *Ecology and Evolution of Communities*, ed. M.L. Cody & J.M. Diamond, pp. 315–341. Cambridge, MA: Belknap Press.

Brown, J.H. & Lieberman, G.A. (1973). Resource utilization and coexistence of seed-eating desert rodents in sand dune habitats. *Ecology* 54: 788–797.

Brown, J.H. & Munger, J.C. (1985). Experimental manipulation of a desert rodent community: food addition and species removal. *Ecology* 66: 1545–1563.

Brown, J.H. & Kurzius, M.A. (1987). Composition of desert rodent faunas: combinations of coexisting species. *Annales Zoologici Fennici* 24: 227–237.

Brown, J.H. & Heske, E.J. (1990). Temporal changes in a Chihuahuan Desert rodent community. *Oikos* 59: 290–302.

Brown, J.H. & Harney, B.A. (1994). Population and community ecology of heteromyid rodents in temperate habitats. In *Biology of the Heteromyidae*, ed. H.H. Genoways, & J.H. Brown, Special Publication 10, pp. 618–651. American Society of Mammalogists.

Brown, J.H. & Reichman, O.J. & Davidson, D.W. (1979). Granivory in desert ecosystems. *Annual Reviews of Ecology and Systematics* 10: 201–227.

Brown, J.S. (1989). Coexistence on a seasonal resource. *American Naturalist* 133: 168–182.

Chesson, P.L. (1986). Environmental variation and the coexistence of species. In *Community Ecology*, ed. J. Diamond, & T.J. Case, pp. 240–256. NY: Harper and Row.

Cody, M.L. & Diamond, J. (eds). (1975). *Ecology and Evolution of Communities*. Cambridge, MA: Belknap Press.

Cohen, J.E., R.A. Beaver, S.H. Cousins, D.L. DeAngelis, L. Goldwasser, K.L. Heong, R.D. Holt, A.J. Kohn, L.H. Lawton, N. Martinez, R. O'Malley, L.M. Page, B.C. Patten, S.L. Pimm, G.A. Polis, M. Rejmánek, T.W. Schoener, K. Schoenly, W.G. Sprules, J.M. Teal, R.E. Ulanowicz, P.H. Warren, H.M. Wilbur, and P. Yodzis. (1993). Improving food webs. *Ecology* 74: 252–258.

Connell, J. (1961a). Effects of competition, predation by *Thais lapillus* and other factors on natural populations of the barnacle *Balanus balanoides*. *Ecological Monographs* 31: 61–104.

Connell, J. (1961b). The influence of interspecific competition and other factors on the distribution of the barnacle *Chthalamus stellatus*. *Ecology* 42: 710–723.

Connor, E.F. & Simberloff, D. (1979). The assembly of species communities: chance or competition? *Ecology* 60: 1132–1140.

Connor, E.F. & Simberloff, D. (1983). Interspecific competition and species co-occurrence patterns on islands: null models and the evaluation of evidence. *Oikos* 41: 455–465.

Connor, E.F. & Simberloff, D. (1984). Neutral models of species' co-occurrence patterns. In *Ecological Communities: Conceptual Issues and the Evidence*, ed. D.R. Strong, Jr., D. Simberloff, L.G. Abele, & A.B. Thistle, pp. 317–331. Princeton, NJ: Princeton University Press.

Darwin, C. (1859). *The Origin of Species by Means of Natural Selection*. London: Murray.

Dayan, T. & Simberloff, D. (1994). Morphological relationships among coexisting heteromyids: an incisive dental character. *American Naturalist* 143: 462–477.

Diamond, J.M. (1975). Assembly of species communities. In *Ecology and*

Evolution of Communities, ed. M.L. Cody & J.M. Diamond, pp. 342–444. Cambridge, MA: Harvard University Press.

Diamond, J.M. (1984). Distributions of New Zealand birds on real and virtual islands. *New Zealand Journal of Ecology* 7: 37–55.

Diamond, J. & Case, T. (eds.) (1986). *Community Ecology*. NY: Harper and Row.

Diamond, J.M. & Gilpin, M.E. (1982). Examination of the 'null' model of Connor and Simberloff for species co-occurrences on islands. *Oecologia* 52: 64–74.

Fox, B.J. (1987). Species assembly and the evolution of community structure. *Evolutionary Ecology* 1: 201–213.

Fox, B.J. (1989). Small-mammal community pattern in Australian heathland: a taxonomically based rule for species assembly. In *Patterns in the Structure of Mammalian Communities*, ed. D.W. Morris, Z. Abramsky, B.J. Fox, & M.R. Willig, Special Publication 28, pp. 91–103. Lubbock, TX: The Museum, Texas Tech University.

Fox, B.J. & Brown, J.H. (1993). Assembly rules for functional groups in North American desert rodent communities. *Oikos* 67: 358–370.

Fox, B.J. & Brown, J.H. (1995). Reaffirming the validity of assembly rules for functional groups or guilds: a reply to Wilson. *Oikos* 73: 125–132.

Freeman, P.W. & Lemen, C. (1983). Quantification of competition among coexisting heteromyids in the southwest. *Southwestern Naturalist* 28: 41–46.

Frye, R.J. (1983). Experimental field evidence of interspecific aggression between two species of kangaroo rat (*Dipodomys*). *Oecologia* 59: 74–78.

Genoways, H.H. & Brown, J.H. (eds.) (1993). *Biology of the Heteromyidae*. Special Publication 10, The American Society of Mammalogists.

Gilpin, M.E. & Diamond, J.M. (1981). Immigration and extinction probabilities for individual species: relation to incidence functions and species colonization curves. *Proceedings of the National Academy of Sciences*, USA 78: 392–396.

Gilpin, M.E. & Diamond, J.M. (1982). Factors contributing to non-randomness in species co-occurrences on islands. *Oecologia* 52: 75–84.

Gilpin, M.E. & Diamond, J.M. (1984). Are species co-occurrences on islands non-random, and are null hypotheses useful in community ecology? In *Ecological Communities: Conceptual Issues and the Evidence*, ed. D.R., Jr., Strong, D. Simberloff, L.G. Abele, & A.B. Thistle, pp. 297–315. Princeton, NJ: Princeton University Press.

Gotelli, N.J. & Graves, G.R. (1995). *Null Models in Ecology*. Washington, DC: Smithsonian Institution Press.

Grant, P.R. (1986). *Ecology and Evolution of Darwin's Finches*. Princeton, NJ: Princeton University Press.

Grinnell, J. & Orr, R.T. (1934). Systematic review of the californicus group of the rodent genus *Peromyscus. Journal of Mammalogy* 15: 210–220.

Hafner, M.S. (1977). Density and diversity in Mojave Desert rodent and shrub communities. *Journal of Animal Ecology* 46: 925–938.

Hanski, I. (1982). Dynamics of regional distribution: the core and satellite hypothesis. *Oikos* 38: 210–221.

Heske, E.J., Brown, J.H., & Mistry, S. (1994). Long-term experimental study of a Chihuahuan desert rodent community: 13 years of competition. *Ecology* 75: 438–445.

Holden, M.E. (1993). Family Dipodidae. In *Mammal Species of the World: A Taxonomic and Geographic Reference*, ed. D.E. Wilson, & D.M. Reeder, pp. 487–499. 2nd edn. Washington, DC: Smithsonian Institution Press.

Hopf, F.A., Valone, T.J., & Brown, J.H. (1993). Competition theory and the structure of ecological communities. *Evolutionary Ecology* 7: 142–154.

Hutchinson, G.E. (1957). Concluding remarks. *Cold Springs Harbor Symposium in Quantitative Biology* 22: 415–427.

Hutchinson, G.E. (1959). Homage to Santa Rosalia, or Why are there so many kinds of animals? *American Naturalist* 93: 145–159.

Jorgensen, C.D. & Hayward, C.L. (1965). Mammals of the Nevada test site. *Brigham Young University Science Bulletin, Biology* Series, VI: 1–81.

Kelt, D.A., Taper, M.L. & Meserve, P.L. (1995). Assessing the impact of competition on community assembly: a case study using small mammals. *Ecology* 76: 1283–1296.

Kelt, D.A., Brown, J.H., Heske, E.J., Marquet, P.A., Morton, S.R., Reid, J.R.W., Rogovin, K.A., & Shenbrot, G. (1996). Community structure of desert small mammals: comparisons across four continents. *Ecology* 77: 746–761.

Lack, D. (1947). *Darwin's Finches.* Cambridge: Cambridge University Press.

MacArthur, R.H. (1958). Population ecology of some warblers on northeastern coniferous forests. *Ecology* 39: 599–619.

Magurran, A.E. (1988). *Ecological Diversity and its Measurement.* Princeton, NJ: Princeton University Press.

Mares, M.A. (1993). Heteromyids and their ecological counterparts: a pandesertic view of rodent ecology and evolution. In *Biology of the Heteromyidae*, ed. H.H. Genoways & J.H. Brown Special Publication 10, American Society of Mammalogists.

M'Closkey, R.T. (1976). Community structure in sympatric rodents. *Ecology* 57: 728–739.

M'Closkey, R.T. (1978). Niche separation and assembly in four species of Sonoran Desert rodents. *American Naturalist* 112: 683–694.

Morton, S.R., Brown, J.H., Kelt, D.A., & Reid, J.R.W. (1994). Comparisons of community structure among small mammals of North American and Australian Deserts. *Australian Journal of Zoology* 42: 501–525.

Munger, J.C. & Brown, J.H. (1981). Competition in desert rodents: an experiment with semipermeable exclosures. *Science* 211: 510–512.

Owen, J.G. (1998). On productivity as a predictor of rodent and carnivore diversity. *Ecology* 69: 1161–1165.

Paine, R.T. (1966). Food web complexity and species diversity. *American Naturalist* 100: 65–75.

Patterson, B.D. (1987). The principle of nested subsets and its implications for biological conservation. *Conservation Biology* 1: 323–334.

Patterson, B.D. (1990). On the temporal development of nested subset patterns of species composition. *Oikos* 59: 330–342.

Patterson, B.D. & Atmar, W. (1986). Nested subsets and the structure of insular mammalian faunas and archipelagos. In *Island Biogeography of Mammals*, ed. L.R. Heaney & B.D. Patterson, pp. 65–82. NY: Academic Press.

Patterson, B.D. & Brown, J.H. (1991). Regionally nested patterns of species composition in granivorous rodent assemblages. *Journal of Biogeography* 18: 395–402.

Patton, J.L. (1993). Family Geomyidae. In *Mammal Species of the World: A Taxonomic and Geographic Reference*, ed. D.E. Wilson & D.M. Reeder, pp. 469–486, 2nd edn. Washington, DC: Smithsonian Institution Press.

Patton, J.L. & Rogers, D.S. (1993). Cytogenetics. *Biology of the Heteromyidae.* In ed. H.H. Genoways & J.H. Brown, pp. 236–258. Special Publication 10, American Society of Mammalogists.

Pielou, E.C. (1969). *An Introduction to Mathematical Ecology.* NY: Wiley-Interscience.

Polis, G.A. & Strong, D.R. (1996). Food web complexity and community dynamics. *American Naturalist* 147: 813–846.

Price, M.V. (1978). The role of microhabitat in structuring desert rodent communities. *Ecology* 59: 910–921.

Price, M.V. (1983). Ecological consequences of body size: a model for patch choice in desert rodents. *Oecologia* 59: 384–392.

Reichman, O.J. & Brown, J.H. (eds.) (1983). Biology of desert rodents. *Great Basin Naturalist Memoirs* 7.

Riddle, B.R. (1995). Molecular biogeography in the pocket mice (*Perognathus and Chaetodipus*) and grasshopper mice (*Onychomys*): the late Cenozoic development of a North American aridlands rodent guild. *Journal of Mammalogy* 76: 283–301.

Riddle, B.R. (1996). The molecular phylogeographic bridge between deep and shallow history in continental biotas. *Trends in Ecology and Evolution* 11: 207–211.

Rosenzweig, M.L. (1968). Net primary productivity of terrestrial communities: prediction from climatological data. *American Naturalist* 102: 67–74.

Rosenzweig, M.L. (1973). Habitat selection experiments with a pair of coexisting heteromyid rodent species. *Ecology* 54: 111–117.

Rosenzweig, M.L. & Winakur, J. (1969). Population ecology of desert rodent communities: habitats and environmental complexity. *Ecology* 50: 558–572.

Rosenzweig, M.L. & Sterner, P.W. (1970). Population ecology of desert rodent communities: body size and seed husking as bases for heteromyid coexistence. *Ecology* 51: 217–224.

Rosenzweig, M.L., Smigel, B. & Kraft, A. (1975). Patterns of food, space and diversity. In *Rodents in Desert Environments*, ed. I. Prakash & P.K. Ghosh, pp. 241–268. The Hague: Dr W. Junk.

Roughgarden, J. (1995). *Anolis Lizards of the Caribbean: Ecology, Evolution, and Plate Tectonics*. NY: Oxford University Press.

Schoener, T.W. (1969). Size patterns in West Indian Anolis lizards. I. Size and species diversity. *Systematic Zoology* 18: 386–401.

Schoener, T.W. (1970). Size patterns in West Indian Anolis lizards. II. Correlations with the sizes of particular sympatric species-displacement and convergence. *American Naturalist* 104: 155–170.

Schroder, G.D. (1987). Mechanisms for coexistence among three species of *Dipodomys*: habitat selection and an alternative. *Ecology* 68: 1071–1083.

Schroder, G.D. & Rosenzweig, M.L. (1975). Perturbation analysis of competition and overlap in habitat utilization between *Dipodomys ordii* and *Dipodomys merriami*. *Oecologia* 19:9–28.

Shenbrot, G.I., Rogovin, K.A., & Heske, E.J. (1994). Comparison of niche-packing and community organization in Asia and North America. *Australian Journal of Zoology* 42: 479–499.

Simberloff, D. & Boecklen, W. (1981). Santa Rosalia reconsidered: size ratios and competition. *Evolution* 35: 1206–1228.

Simberloff, D. & Dayan, T. (1991). The guild concept and the structure of ecological communities. *Annual Review of Ecology and Evolution* 22: 115–143.

Strong, D.E., Jr., Simberloff, D., Abele, L.G., & Thistle, A.B. (eds.). (1984). *Ecological Communities: Conceptual Issues and the Evidence*. Princeton, NY: Princeton University Press.

Thompson, S.D. (1982a). Structure and species composition of desert heteromyid rodent species assemblages: effects of a simple habitat manipulation. *Ecology* 63: 1313–1321.

Thompson, S.D. (1982b). Microhabitat utilization and foraging behavior of bipedal and quadrupedal heteromyid rodents. *Ecology* 63: 1303–1312.

Valone, T.J. & Brown, J.H. (1995). Effects of competition, colonization, and extinction on rodent species diversity. *Science* 267: 880–883.

Valone, T.J. & Brown, J.H. (1996). Desert rodents: long-term responses to natural changes and experimental perturbations. In *Long-term Studies of Vertebrate Communities*, ed. M.L. Cody & J.A. Smallwood, pp. 555–583. Orlando, FL: Academic Press.

Williams, D.F., Genoways, H.H. & Braun, J.K. (1993). Taxonomy and systematics. In *Biology of the Heteromyidae*, ed. H.H. Genoways & J.H. Brown, Special Publication 10, pp. 38–196. American Society of Mammalogists.

Wilson, J.B. (1995) Null models for assembly rules: the Jack Horner effect is more insidious than the Narcissus effect. *Oikos* 72: 139–142.

Appendix

The null hypothesis of community assembly states that there is a much higher probability that a species establishing itself in a local community will come from a functional group that is not already represented at that site, until all functional groups are equally represented, when the cycle repeats itself. Sites that agree with this rule are considered to be in a 'favored' state, while those that are not in agreement with the rule are considered 'unfavored'. For example, a site with two herbivores, two carnivores, and two omnivores is in agreement with the model, and is in a favored state. Similarly, a site with three herbivores, two carnivores, and three omnivores is a favored state. In contrast, a site containing one herbivore, two carnivores, and three omnivores would be considered unfavored, because the model predicts that a second herbivore should have entered the community before a third omnivore entered. Of course, the pool of species with access to a site may be biased towards or against certain functional groups, in which case these sites will be predisposed towards being unfavored. The model corrects for this by comparing the observed number of communities in favored states to a distribution of expected number of favored states that is generated by an iterative computer model. If the species pool for a given site is biased, this should be reflected in the distribution of expected favored states. By incorporating detailed information on the unique species pools for each site, this concern is minimized. In the present study species were allocated to three functional groups, following the lead of Fox and Brown (1993): bipedal heteromyids (*Dipodomys* and *Microdipodops*); quadrupedal heteromyids (*Chaetodipus, Perognathus*), and cricetines (*Peromyscus, Reithrodontomys, Baiomys*, etc.).

Assembly of a simulated community proceeds as follows. The species that are present at all sites, as well as those in the pool for the sites, are characterized according to their membership in these functional groups. Species are drawn randomly from these functional groups, according to the relative sizes of the groups. For example, if the pool for a site contained three herbivores and a single carnivore, the probability of an herbivore entering the community would be $\frac{3}{4} = 0.75$, while the probability of a carnivore entering the community would be $\frac{1}{4} = 0.25$. These probabilities are re-evaluated after each species addition, so that species are drawn from the pools without replacement. Thus, if an herbivore was drawn in the above case, then the probability that the second species entering the site is an herbivore would be $2/3 = 66.67\%$, while the probability of a carnivore entering would be $1/3 = 33.33\%$ The important point to note here is that species are not the unit of investigation. This model is based not on the individual species, but on characteristics of these species (e.g., diet, morphology, etc.). Hence, the unit of investigation is the ecological attribute that is embodied in the functional group designation, and the model addresses the organization of communities based on species characters.

   In its simplest form, the model operates as follows. For each real site, the number of species present is tallied. Simulated communities are assembled and evaluated to determine if they are in a favored or unfavored state, and the number of communities that are in favored states is recorded. This number represents one estimate of the number of communities expected to be in favored states, and constitutes one iteration of the model. This is repeated many times (2000 in the present paper) to produce a distribution of the expected number of favored states among the sites examined. The observed number of favored states is then compared to this distribution, and the probability that the observed value was obtained at random is determined by the proportion of the expected distribution that is equal to, or more extreme, than the observed. Because this is a one-tailed test (the alternative hypothesis is that the observed communities are more structured than random; this implies that the observed number of favored states should be greater than expected at random) the null hypothesis or random assembly will be rejected if the observed value lies in the upper 5% tail of the distribution.

   The above calculations might be a trivial and unnecessary exercise in computer modeling, because we could produce the exact probabilities of observed assemblages with the multivariate hypergeometric distribution. Clearly, an exact probability would be a better metric than the estimate produced by our iterative model. However, a significant element has been added to our model, which allows us to refine our analysis beyond that which could be attained with the multivariate hypergeometric distribution. In particular, interspecific interactions can be simulated by incorporating an interaction coefficient (θ) into the model. Thus, our probability function is given as (Kelt *et al.*, 1995):

$$P(y\{j\} = i \mid X\{j - 1\}) = \frac{(n\{i\} - X\{i,j - 1\})(1 - \theta)X^{(i,j - 1)}}{\sum_i[(n\{i\} - X\{i,j - 1\})(1 - \theta)^{(i,j - 1)}]} \quad (1)$$

where *i* and *j* index functional groups and species, respectively, *n{i}* is the number of species in functional group *n*,*y{j}* is a random variable, indicating the guild to which the *j*th species will be placed, X*{j−1}* is a vector of local species composition after the *j−1* th species has entered the community, and X*{i, j−1}* is a scalar which gives the number of species in functional group I after the *j−1*th species has entered. The numerator gives the combined probability of a species immigrating and becoming established. The denominator normalizes these probabilities so that they sum to one (see Kelt *et al.*, 1995 for further details).

The coefficient θ measures the effect of species from a functional group that is already present in a community upon the establishment of another species from the same functional group. Positive values of θ decrease the probability of establishment, and reflect competitive exclusion. Negative values of θ increase the probability of establishment, and reflect facilitation. When θ is zero, community composition has no effect on the establishment of a new species.

Finally, this model allows us to calculate two final parameters. By varying the value of θ, the strength of interaction that best fits the observed data can be determined. That is, the value of θ that produces a distribution of expected number of favored states that best agrees with the observed number of favored states can be determined. This will be referred to as $\hat{\theta}$. This constitutes a maximum likelihood estimate of the mean strength of interaction across all sites and all species, and is clearly a general value. However, by calculating this value, the statistical power of our model can then be estimated. Power is the probability that a hypothesis is incorrect and therefore should be refuted. If it is assumed that $\hat{\theta}$ reflects the real strength of interaction among these species, then the proportion of the distribution of expected values (number of favored states) produced with θ=$\hat{\theta}$ that lies in the 5% critical region of the distribution of expected values produced with $\hat{\theta}$ (the null hypothesis) constitutes an estimate of the power with which it can be stated that observed communities are significantly different from the null.

4

Introduced avifaunas as natural experiments in community assembly

Julie L. Lockwood, Michael P. Moulton, and Karla L. Balent

Introduction

In 1975, Jared Diamond hypothesized a series of rules for assembling communities of birds on islands in the south Pacific. In large part, these rules were based on incidence functions that dealt with the probability of species occurrence as a function of species richness (Diamond, 1975). His work sparked a heated debate and the concept of 'assembly rules' was largely put to rest (Connor & Simberloff 1979; Gilpin & Diamond, 1982). Perhaps one of the most serious, and yet least appreciated, problems with applications of assembly rules is that they are, in practice, based principally, if not exclusively, on extant community membership (e.g., Wilson & Roxburgh, 1994; Fox & Brown, 1993). As Diamond (1975) noted, the principal goal in elucidating assembly rules is to discover why some species become members of a community and some do not. How can the true criteria for inclusion in a community be decided without knowledge of criteria for rejection?

In order to derive meaningful assembly rules, one must begin with knowledge of all species once present, but now missing, from communities. In natural communities, a knowledge of this depth is generally lacking, reducing the strength of tests for assembly rules. Communities of non-indigenous bird species represent a companion system in which to study the effects of various selection criteria for community membership. Within such communities there is often a very thorough knowledge of which species failed to become community members enabling evaluation of the influence of specific mechanisms of membership rejection.

The analyses described below were based on a data set that includes the passeriform species introduced on four oceanic islands: Oahu, Tahiti, Saint Helena, and Bermuda. Non-indigenous bird species have been accidentally or intentionally introduced for centuries on these islands (Long, 1981). With the help of ardent bird watchers past and present, very clear records have been obtained of which species were introduced, in what manner, and when, and if they became extinct (Moulton & Pimm, 1983).

The assembly history of each island avian community is reconstructed using this information. With this reconstructed assembly trajectory sensible null models can be built with which to compare the extant assemblage. Two of the tests below were built specifically to search for patterns produced by inter-specific competition: one of the most often cited, and controversial, criterion for membership rejection (for reviews see Connell, 1983; Schoener, 1983). An alternative, and equally plausible criteria for rejection, is that a particular species simply is not intrinsically capable of surviving within the new environment (Simberloff, 1992). Thus, patterns of success are also sought, associated with a species' native range size which is believed to reflect a species intrinsic ability to compensate for environmental variability (Williamson, 1996).

Results of previous studies of introduced birds have been conducted mostly on an island by island basis (Lockwood & Moulton, 1994; Brooke *et al.*, 1995; Moulton, 1993; Lockwood *et al.*, 1993, 1997). Here, a much broader view of the phenomena of introduced avian community assembly is attempted by incorporating a large number of species (99) across four oceanic islands. Each of the three patterns described above is addressed in more detail individually and completely before moving to the next. First, however the species introduced must be described, together with the islands, and the circumstances surrounding the introduction events.

The islands

All four islands are oceanic, but they vary in degree of geographical isolation. Bermuda is perhaps the least isolated recording more than a hundred migrants every year (Wingate, 1973). Oahu and Tahiti are members of archipelagos and thus may receive natural immigrants from other nearby islands, but not from any mainland source. Saint Helena is the most geographically isolated sitting alone over 2000 km off the Cape of Africa in the Atlantic Ocean. It is thus doubtful that Saint Helena is subject to any regularly occurring attempts at natural colonization. Bermuda is temperate and relatively small (51 km²). Tahiti and Oahu are tropical islands with sizes of 1057 km² and 1536 km², respectively. Saint Helena is also tropical but relatively small in size (122 km²).

Each island has had a significant proportion of its native flora and fauna altered through human manipulation (Moulton & Pimm, 1983; Brooke *et al.*,

1995; Lockwood & Moulton, 1994; Lockwood *et al.*, 1993). There remain however, some physiographic differences between islands. For example, Bermuda and Saint Helena are relatively low relief islands in which nearly all native forest has been cleared (Wingate, 1973; Benson, 1950). Tahiti and Oahu are high relief islands that still retain a small, but significant, portion of their native forests (Moulton, 1993; Lockwood *et al.*, 1993).

Through field observations on each island, it seems that most successful species are utilizing suburban or disturbed habitat rather than the remaining native habitat (JLL, MPM, R. Brooke, pers. comm.). This is supported by others conducting avian surveys on each island (Scott *et al.*, 1986; Crowell & Crowell, 1976; Haydock, 1954). Not only are the avian assemblages of human derivation, but so too are the habitats that sustain them. The introduced passeriforms on each island make their living from suburban lawns, parks, and city streets thus reducing the variability in habitat across islands.

The formulation of the species pool

Perhaps the most important aspect of determining a criteria of rejection is determining which species one should include in the species pool. The pool consists of all species likely to have attempted establishment during the assembly history. The first step in formulating the pool is by far the easiest and most enjoyable: determining the composition of the extant community. In all the data sets described below, the extant community was determined by combining field observations by JLL, MPM or R. Brooke with recently published field guides (e.g., Pratt *et al.*, 1987; Wingate, 1973). The second step involves an in-depth exploration into the published literature. We compiled lists of species introduced from published references and references contained within the following: Saint Helena; Brooke *et al.* (1995), Tahiti; Lockwood *et al.* (1993), Bermuda; Lockwood and Moulton (1994), and Oahu; Moulton (1993).

The results of this search revealed that, across all four islands, a total of 99 species have been introduced incorporating 131 total introduction events (see Appendix). Of the 24 species introduced on more than one island, only one, the Common Waxbill (*Estrilda astrild*), was introduced on all four islands. These islands are clearly not geographically close enough to exchange species. Thus, each island represents a relatively distinct assemblage with its own unique avifaunal assembly history.

These references were also the basis for decisions regarding when species were introduced as well as if and when they failed. On some occasions there was a significant lag between published references concerning which species

Table 4.1. *Success rates for each island: the number of successful species out of the number of passeriforms introduced*

Island	Success rate
Saint Helena	5 out of 31 (16%) or
	5 of 17 excluding 1929 introductions (29%)
Tahiti	7 out of 41 (17%)
Bermuda	7 out of 14 (50%)
Oahu	27 out of 45 (60%)

were present on what island. For example, two species on Bermuda were introduced around 1800 but, were not referred to again until 1957 (Bourne, 1957). In this account, Bourne indicated that these species had not established after initial introduction however, he gave no date of last sighting. Thus, each species was counted as being present between 1800 and 1957. Such difficulties do not influence the results as long as each species' time of extinction can be placed relative to all others. In this case, no species introduced between 1800 and 1957 subsequently went extinct.

Of the 131 introduction events, 46 are counted as successful and 85 as failed (35%). However, the individual island success rates varied considerably (see Table 4.1). Saint Helena and Tahiti had similar overall success rates at 16% (5 of 31) and 17% (7 of 41), respectively. A total of 45 species were released on Oahu, 27 of those are currently considered successful (60%). On Bermuda, there were 14 exotic species introductions and half of those were successful (50%). There is no significant tendency for the number of successful species to increase with island size ($P = 0.2668$, $r^2 = 0.538$, see Fig. 4.1), although the largest island had the greatest number of successful introductions.

The first two of our three tests are sensitive to various influences on the formation of the species pool. There was one criterion, however, that was universally followed. Species were excluded if they invaded one of the islands naturally or if only five or fewer individuals were known to have been released (as these species' fates would more likely be determined by stochastic processes). The other influences include the phylogenetic variability seen within the species pool, the timing and method of introduction, and the presence or absence of a native avifauna. Each is dealt with in turn.

Colwell and Winkler (1984) point to a potential bias in the formation of the species pool due to the amount of phylogenetic variability incorporated: They call this the J.P. Morgan Effect. If species with extreme morphologies are included in the pool, the remaining species will appear clumped in any

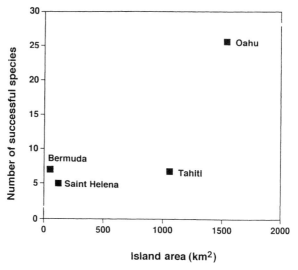

Fig. 4.1. The number of species successfully colonizing islands included in our data set does not correspond to island size. The largest island does hold the greatest number of successful species, however.

morphologically defined space. Since morphological characters are conserved within phylogenetic lineages, the likelihood of including an extreme morphology correlates with the degree of phylogenetic variability seen within the species pool (Colwell & Winkler, 1984). An extreme species (or group of species) included within a group of very similar species serves to obscure any pattern present.

In Lockwood *et al.* (1997) it was argued that one way to minimize the chance for a J.P. Morgan effect would be to look only within the group of finches introduced to Saint Helena. Finches make up the overwhelming majority of all species introduced (74%) and species that were successful (84%: Lockwood *et al.*, 1997). Clearly, within this group competition will be the strongest as they all feed and nest in very similar fashion (Clement *et al.*, 1993). The presence of distantly related species in the pool may obscure any competitively generated pattern.

The introduction events ranged in time from 1776 to the mid-1980s, with most occurring during the late 1800s to early 1940s. Recently introduced species (i.e., within the past 20 years) are often difficult to place as either successful or failed. This is especially true if some species are currently present in small numbers or are restricted to a particular habitat. We should perhaps consider these species 'the walking dead' (Simberloff, 1992) rather than successful colonizers. This is the case with several species introduced to Oahu since 1960. However, most other species considered here were released several years before the present (e.g., the last species released on Saint Helena occurred 67 years ago). The remaining few are very clearly successful (e.g., the Great Kiskadee was released on Bermuda in 1957 and is currently one of the most common birds observed there; Wingate, 1973, JLL).

The modes of introduction varied across islands; however, they can be classified into the following categories: ship or airplane stowaways, cage escapees, and purposeful introductions (Long, 1981; Lever, 1987). No individual or group introduced avian species on more than one of these islands. However, on two islands (Tahiti and Saint Helena) one individual was responsible for the release of the majority of species included in the species pool. The circumstances surrounding each introduction event are unique and influenced how we constructed each pool differently.

On Tahiti, Eastham Guild released 85% (35 of 41) of all species introduced (Guild, 1938, 1940; Lockwood *et al.*, 1993). Only two of these species survived, although we know Guild exerted considerable effort to ensure proper release (Guild, 1938, 1940). Guild kept individuals in an aviary for some time after their arrival, allowing him to remove diseased or sick birds. Further, Guild and other island residents established feeding stations to enhance the bird's chances for success. For these reasons, it was appropriate to include in the pool all the species Guild released.

H. Bruins-Lich released 45% (14 of 31) of all species introduced on Saint Helena (Haydock, 1954; Brooke *et al.*, 1995). The argument has previously been made that these species possibly should be excluded from the pool (Brooke *et al.*, 1995) as there is very little information on the release methods. Further, none of the 14 species he released was successful. Thus, unlike Guild, the possibility cannot be dismissed that he may have, in some way, inadvertently doomed all of these species to failure. For this reason, our analyses were conducted with and without these 14 species.

On all islands except Bermuda, the native passeriform fauna was either decimated before introductions occurred (Brooke *et al.*, 1995) or severely restricted in range or numbers (Lockwood *et al.*, 1993; Moulton, 1993). Thus, on these three islands interactions between native and exotic passeriforms is minimal at best (Brooke *et al.*, 1995; Lockwood *et al.*, 1993; Moulton, 1993). However, on the island of Bermuda three native passeriforms remain that are widespread and abundant (Wingate, 1973; Lockwood & Moulton, 1994). These are the Gray Catbird (*Dumetella carolinensis*), Eastern Bluebird (*Sialia sialis*), and the White-eyed Vireo (*Vireo griseus*). Native–exotic interactions cannot be discounted when looking for pattern within this community. The three native species on Bermuda are included where appropriate.

Priority effect

If the group of species that successfully colonized a particular island did so when significantly fewer other introduced species were present, this community is said to exhibit a priority effect (Moulton, 1993). Simply, species that arrive first are less likely to encounter competitors which may preclude their establishment. Once establishment is ensured, these same species are more likely to become competitively dominant over later arrivals (Alford & Wilbur, 1985). This pattern would be expected if interspecific competition was a predominate criterion for membership rejection.

Methods

Moulton (1993) details the procedure for testing for a priority effect. Briefly, we rank each species introduced to a particular island according to its date of introduction. For each species, we calculated the number of other introduced species present on the island at the time of introduction. Using a Mann–Whitney U-test, the mean number of other species present for failed introductions is compared to the mean for the group that was successful. Each island is considered separately.

Results

Three of our four island avifaunas exhibits a priority effect (Table 4.2). The most pronounced patterns are within the species introduced on Saint Helena and Tahiti. On Tahiti, successfully introduced species initially faced an average of 13.14 other species, whereas those that failed faced an average of 38.0 ($P > \chi^2 = 0.0032$). The average number of other species faced by successfully introduced species on Saint Helena was 5.4, and for unsuccessful species 14.15 ($P > \chi^2 = 0.0009$: Brooke *et al.*, 1995). As noted above, there is reason to exclude the 1929 introductions on Saint Helena. Their inclusion certainly inflates the priority effect. However, even if these species are excluded, the priority effect on Saint Helena still holds ($P > \chi^2 = 0.008$).

The Bermuda avifauna tends toward a priority effect when the native species are included as the first to arrive, which undoubtedly they were ($P > \chi^2 = 0.057$, mean successful = 5.6, mean failed = 9.14). However, it is not known how many other species the natives encountered when they first arrived. One could argue that the natives should be excluded from the analysis and only counted as 'other species present' in our calculations. If the natives are 'fixed' into the analyses in this way, there is clearly no evidence

Table 4.2. *Results of our search for a priority effect using the passeriform assemblages on Tahiti, Saint Helena, Bermuda, and Oahu*

Island	Species pool used	Results
Tahiti	All passeriforms	$P = 0.0032$, Yes
Saint Helena	All passeriforms	$P = 0.0009$, Yes
	Passeriforms introduced before 1929	$P = 0.008$, Yes
Bermuda	All passeriforms including native species	$P = 0.057$, tendency
	All passeriforms with native species 'fixed'	$P = 0.252$, No
Oahu	All passeriforms introduced before 1960	$P = 0.009$, Yes
	All passeriforms introduced before 1981	$P = 0.235$, No

Three out of four island assemblages show statistical proof of a priority effect.

for a priority effect ($P > \chi^2 = 0.252$, mean successful = 7.14, mean failed = 9.14).

Moulton (1993) conducted tests for the priority effect among the introduced passeriforms of Oahu. Originally, Moulton and Pimm (1983) excluded species introduced to Oahu after 1960 as they believed the current status of these species was uncertain. If only species introduced prior to 1960 are examined, as is in Moulton and Pimm (1983), a priority effect ($P > \chi^2 = 0.009$, mean successful = 12, mean for failed = 19) is observed.

After more than ten years of continued monitoring of these questionable species beyond 1983, a much better picture is available of which species have become established and which have not (Moulton, 1993). When the list was updated to include introductions up to 1981 (thus still excluding very recent introductions due to their questionable status), there is no evidence for a priority effect ($P < \chi^2 = 0.235$, mean for successful = 17.5, mean for failed = 20).

Discussion

In three out of four independent assemblies, the presence of a priority effect can be statistically shown. This result is consistent with the hypothesis that interspecific competition was a predominate criterion of membership rejection in each assembly. Species which establish early likely encounter fewer competitors and thus have a greater chance of being competitively dominant over later arrivals (Alford & Wilbur, 1985).

With this in mind, it is not surprising that evidence for a priority effect within the avifaunal assemblages on Oahu and Bermuda is not found consistently. In each case, there is a lack of sufficient knowledge concerning which species were truly 'early' and which of the 'late' arrivals failed to establish. It is not known which set of species may have been present when the three Bermuda natives first colonized the island. Similarly, it is not known how to classify the outcomes of species introduced recently on Oahu even though the set of species they likely encountered is known. Our results depend on how this lack of knowledge is compensated for. It is thus clear that insufficient information on the assembly history, and thus about the circumstances of membership rejection (i.e., competition or not), influences our perception of the assembly process.

Morphological overdispersion

Morphological overdispersion will result if the species that successfully colonized a community are more spread-out in morphological space and more evenly positioned than what is expected by chance (Lockwood *et al.*, 1993; Moulton & Pimm, 1987; Ricklefs & Travis, 1980). Overdispersion is also consistent with the hypothesis that interspecific competition determined community membership.

Methods

Several morphological variables (beak width, depth, and length and wing length) were measured on all species introduced onto each island from museum specimens housed at the Los Angeles County Museum of Natural History or the United States National Museum of Natural History. Each morphological variable was chosen because of a believed association with some aspect of avian ecology (Ricklefs & Travis, 1980; Grant, 1986). There are two possible approaches when deciding how many morphological variables to incorporate into such an analysis. The first is to include only one variable, and preferably one that reflects a multitude of ecological characteristics (e.g., body size). The second is to include several variables in the hopes of homogenizing any variability seen within any one correlation. Thus, if a species whose body size is not highly linked to its food preference, maybe its beak shape is, and in a multivariate analysis this would not go unnoticed (Ricklefs & Travis, 1980; Karr & James, 1975; Findley, 1973, 1976).

These raw variables are then log-transformed to normalize the inherent variability of such measurements and a principal components analyses is conducted using the covariance matrix (Ricklefs & Travis, 1980). The first two principal component scores invariably accounted for the majority of the variance seen within the raw data in our analyses (i.e., > 90%). The first principal component typically reflected body size and accounted for the majority of the variability within the data. The second principal component typically reflected the size and shape of the beak relative to the body size and accounted for significantly less variance.

Principal components analysis was used to produce a set of uncorrelated trait axes, upon which species can be plotted to show morphological similarity. The minimal spanning tree (MST) is used as a measure of dispersion within this morphological space and thus becomes our test statistic. The MST is the minimum sum of the lengths of n-1 line segments that connect n points such that no loops are created (Ricklefs & Travis, 1980). The standard deviation (SDEV) of the MST line segments is used as a measure of how evenly positioned the species are in that space.

An MST is drawn for the extant set of species on each island and then compared to a null distribution of MST and SDEV values derived from Monte-Carlo simulations (Ricklefs & Travis, 1980). Thus, if an introduced community comprises ten species and it is known that 20 species were introduced, MSTs and SDEVs are calculated for 1000 random selections of ten species from the pool of 20. An observed community is considered to be morphologically overdispersed if a large proportion (i.e., 90–95%) of null community MST values are smaller and, simultaneously, the SDEV values are larger (i.e., more evenly spaced) than the observed. Again, each island avifaunal assemblage was considered separately.

The observed passeriform communities on Bermuda, Tahiti and Saint Helena were all compared to 1000 randomly constructed communities (Lockwood *et al.*, 1993, 1996; Lockwood & Moulton, 1994) when possible. The set of successful finches introduced to Saint Helena before 1929 were compared to all possible alternative outcomes. The extant assemblage on Oahu was compared only to 200 random communities (Moulton, 1993). For a more detailed account of methods, see Lockwood *et al.* (1993).

Results

Tests for morphological overdispersion among the successful passeriforms on all four islands have previously been conducted (Lockwood *et al.*, 1993, 1997; Moulton, 1993; Lockwood & Moulton, 1994). Here only a brief review of our results are provided (see Fig. 4.2 for all morphological spaces and Table 4.3 for a summary of results). Oahu, Tahiti and Bermuda all exhibited pronounced overdispersion, while simultaneously being more evenly spaced among all successful passeriforms introduced (Oahu: $P = 0.035$, Tahiti: $P = 0.002$). On Bermuda, this is true regardless of whether or not the native species

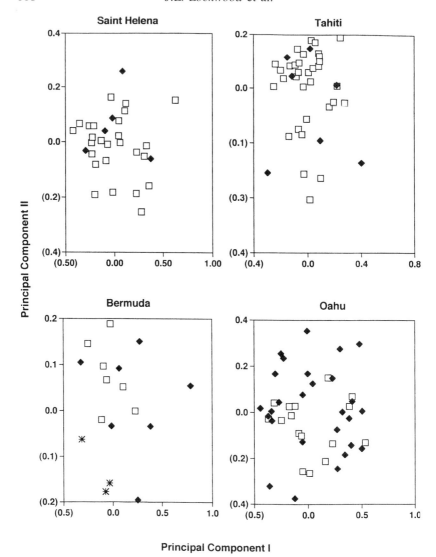

Fig. 4.2. The morphological spaces defined by the first two principal component scores for all island avifaunal assemblages. Principal component I indicates overall body size, whereas principal component II indicates the size and shape of the beak relative to body size. Because analyses used different species assemblages, the axes are only broadly comparable. Graphs are redrawn from: Lockwood and Moulton (1994), Lockwood *et al.* (1993, 1997), and Moulton and Pimm (1987). Closed diamonds represent successful species; open squares represent failed species; stars represent native species.

Table 4.3. *The results of our search for the pattern of morphological overdispersion within the four island passeriform assemblages*

Island	Species Pool	Results
Bermuda	All passeriforms including natives	$P = 0.006$, Yes
	All passeriforms 'fixing' natives	$P = 0.014$, Yes
Tahiti	All passeriforms	$P = 0.002$, Yes
Oahu	All passeriforms	$P = 0.035$, Yes
Saint Helena	All passeriforms	$P = 0.139$, No
	All passeriforms introduced before 1929	$P = 0.301$, No
	Only finches	$P = 0.092$, No
	Only finches introduced before 1929	$P = 00.076$, Tendency

Again, three of the four assemblages show statistical evidence for the pattern of morphological overdispersion.

were fixed in the morphological space ($P = 0.006$) or considered part of the species pool ($P = 0.014$).

On Saint Helena overdispersion was searched for among all introduced passeriforms and among only the group of finches to compensate for a potential J.P. Morgan Effect (see above). It is shown that the group of introduced finches are not morphological overdispersed ($P = 0.092$). As noted above, there are compelling reasons to exclude the 14 species released in 1929. The data suggest that the finches introduced prior to 1929 are overdispersed ($P = 0.076$; an exact statistic from a complete permutation test). However, the set of all species introduced are also not overdispersed ($P = 0.139$), nor are the surviving species introduced before 1929 ($P = 0.301$).

Discussion

In three of four independent assemblies, the pattern of morphological overdispersion is clearly seen. Overdispersion, like the priority effect, is consistent with the hypothesis that interspecific competition was a predominate criterion for species rejection. Thus, using two methodological approaches, the same general conclusion is reached: competition between existing community members and those that are newly released often determines the fate of the new colonizers.

Morphological overdispersion will only result if two criteria are met. First, interspecific competition must be strong enough to drive one of the competitors to extinction. Second, competition must be widespread enough to effect all the species included in the species pool (Moulton & Pimm, 1987). Thus, determining the composition of the species pool takes on added importance. A balance must be struck between including all species in the community and only those that are likely to compete intensely and extensively enough to perhaps produce pattern. This balance is sometimes tenuous as is the case with the introduced avifauna of Saint Helena. It is not clear if the eight phylogenetically (and thus morphologically) distinct species released on Saint Helena significantly altered the chances that the set of finches released would successfully colonize. Clearly, how that question was answered influenced the results obtained.

Many problems are solved simultaneously by defining the species pool as all the passeriforms introduced onto an oceanic island (i.e., it is known who, how, and why species attempted colonization on the island). This is a luxury that those working within natural systems do not enjoy. Indeed, the lack of predators, native species and natural colonizers on oceanic islands makes the definition of an island community all the easier. However, it does not solve all problems of definition. Our results indicate that the definition of the species pool, and the extent of the community itself, can play a considerable role in how a rejection criteria is determined and therefore the assembly process itself.

Size of native range

Finally, we test a hypothesis set forth by Moulton and Pimm (1986) and Daehler and Strong (1993) that species which have expansive native ranges will more likely successfully colonize a new locale after introduction. Species with larger range sizes may exhibit greater inherent ecological or physiological flexibility or may be better dispersers (Williamson, 1996; Brown, 1995). These species thus may be predisposed to be successful invaders no matter where released or what other species may be present. This is consistent with the 'All or None' hypothesis of Simberloff and Boecklen (1991), which has been put forth as an alternative explanation for patterns observed above. Thus, we explore the same data set to test this hypothesis as well.

Methods

Following the methods of Moulton and Pimm (1986), native range sizes were estimated using maps published in Long (1981). A grid was placed over each map and the number of intersecting points within the native range specified by Long (1981) were counted. This number was then divided by the number of intersections within the continent of Australia to standardize range sizes between maps (there were two map sizes). Thus, relative range sizes were calculated and not exact range sizes. All relative range sizes are presented as the percentage of the area of Australia each species' range incorporates. Thus, a relative range size of 50% indicates that this species' actual range is roughly half the size of Australia. All species were ranked according to their relative range size and grouped as either successful or failed. Mean range sizes are compared for each group using a Mann–Whitney U test. All introduction events across all islands are considered. Thus, if the same species was introduced on two or more islands, each introduction was counted as an independent event. Effects of range size were also tested for on an island-by-island basis.

Results

There were 131 independent introduction events (99 species) recorded as occurring on the four islands. However, there were 12 species for which Long (1981) provided no native range maps. Any introduction events associated with these species were thus excluded. Of the remaining 118 introduction events (87 species), 41 were successful and 77 failed. The average relative range size of the 41 successful species was 96%, whereas the average relative range size for failed species was 109%. There was no statistical difference between these two means ($P > \chi^2 = 0.617$).

It may be that the species comprising one or more island communities do show increased chances for success based on their native range sizes if the island communities are considered independently. Our result above may mask effects present at the island level. Thus, the test was conducted on an island-by-island basis. None of the island communities showed significant correlations between successful establishment and large native range (see Table 4.4).

Discussion

To some degree, all species that strive for membership in a community face a novel situation. The species they encounter will likely not be the same as those present in the community from which they originated. The variability in abiotic conditions will also not be the same. This is certainly more pronounced when human-mediated introduction are considered. These species are often transported long distances and then released (Long, 1981; Lever, 1987). Thus, it may be expected that patterns regarding native range size will be more pronounced within species introduced by human actions.

Our results, however, indicate that range size in itself is not a good predictor of successful establishment within an assembling community. There may be species-specific attributes that will influence establishment within a

Table 4.4. *Island by island analyzes of the effect of native range size on the probability of successful establishment after introduction*

Island	Average range size of successful species	Average range size of failed species	Associated probability
Saint Helena	59%	144%	$P = 0.1054$
Tahiti	69%	49%	$P = 0.2912$
Bermuda	235%	123%	$P = 0.1792$
Oahu	104%	74%	$P = 0.5541$

None of the islands showed an effect.

particular context. Indeed, there is a growing subdiscipline surrounding the topic (Drake *et al.*, 1989; Williamson, 1996; Kareiva, 1996). It is not clear that any one, or even a small suite, of these characteristics will consistently determine a species fate when attempting to enter a community (Williamson & Fritter, 1996; Gilpin, 1990). Our results indicate that species-specific attributes, when represented by a 'macro-variable' such as range size, are not necessarily good predictors of success within a unique assembly (within an island community) or across several independent assemblies (across several island communities).

Summary and conclusion

The determination of 'rules' that govern the assembly of natural communities is riddled with several methodological pitfalls. The most obvious of which is the inability to determine a true criteria of rejection when looking only within an extant community (Drake, 1991). However, our results indicate that even when companion systems of introduced avifaunas are used, where species rejections can be documented, a clear picture of those rejecting criteria cannot always be gained. There remain questions concerning introduction timing and the definition of the species pool and extent of community. Most analyses of natural communities uniformly ignore such questions.

Our results, some new and some summarized here, indicate that interspecific competition can play a deciding role in which species will become community members. They also point to a deficiency in the predictive ability of species-specific attributes in determining success. Any criteria, if it is to prove useful as an assembly rule, must provide an acceptable predictive ability across many assembly scenarios. As yet, it is believed neither of these two hypothesized rules completely fits the bill.

Acknowledgments

We wish to thank Dr Richard Banks and Dr Gary Graves at the National Museum of Natural History, Washington, DC and Kimball Garrett of the Los Angeles County Museum of Natural History, Los Angeles, CA for allowing JLL and MPM access to their avian collections. This work incorporates the knowledge of several individuals including R.K. Brooke, D. Wingate, S.K. Anderson, and S.L. Pimm.

References

Alford, R.A. & Wilbur, H.M. (1985). Priority effects in experimental pond communities: competition between *Bufo* and *Rana*. *Ecology* 66: 1097–1105.

Benson, C.W. (1950). A contribution to the ornithology of St Helena, and other notes from a sea-voyage. *Ibis* 92: 75–83.

Bourne, W.P. (1957). The breeding birds of Bermuda. *Ibis* 99: 94–105.

Brooke, R.K., Lockwood, J.L. & Moulton, M.P. (1995). Patterns of success in passeriform bird introductions on Saint Helena. *Oecologia* 103: 337–342.

Brown, J.H. (1995). *Macroecology*. Chicago, IL: Chicago University Press.

Clement, P., Harris, A & Davis, J. (1993). *Finches and Sparrows: An Identification Guide*. London: Christopher Helm.

Colwell, R.K. & Winkler D.W. (1984). A null model for null models in biogeography. In *Ecological Communities: Conceptual Issues and the Evidence*, ed. D.R. Strong, Jr., D. Simberloff, L.G. Abele, & A.B. Thistle, pp. 344–359. Princeton, NJ: Princeton University Press.

Connell, J.H. (1983). On the prevalence and relative importance of interspecific competition: evidence from field experiments. *American Naturalist* 122: 661–696.

Conner, E.F. & Simberloff, D. (1979). The assembly of species communities: chance or competition? *Ecology* 60: 1132–1140.

Crowell, K.L. & Crowell, M.R. (1976). Bermuda's abundant, beleaguered birds. *Natural History* 85: 48–56.

Daehler, C.C. & Strong, Jr. D.R., (1993). Prediction and biological invasions. *Trends in Ecology and Evolution* 8: 380.

Diamond, J.M. (1975). Assembly of species communities. In *Ecology and Evolution of Communities*, ed. M.L. Cody & J.M. Diamond, pp. 342–444. Cambridge, MA: University Press.

Drake, J.A. (1991). Community assembly mechanics and the structure of an experimental species ensemble. *American Naturalist* 137: 1–26.

Drake, J.A., Mooney, H.A., di Castri, F., Groves, R.H., Kruger, F.J., Rejmanek, M. & Williamson, M. (1989). *Biological Invasions. A Global Perspective*. Chichester, UK: John Wiley.

Findley, J.S. (1973). Phenetic packing as a measure of faunal diversity. *American Naturalist* 107: 580–584.

Findley, J.S. (1976). The structure of bat communities. *American Naturalist* 110: 129–139.

Fox, B.J. & Brown, J.H. (1993). Assembly rules for functional groups in North American desert rodent communities. *Oikos* 67: 358–370.

Gilpin, M.E. (1990). Ecological prediction. *Science* 248: 88–89.

Gilpin, M.E. & Diamond, J.M. (1982). Examination of the 'null' models of Connor and Simberloff for species co-occurrences on islands. *Oecologia* 52: 64–74.

Grant, P.R. (1986). *Ecology and Evolution of Darwin's Finches*, pp. 77–96. Princeton, NJ: University Press.

Guild, E. (1938). Tahitian aviculture: acclimation of foreign birds. *Avicultural Magazine* 3: 8–11.

Guild, F. (1940). Western bluebirds in Tahiti. *Avicultural Magazine* 5: 284–285.

Haydock, E.L. (1954). A survey of the birds of St Helena. *Ibis* 6: 818–821.

Kareiva, P. (1996). Special feature: developing a predictive ecology for non-indigenous species and ecological invasions. *Ecology* 77: 1651–1697.

Karr, J.R. & James, F.C. (1975). Eco-morphological configurations and convergent evolution in species communities. In *Ecology and Evolution of Communities*, ed. M.L. Cody & J.M. Diamond, pp. 258–291. Cambridge, MA: Belknap Press of Harvard University Press.

Lever, C. (1987). *Naturalized Birds of the World*. NY: Longman Scientific and Technical.

Lockwood, J.L. & Moulton, M.P. (1994). Ecomorphological pattern in Bermuda birds: the influence of competition and implications for nature reserves. *Evolutionary Ecology* 8: 53–60.

Lockwood, J.L., Moulton, M.P. & Anderson, S.K. (1993). Morphological assortment and the assembly of communities of introduced passeriforms on oceanic islands: Tahiti versus Oahu. *American Naturalist* 141: 398–408.

Lockwood, J.L., Moulton, M.P. & Brooke, R.K. (1997). Morphological dispersion of the introduced passeriforms of Saint Helena. *Ostrich* 67: 111–117.

Long, J. (1981). *Introduced Birds of the World*. London: David & Charles.

Moulton, M.P. (1993). The all-or-none pattern in introduced Hawaiian passeriforms: the role of competition sustained. *American Naturalist* 141: 105–119.

Moulton, M.P. & Pimm, S.L. (1983).The introduced Hawaiian avifauna: biogeographical evidence for competition. *American Naturalist* 121: 669–690.

Moulton, M.P. & Pimm, S.L. (1986). Species introduction to Hawaii. In *Ecology of Biological Invasions of North American and Hawaii*, ed. H.A. Mooney & J.A. Drake, pp. 231–249. NY: Springer-Verlag.

Moulton, M.P. & Pimm, S.L. (1987). Morphological assortment in introduced Hawaiian passerines. *Evolutionary Ecology* 1: 113–124.

Pratt, H.D., Bruner, P.L. & Berrett, D.G. (1987). *A Field Guide to the Birds of Hawaii and the Tropical Pacific*. Princeton, NJ: Princeton University Press.

Ricklefs, R.E. & Travis, J. (1980). Morphological approach to the study of avian community organization. *Auk* 97: 321–338.

Schoener, T.W. (1983). Field experiments on interspecific competition. *American Naturalist* 122: 24–285.

Scott, J.M., Mountainspring, S., Ramsey, F.L. & Kepler, C.B. (1986). Forest Bird communities of the Hawaiian Islands: their dynamics, ecology and conservation. Las Cruces, NM: Cooper Ornithological Society.

Sibley, C.G. & Monroe, Jr. B.L. (1990). *Distribution and Taxonomy of Birds of the World*. New Haven, CT: Yale University Press.

Simberloff, D. (1992). Extinction, survival, and effects of birds introduced to the Mascarenes. *Acta Oecologica* 13(6): 663–678.

Simberloff, D. & Boecklen, W. (1991). Patterns of extinction in the introduced Hawaiian avifauna: a re-examination of the role of competition. *American Naturalist* 138: 300–327.

Williamson, M. (1996). *Biological Invasion*. London: Chapman and Hall.

Williamson, M. & Fritter, A. (1996). The varying success of invaders. *Ecology* 77: 1661–1666.

Wilson, J.B. & Roxburgh, S.H. (1994). A demonstration of guild-based assembly rules for a plant community, and determination of intrinsic guilds. *Oikos* 69: 267–276.

Wingate, D. (1973). *A Checklist and Guide to the Birds, Mammals, Reptiles, and Amphibians of Bermuda*. Hamilton, Bermuda: Bermuda Audubon Society.

Appendix Table *All species introduced on Oahu, Tahiti, St Helena or Bermuda*

Species	Oahu	DI	Tahiti	DI	St Helena	DI	Bermuda	DI	Relative range size
Pitangus sulphuratus							S	1950	1.76923076
Corvus brachyrhynchos							S	1876	2.30769230
Rhipidura leucophrys	F	1926							1.30769230
Grallina cyanoleuca	F	1922							1.23076923
Dumetella carolinensis							N	native	
Sialia sialis							N	native	
Sialia mexicana			F	1938					0.46153846
Turdus merula					F	1820			2.84615384
Turdus philomelos					F	1820			2.64615384
Cyanoptila cyanomelana	F	1929							
Erithacus rubecula					F	1820			2.07692307
Luscinia akahige	F	1929							
Luscinia komadori	F	1931							
Copsychus malabaricus	S	1939							1.23076923
Copsychus saularis	F	1932							
Sturnus vulgaris	S	1872			F	1852	S	1950	3
Acridotheres tristis	S	1960	S	1908	S	1820			0.40909090
Gracula religiosa					F	1829			0.61538461
Mimus gilvus					F	1853			0.46153846
Mimus polyglottos	S	1928	F	1938			F	1983	1.07692307
Parus varius	F	1928							0.23076923
Pycnonotus cafer	S	1966							0.53846153
Pycnonotus jocosus	S	1965	S	1979					0.46153846
Zosterops japonicus	S	1929							0.30769230
Zosterops lateralis			S	1938					0.30769230

Species									
Cettia diphone	S	1929							0.53846153
Garrulax caerulatus	S	1947							0.38461538
Garrulax canorus	S	1900							0.30769230
Leiothrix lutea	S	1928	F	1938					4.53846153
Alauda arvensis	S	1847			F	1820	S	1870	5.90909090
Passer domesticus	S	1871			F	1870	F	1800	3.45454545
Passer montanus									0.23076923
Sporopipes squamifrons					F	1929			0.15384615
Ploceus capensis					F	1929	F	1970	1.15384615
Ploceus velatus					S	1850			0.07692307
Foudia madagascarensis					F	1929			0.38461538
Euplectes albonotatus					F	1929			0.30769230
Euplectes progne					F	1929			1.15384615
Euplectes orix			F	1938	F	1820	F	1970	1.15384615
Pytilia melba					F	1929	F	1970	1.69230769
Lagonosticta senegala	F	1965	F	1938					1.38461538
Estrilda astrild	S	1981	S	1908	S	1820	F	1970	0.46153846
Estrilda caerulescens	S	1965							0.53846153
Estrilda erythronotus			F	1938	F	1929			0.6153846
Estrilda melpoda	S	1965			F	1929	S	1970	0.38461538
Estrilda melanotis			F	1938					0.46153846
Estrilda troglodytes	F	1965	F	1938					0.92307692
Uraeginthus angolensis	F	1965	F	1938					0.07692307
Uraeginthus bengalus	F	1965	F	1938	F	1929			0.53846153
Uraeginthus cyanocephala	F	1965							0.46153846
Uraeginthus granatinus			F	1938	F	1929			
Amandava amandava	S	1900	F	1938	F	1929			1.30769230
Amandava subflava			F	1938					
Stagonopleura guttata			F	1938					
Erythrura psittacea			F	1938					0.07692307

Appendix Table (*continued*)

Species	Oahu	DI	Tahiti	DI	St Helena	DI	Bermuda	DI	Relative range size
Erythrura cyaneovirens			F	1938					0.07692307
Erythrura prasina			F	1938					0.30769230
Erythrura trichroa			F	1938					0.0769230
Neochmia modesta			F	1938					
Neochmia ruficauda			F	1938					0.23076923
Neochmia temporalis			S	1899					
Poephila acuticauda			F	1938					0.0769230
Taeniopygia bichenovii			F	1938					
Chloebia gouldiae			F	1938					0.07692307
Padda oryzivora	S	1964			S	1790			0.23076923
Lonchura castaneothorax			S	1899					1.38461538
Lonchura cucullata			F	1938					1
Lonchura fringilloides			F	1938					1.15384615
Lonchura malabarica	S	1984							0.53846153
Lonchura malacca	S	1936	F	1938					0.76923076
Lonchura punctulata	S	1883	F	1938					1.76923076
Vidua macroura	F	1962							0.76923076
Vidua paradisea					F	1929			0.38461538
Vidua regia					F	1929			2.38461538
Fringilla coelebs					F	1820			0.76923076
Serinus atrogularis					F	1929			0.07692307
Serinus canaria							F	1968	0.15384615
Serinus canicollis					F	1850			0.30769230
Serinus flaviventris					S	1776			0.30769230
Serinus leucopygius	F	1965							

Species	Oahu	Tahiti	Saint Helena	Bermuda	
Serinus mozambicus	S 1964	F 1938	F 1929		1.07692307
Carduelis carduelis				S 1800	2.40909090
Carduelis chloris		F 1938	F 1870	F 1800	1.84615384
Carduelis tristis		F 1938			1.30769230
Carpodacus mexicanus	S 1870				0.76923076
Paroaria coronata	S 1928				0.30769230
Paroaria dominicana	F 1931				0.15384615
Tachyphonus rufus		F 1938			1.23076923
Ramphocelus bresilius		F 1938			0.15384615
Ramphocelus carbo		F 1938			0.92307692
Ramphocelus dimidiatus		S 1938			0.07692307
Thraupis episcopus		F 1938			0.84615384
Tangara arthus		F 1938			0.23076923
Tangara larvata		F 1938			
Cyanerpes cyaneus	S 1965				1.23076923
Sicalis flaveola	S 1974				0.76923076
Tiaris olivacea	F 1934				0.23076923
Passerina cyanea	F 1941				0.61538451
Passerina leclancherii	F 1931				0.07692307
Sturnella neglecta				S 1800	1.07692307
Cardinalis cardinalis	S 1929				0.53846153
Vireo griseus				N native	
Total introductions	45	41	31	17*	
Total successes	27	7	5	7	

DI = date of introduction. F = failed. S = successful.
Oahu dates taken from Moulton (1993), Tahiti from Lockwood *et al.* (1994), Saint Helena from Brooke *et al.* (1995), and Bermuda from Lockwood and Moulton (1994).
* includes native species.
Nomenclature follows Sibley and Monroe (1990).

5

Assembly rules in plant communities

J. Bastow Wilson

Introduction

'To do science is to search for repeated patterns' (MacArthur, 1972). Therefore, to do community ecology must be to search for repeated community
patterns. Two basic kinds of community pattern can be envisaged, with
different causes:

(a) *Environmentally mediated patterns*, i.e., correlations between species
due to their shared or opposite responses to the physical environment.
Ecologists have long tried 'to find out which species are commonly associated together upon similar habitats' (Warming, 1909). Modern methods
allow more subtle questions to be examined, such as the shape of environmental responses (e.g., Bio *et al.*, 1998), the niche widths (e.g., Diaz
et al., 1994), and how repeatable the associations of species are (e.g.,
Wilson *et al.*, 1996c). However, the simple existence of environmentally
mediated patterns is now too obvious to need demonstrating; Warming
(1909) described it as 'this easy task'.
(b) *Assembly rules*, i.e., patterns due to interactions between species, such as
competition. These patterns, when we can find them, are fascinating evidence that competition, allelopathy, facilitation, mutualism, and all the
other biotic interactions that we know about in theory, actually affect
communities in the real world.

Of course, to make this distinction, it has to be known what processes have
caused each pattern – physical environment or biotic interactions – but that
is our task as community ecologists. Both types of process may occur. This
combination of processes can be seen as initial exclusion of species that are
unable to tolerate the physical environment (i.e., environmental filtering), followed by the operation of assembly rules (i.e., biotic filtering; Keddy, 1992).

Table 5.1. *Types of assembly rule*

1 Rules based on particular named species
2 Rules based on presence/absence:
(a) variance in richness
(b) local vs. regional richness
(c) large-scale distributions.
3 Rules based on plant characters:
(a) texture convergence
(b) limiting similarity
(c) guild proportionality.
4 Rules based on species abundance:
(a) biomass constancy
(b) abundance-based guild proportionality
(c) RAD (relative abundance distribution):
(i) evenness
(ii) shape of the RAD.

The main aim here is to review the kinds of assembly rules that might exist for plants, and to describe attempts to find examples (Table 5.1), but first the definition of 'assembly rule', and some basic problems in assembly rule work will be considered.

Definition of 'assembly rule'

An assembly rule definition will be followed that is based on that of Wilson & Gitay (1995a): 'ecological restrictions on the observed patterns of species presence or abundance that are based on the presence or abundance of one or more other species or groups of species (not simply the response of individual species to the environment)'. The inclusion of 'ecological' indicates that such a rule is intended to express the results of processes in ecological time, not evolutionary processes such as character displacement.

Diamond (1975), in coining the term 'assembly rule', had defined rules in terms of individual species, e.g., '*Macropygia amboinensis* and *Macropygia mackinlayi* cannot both be present on any one island'. However, almost every author since has envisaged rules of a less specific nature, making it easier to generalize rules between communities. Some authors have bent the original concept into rules that specify the *processes* by which communities are assembled (Cole, 1983; Hunt, 1991); I retain the majority view that assembly rules describe *patterns* that are evidence of such processes (e.g., Lawler, 1993; Graves & Gotelli, 1993).

This definition, Diamond's usage and any other definitions, do not include, as Keddy (1992) seems to, environmentally mediated patterns. In ecology terms are often used, for effect, so far outside their original intent that they lose any meaning. If this happens to 'assembly rule' with the inclusion of the environmental sieve, as Keddy seems to advocate, another term will have to be invented to cover real assembly rules, based only on species interactions. That would be a nuisance.

Problems associated with assembly-rule work

Claims for demonstration of assembly rules have often caused considerable debate (Weiher & Keddy, 1995). Some ecologists appear reluctant to accept any evidence for their existence. One reason for this is belief that only experimental evidence is useful in determining the importance of species interactions in communities (Hastings, 1987; Goldberg, 1995; for a contrary view see Wilson, 1995b and Wilson & Gitay, 1995a). Others are unsure that species interactions are a significant force in structuring natural communities, and are therefore wary of any evidence that they are (Simberloff, 1982; Wilson, 1991b).

Such caution and care is appropriate. If only it were shown more often in ecology! Yet, it can occasionally be taken to extremes. In most guild work, guilds are hypothesized, and hopefully then tested. This is a standard approach in science. However, de Kroon and Olff (1995) argue that, in assembly-rule work, the guild hypothesis has to be proved before it can be tested.

Evidence has to be examined very carefully, but the baby not thrown out with the bathwater. Stone *et al.* (1996) produced valid criticisms of the assembly-rule work of Fox & Brown (1993), but they entitled their paper: 'Community-wide assembly patterns unmasked ...', as if the mistakes in Fox & Brown's analyzes invalidated all assembly-rule work.

There are a number of particular problems in assembly-rule work.

Problem 1: The difficulty in framing valid null models

In assembly-rule work, the null models have to be carefully chosen. When we look for a correlation between two variates, or for a difference between two groups (as for a *t*-test), it is fairly obvious what the data would look like if there were no pattern, i.e., under a null model. With ecological communities, random assemblages may show surprising patterns (Lockwood *et al.*, 1997), so it is far from clear what an assemblage of species would look like were there no assembly rules. As Andrew D.Q. Agnew (pers. comm.) once asked: 'What does a plant community look like when it isn't there?'.

Occasionally, assembly rules can be tested using standard statistical tests (e.g., r or t). Usually, however, a special null model has to be devised, and a corresponding randomization test used to determine significance. There are two aspects of any such explicit null-model test: the test statistic (i.e., the value which is compared between the observed data and the randomizations) and the null model (i.e., in practical terms how the randomization is done).

Choosing a test statistic represents no problems of validity: any test statistic, calculated on the observed and randomized data alike, is valid. The danger is that, by choosing an inappropriate test statistic, evidence for an assembly rule might be missed. From this point of view the choice of test statistic is crucial.

Moreover, it is all too easy to use a null model which looks plausible, yet yields answers that give no evidence on the existence of assembly rules. There are three basic safeguards against this:

(a) The method should be tested on random datasets. A valid method should produce significance at the 5% level in 5% of the random datasets, perhaps fewer but certainly not more. This check can expose a lot of mistakes (e.g., Wilson, 1987; Wilson, 1995c; Drobner *et al.*, 1998). However, the method of producing random datasets is important. If it is too similar to the method used in the null-model randomization, the test will give only 5% significances by definition, and the test is of little value.

(b) The null model should include every feature of the observed data except the one it is intended to test (Tokeshi, 1986).

There is a danger of making a null model that includes some of the effects one is trying to test, the 'Narcissus effect' (Colwell and Winkler 1984). The Narcissus effect is bound to occur to some extent. For example, a pool of extant species is considered, even though some species may have become extinct because of competition. This gives a somewhat conservative test.

The opposite problem (Wilson, 1995c) is a model that fails to include obvious features, i.e., that ignores Tokeshi's principle. For example, in comparing islands it is pretty certain that some islands (large islands, probably) will contain more species than others (small islands). If it is not intended to test this feature, the island species-richnesses must be built into the null model. Otherwise, a departure from the null model will probably just be disclosing the obvious fact that large islands have more species. The danger is that, if one does not notice what has happened, this will spuriously appear to be an assembly rule, the trap into which Fox and Brown (1993) fell. This is termed the 'Jack Horner effect'

(Wilson, 1995c), a test in which the only conclusion from the analysis is an already obvious fact, in the analogy that plum puddings contain plums[1], in ecology that, for example, large islands have more species.

A danger of the Narcissus effect (a conservative test, in effect an unavoidable Type II error) should be accepted, rather than the Jack Horner effect (a spurious 'assembly rule', a Type I error), especially when so many skeptics are watching.

(c) As far as possible, the null model should represent an explicit ecological process by which the community could conceivably have been assembled (Manly, 1991). For example, Partel *et al.* (1996) used a null model phrased in terms of the position of points on a graph, not in terms of the occurrence of species. They made the mistake of thinking that under an ecological null model all possible values are equally likely, which is rarely the case. This makes it very unclear what their model is testing.

Many null models are phrased in terms of the scattering of species across quadrats or islands (e.g., Connor & Simberloff, 1979; Wilson *et al.*, 1987; almost all of them really). This is unrealistic, because it is individuals that are scattered onto an island (even then as part of population processes). Yet, a model of species scattering generally has to be accepted, because to introduce population processes would cause far more problems than it would solve. At least a model of species scattering is one ecological step below the test statistic.

Problem 2: What assembly rule to test for

It is not known what assembly rules might be operating in any community. Because ecologists are not plants, the contents of their rule book is unknown to us. Therefore, it is not known what to test for (Wilson 1991b, 1994). Andrew Agnew (pers. comm.) followed up his question: 'What does a community look like when it isn't there?' by asking: 'What does a community look like when it *is* there?'

Problem 3: Environmental patchiness

The responses of species to the physical environment are of major interest in ecology (e.g., Austin *et al.* 1990; Bio *et al.*, 1998), but when we are seeking assembly rules they are a nuisance, obscuring or mimicking the assembly rules that we are looking for. Patchiness in successional stage has similar

---

[1]'plum' is used here in the sense of dried grape.

effects (Zobel *et al.*, 1993). There is environmental variation everywhere. The danger is that an analysis intended to test for assembly rules may merely demonstrate that different species grow in different environments, which we already know well (Warming, 1909).

A mitigating factor is that environmental differences, because they represent an increase in heterogeneity, usually lead to aggregated distributions, to high variation (e.g., clustered gradient optima, differences in plant characters), whereas most assembly rules, by their definition, lead to regular distributions, to low variation (e.g., dispersed gradient optima, similarity in plant characters). Therefore, environmental variation often represents noise in our analyzes. However, environmental variation can sometimes mimic assembly rules, i.e., cause departures from the null model in the same direction as would our assembly rule (e.g., different but equal species pools: Watkins & Wilson, 1992).

The problem arises at all scales. There can be environmental variation down to the smallest scale we can conceive (Watkins & Wilson, 1992; Wilson *et al.*, 1992), and usually is. Heterogeneity is obvious if the sampling covers several habitats (e.g. Weiher *et al.*, 1998). On a yet larger scale, there is biogeographic heterogeneity (e.g., Fox & Brown, 1993; Armbruster *et al.*, 1994).

The biggest danger is that the heterogeneity includes two or more species pools. A null model that randomizes over all the quadrats (e.g., Fox & Brown, 1993; Weiher *et al.*, 1998) is then obviously false before we start: all species can not occur equally in all sites. The test therefore becomes a test between species pools. Such a test is largely of the fundamental response of species to the environment, i.e., a test of evolutionary patterns, and therefore arguably not a test of assembly rules anyway. But, the greater problem is that there are usually several quadrats in each micro-habitat/habitat/region, so that we do not have as many independent species pools as the analysis implies. This is related to the problem of spatial autocorrelation (see below).

The problem can be reduced by examining areas that are more uniform in environment, but no area is ever completely uniform. Other solutions to the problem are patch models (Fig. 5.1), in which the null-model predictions are based on a small patch around the target quadrat (Watkins & Wilson, 1992; Bycroft *et al.*, 1993; Armbruster *et al.*, 1994; Wilson & Watkins, 1994; Wilson & Gitay, 1995a), and accumulating variation within areas/environments/communities, but not between (e.g., Wilson & Roxburgh, 1994; Wilson *et al.* 1995a).

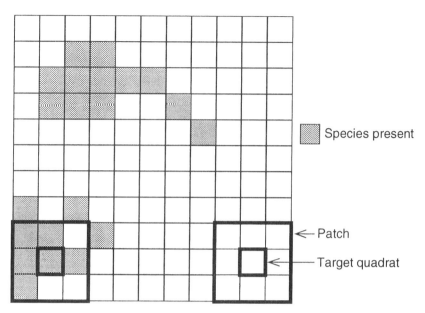

Fig. 5.1. A patch model. The null-model prediction for each target quadrat is based on the frequency or abundance of the species in a patch of quadrats (here nine quadrats) centered on the target quadrat. With data comprising a line of adjacent quadrats, the patch will be a line of quadrats centered on the target quadrat, typically seven.

Problem 4: Spatial autocorrelation

As in almost all ecological data, there are problems with spatial autocorrelation: samples taken nearby are likely to be more similar. This does not bias towards finding a certain result, but it affects significance testing, because it means that the samples are not independent. The problem arises on all scales. Samples only a few centimetres apart can be similar when the sampled area spans a metre, but so can samples only a few miles apart when the sampled area spans a hundred miles.

There must be thousands of demonstrations in the literature that a particular vegetational gradient correlates significantly with soil water status, pH, elevation, or whatever. It can be guessed that more than 99% of these tests are invalid because of spatial autocorrelation. No one minds, because no one doubts that vegetational gradients are related to water, pH and elevation. When assembly rules are proposed, many people do mind. They should.

Four solutions are known:

(a) Patch models (Watkins & Wilson, 1992). These examine only a small group of adjacent quadrats at a time (Fig. 5.1). There is therefore little opportunity for nearby quadrats to be more similar, because all the quadrats in the patch are close to the target quadrat. In fact, such methods tend to be a bit conservative in the face of spatial autocorrelation (Watkins & Wilson, 1992); at least this conservatism removes the problem of obtaining spurious significance because of spatial autocorrelation (Wilson, 1995d). Armbruster *et al.* (1994) used what is essentially a patch model, but on a biogeographic scale.

(b) Accumulate variation within areas (e.g., Wilson & Roxburgh, 1994; see above).

(c) Rotation/reflection and random shifts methods (Palmer & van der Maarel, 1995). These retain the exact spatial pattern for each species, but randomize the orientation or position of the pattern between species. They are therefore immune to spatial autocorrelation. They are more restricted in the type of sampling scheme for which they can be used, ideally a contiguous grid or a circular transect. However, their major drawback is that they randomize across any environmental patchiness, giving the possibility of spurious 'assembly rules' (Problem (c) above).

(d) The 'random patterns' test (Roxburgh & Chesson, 1998).

Problem (e): Subtle effects

Problems (c) and (d) above are especially severe because the effects that we seek in assembly-rule work are generally quite subtle, tendencies rather than strict rules (Fox & Brown, 1993; Wilson & Roxburgh, 1994), easily overwhelmed by sampling error and by environmental noise.

Presence/absence rules

Some assembly rules are based on the presence and absence of species in samples.

Variance in richness

The simplest type of presence/absence assembly rule is one in terms of species richness. Much theory (Fig. 5.2) is based on the idea that species too similar in niche cannot coexist (e.g., Pacala & Tilman, 1994). If this is true, the number of species that can coexist locally should be limited, because there is a limited number of niches (Ricklefs, 1987). This would give low variance in species richness, compared to a null model. This effect has proved surprisingly difficult to find, but it can be found, at least at a small scale.

Limiting similarity

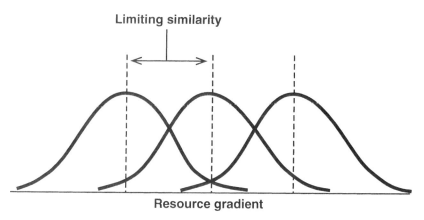

Resource gradient

Fig. 5.2. The theory of limiting similarity, leading to the concept of niche limitation (cf. May & MacArthur, 1972).

Watkins & Wilson (1992) sampled a number of lawns, examining richness in $\frac{1}{2}'' \times \frac{1}{2}''$ quadrats. For six of these sites, the observed frequency histogram of species richness was narrower than expected on a random basis, i.e., species richness was more constant, an assembly rule (Fig. 5.3). It is possible that some of this effect is due to physical constraints on individual module packing (Watkins & Wilson, 1992; Palmer & van der Maarel, 1995), e.g., packing of leaves (some people refer to plant packing; however, the non-packable unit can vary from a leaf to a clone). Yet, up to five species (average 1.6) can be found at a point in this lawn, so quite a few can be packed into a $\frac{1}{2}'' \times \frac{1}{2}''$ quadrat. And plant canopies are normally only 3–5% occupied by plant modules. At least it's not a simple module packing effect.

Even stronger effects of this type can be found at the smallest scale, that of a point (Wilson *et al.*, 1992, 1996b). It seems likely that a considerable part of this effect is due to constraints on module packing. However, that is not the whole story. For example, one treatment of Wilson *et al.* (1992) had received selective herbicide. By the first census, 6 weeks after herbiciding, leaf module density had already been restored to that of the control, but species packing was significantly less tight (i.e., RV_r was higher: Fig. 5.4).

As usual in assembly-rule work, spatial heterogeneity in environment may be a confounding factor. One way to overcome this is to record the same quadrat thorough time. Wilson *et al.* (1995b) did this on limestone grassland in Sweden. Variance in richness, when adjusted for overall year-to-year variance in richness, was significantly less than null-model expectation at two sites (Table 5.2).

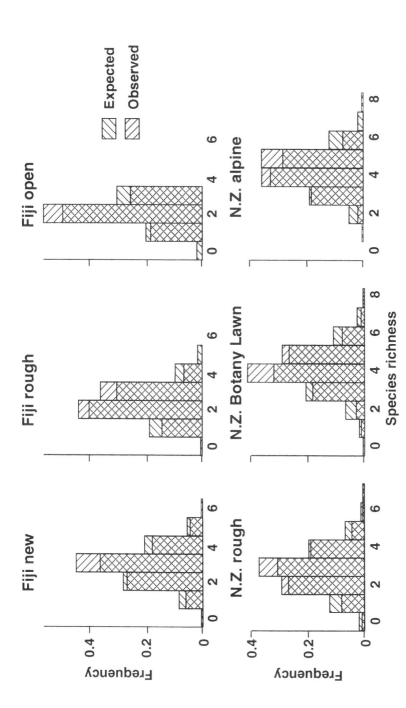

Fig. 5.3. The distribution of species richness in ½" × ½" quadrats on six lawns, compared to that expected under a null model. (After Watkins & Wilson, 1992.)

Table 5.2. *Temporal variance in richness, 1985–1989, in three sites of limestone grassland on Öland, Sweden, sampled with 3×3 cm quadrats, and adjusted for overall year-to-year changes in richness*

Site	RV_r	P
Gettlinge	1.036	ns
Kleva	0.786	0.018
Skarpa Alby	0.785	0.017

RV_r is an index of variance in richness: a value of 1.0 indicates variance equal to that expected at random, a value < 1.0 indicates variance lower than this (a possible assembly rule). P indicates the significance of the departure from the null-model value of 1.0.

Local vs. regional richness

The work above sought a species-richness assembly rule at a fine scale. The opposite approach is to work at the continental level, examining different species pools. Connell & Lawton (1992) pointed out that niche limitation could be seen by plotting local species richness against the size of the regional pool. If there were niche limitation, the curve would flatten out (Fig. 5.5). There are two problems with this approach. First, there has been uncertainty as to how to test for this pattern (Cresswell *et al.* 1995; Partel *et al.*, 1996; Caley & Schluter, 1997), the best idea to date being to test for significant concave-down non-linearity in regression. The second problem is that the points must be reasonably independent, i.e., based on independent species pools. Whereas at the local level one can sample 100, 1000 or 10000 individual quadrats, it is difficult to find many more than ten continents on Earth with independent species pools (Lawton *et al.*, 1993 obtained data for six pools). Inflating the number of continents by including several taxa does not help (Caley & Schluter, 1997); even if the taxa can be considered as independent, there is no reason to suppose that saturation would occur at similar pool and richness values in different groups of organisms. Unless a method of analysis can be found that copes with overlapping pools, this approach will not reach its potential until data can be included from other planets.

Large-scale distributions

Most large-scale patterns can be related to the environment, the 'easy task' of Warming (1909). However, there are features of zonations that could

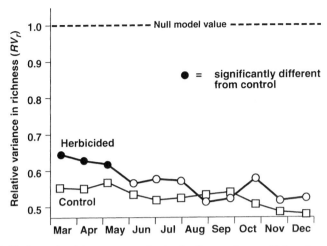

Fig. 5.4. Variance in richness at a point in the Botany Lawn, University of Otago, New Zealand. An RV_r value < 1.0 indicates a trend towards constant species richness. (After Wilson *et al.*, 1992.)

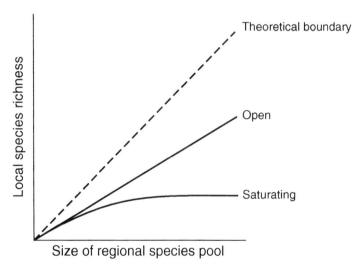

Fig. 5.5. The relation between local species richness and the size of the regional species pool. 'Theoretical boundary' indicates that there logically cannot be more species locally than in the regional pool. 'Open' shews a community-type with no limitation to local species richness. 'Saturating' shews a community-type where the local community becomes saturated with species (cf. Connell & Lawton, 1992).

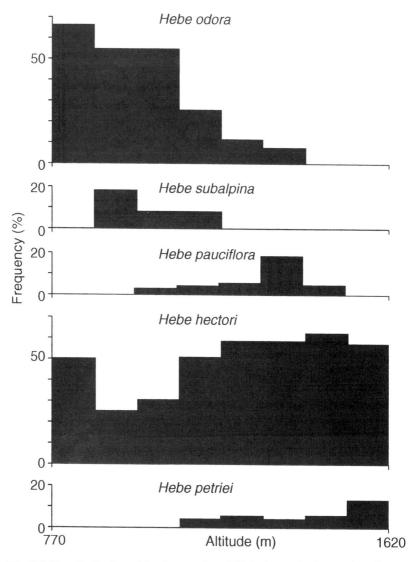

Fig. 5.6. The distribution of the five species of *Hebe* (*sensu lato*) occurring along an altitudinal gradient on a mountain range in south-west New Zealand. (After Wilson & Lee, 1993.)

indicate assembly rules. For example, some have envisaged communities as sets of co-evolved, co-adapted species. Dice (1952) wrote: 'All the species which are members of a given association ... are adjusted more or less perfectly to one another ... [due to] interco-ordinated evolution'. More recently,

Gilpin (1994) envisaged 'cohesive', 'self-organized' communities; if they met, they would not mix, but would 'battle as coordinated armies', each community tending to repel an invader from the enemy. Surely that would be an assembly rule! One would then get sudden boundaries between communities along a gradual environmental gradient, i.e., the species boundaries would be clustered.

Of the attempts that have been made to test this, probably the best is that of Auerbach and Shmida (1993). However, this whole approach founders on the impossibility of defining the environmental scale, and therefore of determining whether or not the environment changes gradually (Wilson, 1994).

The problem can be overcome by asking particular questions that do not depend on the environmental scale, e.g., whether congeners tend to exclude each other, because they occupy similar alpha niches (Boutin & Harper, 1991). Wilson and Lee (1993) found slight evidence for this (Fig. 5.6). Another such question is whether species replace each other along the gradient. Dale (1984) asked this question in a way that did not depend on the environmental scale used (i.e., it was a non-parametric method), and found significant evidence for a pattern, to my surprise (Wilson, 1994).

Assembly rules based on plant characters

Results based on species richness are hard to interpret, because of possible effects of module packing. The most interesting rules are therefore those based on the characters of the plants: limiting similarity, texture convergence and guild proportionality (Leps, 1995; Wilson 1995a; Weiher & Keddy, 1995).

These approaches have the advantage that in the null-model analyses the number of species per quadrat is generally held equal to that observed. Therefore, unlike rules based on species richness, there can be no simple effect of any constraint on module packing. For example, in guild proportionality analyses, for a two-species quadrat there are two species in that quadrat in both the observed and randomized data, the question is only the guild membership of the two species present. Any difference between the observed and randomized data cannot be due to there being only two species in that quadrat, because this feature does not differ between the observed and the randomized.

Limiting similarity

The limiting similarity approach looks to see whether pairs of species, that are judged to be similar in niche, co-occur less often that one would expect on a random basis. Looking at it the other way around: do species that co-occur have less niche overlap than expected at random?

Cody (1986) examined the leaf sizes of *Leucadendron* species in Cape Province, South Africa. He determined morphological overlap between two species as an intersection of their respective male–female morphological ranges, an intuitively obvious if logically puzzling procedure. Cody states that morphologically overlapping species co-occurred at a site significantly less often than expected at random. The basis of the null model is not entirely clear, but this represents a pioneering attempt to examine limiting similarity.

There have been many attempts to find evenly spaced (i.e., staggered) flowering times within a community (e.g., Pleasants, 1980), although with many methodological arguments (e.g., Fleming & Partridge, 1984; Pleasants, 1990). Ranta *et al.* (1981) used a limiting similarity approach to this problem. They examined the plant species in two boreal fields, and clustered them into groups with similar pollinator visitors. In one field, species within two out of seven groups were significantly more evenly spaced in flowering time than expected at random (Fig. 5.7). In another field three out of six groups showed significantly even spacing. So, species that share pollinators can coexist more readily if they differ in flowering time. Moreover, pairs of species that overlapped in both pollinator visitors and flowering time usually differed in a character such as plant height or flower color (though there was no statistical test for this). These characters would encourage different groups of pollinators, and therefore reduce competition for pollinators. It is not quite clear whether the result is due to sorting in ecological time, and thus qualifies as an assembly rule, or whether it represents 'the ghosts of competition past' (Connell, 1980). Probably there are elements of both.

Weiher *et al.* (1998) tested for limiting similarity in the species of several

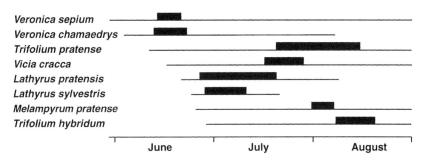

Fig. 5.7. The distribution of the flowering times of species in a field at Puumala, Finland, group PA (the groups are based on similarity in pollinator visitors, using a cluster analysis). (After Ranta *et al.*, 1981.)

wetland habitats. For three characters, they found a significant tendency towards even spacing, in that different quadrats tended to contain species that spanned a wide range of values. All three characters were related to plant size. This pattern was confirmed by guild-proportionality analyses. So, quadrats tended to have some large species and also some small species. This concept is related to guild proportionality in stratification (Wilson, 1989; Wilson *et al.*, 1995a; Wilson & Roxburgh, 1994). The problem with the analyzes of Weiher *et al.* is that species occurrences were explicitly sampled and randomized over several habitats. The problems mentioned above therefore arise, that this is mainly a test of species pools rather than rules for assembly within a community, and that although there were 115 quadrats, there were not 115 independent species pools.

In my view, a very model of how to undertake assembly-rules work is the study by Armbruster *et al.* (1994) of *Stylidium* species in Western Australia. These plants have a complicated pollination system involving bees and flies, different flower shapes allowing different pollinators. Armbruster *et al.* used a patch model to allow for biogeographic variation. Their null model indicates how often pairs of species, that are similar in flower type, would co-occur on a random basis (Fig. 5.8). In fact, the observed data show very few such co-occurrences of similar pairs. Using a one-tailed test, Armbruster found the result significant (Fig. 5.8). A two-tailed test should probably have been used, by which $P > 0.05$. There was strong significance when character displacement was included in the test, but, as evolutionary change, character

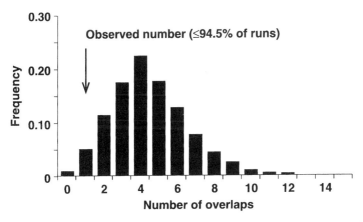

Fig. 5.8. Frequency distribution, under an ecological null model, of the number of co-occurring *Stylidium* species pairs in an area of Western Australia that overlap in pollination niche, and the observed number of such overlaps. (After Armbuster *et al.*, 1994.)

displacement does not seem to qualify as an assembly rule. Armbruster (1986) had done a similar test using the tropical Euphorbiad *Dalechampia*. Again, species assembly was significant using a one-tailed test, but not using a two-tailed one. However, this approach looks promising.

Armbruster (1995) suggested that character displacement might occur more often in reproductive characters, but that since the limitations to coexistence which are the basis of assembly rules would operate over short distances, assembly rules are more likely to be related to the vegetative characters of the species.

Texture convergence

Texture refers to the range of plant characters in a community, irrespective of taxon (Barkman, 1979). The characters considered are generally those believed to be indicative of niche, e.g., leaf thickness, leaf angle, NPK content, chlorophyll content, respiration rate, rooting pattern, etc. For example, a grassland has a different texture from a shrubland, because of differences in leaf shape, woodiness, etc. An assembly rule in this context is observed when biotic interactions cause convergence: similar texture in different sites, even those on different continents.

Many workers have assumed that community convergence would have to be the result of evolution, and have felt obliged to apologize for their lack of information on ancestors (e.g., Orians & Paine, 1983; Schluter, 1986; Wiens, 1991a). Weiher and Keddy (1995) took the opposite extreme, assuming that all 'trait overdispersion' was caused immediately by competition, i.e., by ecological sorting.

If ecological sorting is occurring, then when two species are present that are too similar, one of them will suffer competitive exclusion (Fig. 5.9(*a*)). This assumes that there is a pool of species, of which some are eliminated, but indeed the regional species pool, even after physical filtering, is generally larger than the local pool, depending of the spatial scale examined (Partel *et al.*, 1996). Such ecological sorting will occur during colonization of empty sites. It will continue to operate as species from the regional pool continue to invade, either failing to establish due to suppression by superior competitors vying for the same niche space, or causing functionally similar species already present to succumb to competitive exclusion. If there are niches in a community for a range of functional types, with more or less one species per niche, the result would be expected to be convergence between comparable communities in different areas. The same pressures will act via selection in *evolutionary* time to cause *evolutionary* convergence between regions (Fig. 5.9(*b*)).

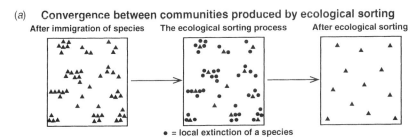

Fig. 5.9. Ecological and evolutionary processes leading to convergence. The square area represents functional space. Ecological and evolutionary processes can both cause an even distribution of species in functional space. If this occurs independently in different continents/sites/patches, the communities will converge.

Thus, convergence, in the sense of communities with a more similar distribution of species in niche space than expected on the basis of random assortment from species pools (e.g., Wilson *et al.*, 1994; Smith *et al.*, 1994), can be produced by either ecological or evolutionary processes. Comparing continents, it is difficult to distinguish between ecological and evolutionary convergence. However, at the community level evolutionary convergence might be expected to be rare, because most species occur in several different associations, and cannot coevolve simultaneously to fit in with each set of associates (Goodall, 1966). The caveat to this argument is ecotypic adjustment to different associates, in effect character displacement.

Most studies of texture convergence have compared Mediterranean shrublands. The problem has been the absence of a null model, which has led some to the defeatist view that such work is impossible (Blondel *et al.*, 1984; Keely, 1992). However, Wilson *et al.* (1994) developed a suitable null model (related to earlier work by Schluter, 1986), and used it to look for convergence in carrs (i.e., wooded fens) in Britain and New Zealand. They measured five functional characters related to light capture (Fig. 5.10). Species presence/absence data revealed no convergence. However, when species were weighted by their abundance, convergence was seen in some variates.

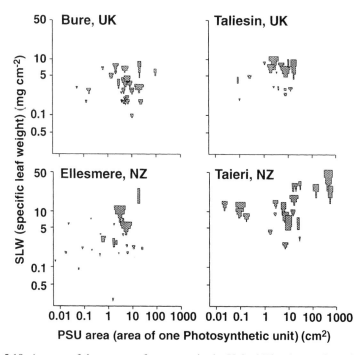

Fig. 5.10. Aspects of the texture of two carrs in the United Kingdom and two in New Zealand. The size of the symbol indicates the abundance (photosynthetic biomass) of the species, and the steps the hierarchy of modules, and the shape leaf shape. (After Wilson *et al.*, 1994.)

Some studies with animals have also failed to find texture convergence (e.g., Wiens, 1991b). In testing assembly rules it is usually necessary to contend with environmental noise, thus convergence may sometimes not be demonstrable because the environments of the continents are not similar enough. With texture convergence, there is the additional problem of historical noise, i.e., the different evolutionary and biogeographic history of different continents may have resulted in species pools that are too different for convergence to have been completed. A negative texture-convergence result could also be because our analyses are not yet sophisticated enough to see convergence above the historical and environmental noise. A recent search for texture convergence in the non-*Nothofagus* component of *Nothofagus* forest in Argentina, Chile, New Zealand, Australia and Tasmania, has demonstrated that, if one looks at only the shape of the distribution, to control for environmental variation, significant convergence is demonstrable (B. Smith & J.B. Wilson, unpublished data).

Texture convergence might occur on a much smaller scale too. For example, Smith *et al.* (1994) compared stands of *Nothofagus* forest a few hundred meters apart. They obtained some evidence for convergence, although they were aware of problems in testing convergence using abundance information when there is species overlap. This problem has now been solved (J.B. Wilson & B. Smith, unpublished data), and comparisons of texture in patches of communities a few meters apart are currently in process (A.J. Watkins & J.B. Wilson, unpublished data). Here, ecological convergence, not evolutionary is clearly being examined. The question is: if two patches are compared, and, in the second patch, a species is present in lower abundance, is that compensated by greater abundance in a species with similar functional characters?

Guild proportionality

The texture approach characterizes a community by the range of functional characters within it. An alternative is to divide the species into arbitrarily discrete guilds (= functional types, = functional groups), according to their functional characters. Convergence can then be sought as a proportional representation from different guilds that is constant between patches of a community. Wilson (1989) attempted this for a gymnosperm/angiosperm rainforest, but found little evidence for an assembly rule. Wilson *et al.* (1995a) used the same method on *Nothofagus* rainforests, and found a significant tendency towards constant proportions of species from the ground and herb guilds.

Wilson and Roxburgh (1994) sampled a species-rich, stable lawn, and found regularity in the proportion of species from grass and forb guilds, a rule which held whether bryophytes were counted as forbs, or omitted (Fig. 5.11(*a*), (*b*)). A similar assembly rule was found in guilds based on height strata (Fig. 5.11(*c*)).

Ecologists skeptical of assembly rules have an important point, that it is very easy to find spurious rules. The grass : forb rule of Wilson and Roxburgh could be reproduced by confusion in identification between different species of grass. There are sampling errors in all work, but one has to be especially wary of errors that would mimic the effect being sought. Our result was submitted for publication only several analyzes had convinced us that this was not the cause of our result (Wilson, 1995a).

Fox and Brown (1993), seeking an assembly rule very similar to guild proportionality (they examined numbers, not proportions) fell into the 'Jack Horner' trap (Wilson, 1995c), because they failed to include, in their null model, differences in frequency between species. The departure from their

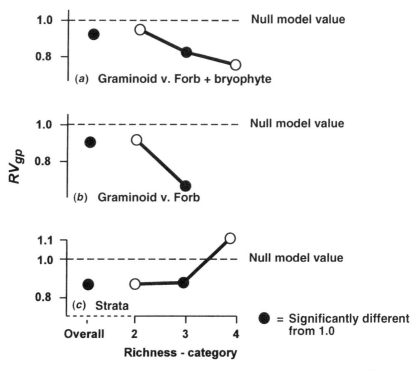

Fig. 5.11. Variation in guild proportions on the Botany Lawn, University of Otago, New Zealand, compared to that expected under a null model. An RV_{gp} value < 1.0 indicates a trend towards constant guild proportions (i.e., guild proportionality). (After Wilson & Roxburgh, 1994.)

null model can therefore be seen as just a demonstration that some species are more frequent than others – hardly an assembly rule (Wilson, 1995c). Fox and Brown (1993) also failed to incorporate in their null model that some species were absent from some sites because their geographic range did not include the whole area of sampling (Stone *et al.*, 1996).

Sometimes, taxonomy has been used as a rough guild classification. Mohler (1990) made a comparison with two subgenera of *Quercus*. For 12 of the 14 regions that he examined there was a significant tendency for the two most dominant oak species to be from different subgenera. This was not related to consistent pairing of particular species. There are some problems with his data and his null model, but the result is still impressive. Mohler examined various explanations: disease/pest pressure, niche differences in fruiting phenology through mast fruiting, dispersal differences, etc., but could not find any clear single explanation. The result is most easily explained by assuming more niche similarity within subgenera than between, but it is not known what niche axis is involved.

Similar questions, of guild proportionality, have been asked at coarser guild levels, such as the proportions of predator and prey (including plant) species in communities. Several claims have been made for constancy in these proportions, which would be an impressive assembly rule (e.g., Jeffries & Lawton, 1985; Gaston *et al.*, 1992). Pimm (1991) gave the supposed assembly rule of near-constant predator : prey ratios as an example of ecological theory that could be used to guide management decisions in nature conservation. Unfortunately for this widely touted 'rule', a null-model comparison shews that predator : prey ratios are usually about as constant as expected at random; Wilson (1996) was able to find no case of predator : prey ratios significantly more constant than expected at random.

Intrinsic guilds

As discussed above, one of the problems in seeking assembly rules is that it is not known what rules to seek, what is in the plants' rule book. This is a particular problem with rules based on plant characters. There is ignorance not only of the assembly rules operating but also of which characters the rules apply to. All limiting-similarity work, all texture work, and almost all guild work, has had to start with an assumption of which characters might be important, and should be measured.

In guild work, this problem can be overcome. If guilds are defined as groups of species which compete more with each other than with species in other guilds (the guild definition of Pianka, 1980, 1988), and if such guilds exist, and if competition limits species coexistence, then there must be guild proportionality. That is to say, if there are guilds limiting coexistence, we can find them (Wilson & Roxburgh, 1994). An *a priori* hypothesis of what the guilds are is not needed, because 'the plants can be asked' what guilds they are in. To do this, the guild classification that maximizes the degree of guild proportionality is found (of course, to avoid circularity the classification has to be determined on one part of the data, and tested on independent data). This allows the intrinsic guilds to be determined, i.e., the guilds intrinsic to the community, the guilds that the plants see, not those imposed by the ecologist.

Correlations then might be found between the intrinsic guilds and the characters that are known. Or no correlations might be seen, because the significant characters are ones that have not been measured, or because coexistence depends on complicated combinations of characters. However,

Table 5.3. *Intrinsic guilds in the Botany Lawn, University of Otago, New Zealand, sampled at point scale*

Guild A	Guild B
Agrostis capillaris	Acrocladium cuspidatum
Anthoxanthum odoratum	Cerastium fontanum
Bellis perennis	Cerastium glomeratum
Holcus lanatus	Eurhynchium praelongum
Hydrocotyle moschata	Festuca rubra
Linum catharticum	Hydrocotyle heteromeria
Poa pratensis	Hypochaeris radicata
Trifolium dubium	Prunella vulgaris
Trifolium repens	Ranunculus repens
	Sagina procumbens

Table 5.4. *Intrinsic guilds in a Welsh salt marsh, sampled at the scale of a 2 cm diameter circle*

Guild A	Guild B
Festuca rubra	Plantago lanceolata
Puccinellia maritima	Salicornia europaea
Triglochin maritima	Spergularia media
	Suaeda maritima

Variable in guild membership:
Armeria maritima   Juncus maritimus
Glaux maritima   Aster tripolium

the intrinsic guilds are, within our assumptions, still the truth[1], even though there might be no idea of what plant interaction mechanism is causing them.

Using this method on a lawn, Wilson and Roxburgh (1994) found significant guild structure, and it seemed to be related to leaf shape and orientation (Table 5.3). Wilson and Whittaker (1995) used the method on a salt marsh, and again found significant guild structure related to leaf shape and orientation (Table 5.4). It is too soon to tell, but the possibility of a general assembly rule emerges, if comparable guilds can be found in other communities.

One advantage of the intrinsic guild approach is that it can fail. Other methods of classifying species into guilds are bound to give guilds. With an intrin-

[1] i.e., $P < 0.05$, which counts as the truth!

sic guild search, if there are no guilds or there is no competitive exclusion, the search will fail, and it often does (Wilson *et al.*, 1995a,b,c).

Abundance-based assembly rules

Assembly rules can also be based on quantitative information, i.e., on species abundances.

Biomass constancy

One obvious abundance-based rule, trivial almost, would be a constancy (between patches of a community) of total biomass because of competition: when the abundance of one species is higher, that of another or of others is lower (Fig. 5.12). Wilson and Gitay (1995a), in a Welsh dune slack, examined variance in total biomass between quadrats, and compared it with that expected under a null model in which the biomasses of the species were allocated at random. Habitat heterogeneity makes it difficult to demonstrate such a rule, but they did so by using a patch model. It *is* possible to obtain evidence for competition from snapshot field data (cf. Goldberg, 1995; Wilson, 1995a).

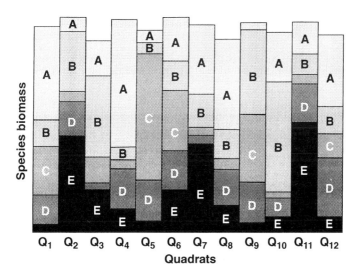

Fig. 5.12. The concept of biomass constancy: when the biomass of one species is higher, that of another or others is lower (artificial data).

Abundance-based limiting-similarity / texture / guilds

The limiting similarity, texture convergence, guild proportionality and intrinsic guild approaches could, in theory, all be based on abundance information, ideally biomass. No such approaches with limiting similarity are known. Plant texture convergence work has included abundance information from the beginning (Wilson *et al.*, 1994).

Guild proportionality could be approached with abundance data. The question 'is there a constant proportion of species from each guild' then becomes 'is there a constant proportion of biomass from each guild'. Wilson *et al.* (1996a) calculated abundance-based guild proportionality within plots of the Park Grass experiment. There had been no evidence of guild proportionality using presence/absence data, and there was none using biomass data.

Preliminary searches for intrinsic guilds in semi-arid grassland in New Zealand, using abundance (biomass) information, have given negative results (e.g., all species in one guild: Table 5.5).

One would expect that the addition of information on abundances would make it easier to see structure, but that does not seem to be the case so far.

Relative abundance distribution (RAD)

A notable feature of any community is the distribution of relative abundances of the species (RAD), irrespective of species identities. To display an RAD, we normally use a ranked-abundance plot: the log of abundance against the rank order of abundance (Fig. 5.13). An assembly rule based on the RAD could be very general, because the RAD of any community can be found, and RADs can be compared worldwide. It is not necessary to know anything about the species involved, except that they are distinct species. The RAD is also the sort of information that is observed instinctively: if in a community 95% of the biomass is one species, this is commented on. If there are many species with about equal abundance, this is commented on.

Evenness

The most obvious feature of the RAD is evenness, in effect the lack of slope on the ranked-abundance plot (although not calculated that way: Smith & Wilson, 1996).

Grime (1973, 1979) predicted a humped-back curve for richness – an increase and then a decrease as total community biomass increased. Using the same logic that has been used for the humped-back richness model

Table 5.5. *The result of an intrinsic guild search in a New Zealand semi-arid grassland, sampled at the scale of 10 cm × 10 cm*

Guild A	Guild B
Agrostis sp.	
Aira caryophyllea	
Anthoxanthum odoratum	
Breutelia affinis	
Carex breviculmis	
Cladonia sp.	
Echium vulgare	
Erodium cicutarium	
Erophila verna	
Geranium sessiliflorum	
Gypsophila australis	
Hieracium pilosella	
Hypericum perforatum	
Leucopogon fraseri	
Minuartia hybrida	
Myosotis discolor	
Neofuscella sp.	
Pimelea sericeo-villosa	
Poa maniototo	
Polytrichum juniperinum	
Raoulia apicenigra	
Raoulia australis	
Rumex acetosella	
Rytidosperma buchanani	
Stellaria gracilenta	
Trifolium arvense	
Triquetrella papillata	
Veronica verna	
Vulpia dertonensis	
Xanthoparmelia amphixantha	
Xanthoparemelia cf. tasmanica	

(Drobner *et al.* 1998) predicted a monotonic decrease for evenness. They examined this in a number of communities in a high-rainfall area of New Zealand, and obtained the relation predicted (Fig. 5.14). However, a very similar result can be found using random numbers (Drobner *et al.*, 1998). The effect seems to be due to the roughly 'geometric' shape of most RADs, and the fact that evenness calculations are based on the proportional relations of the species, but total biomass is the sum of their absolute values. Here is another illustration of the need to examine apparent assembly rules carefully, and to test the method with random data.

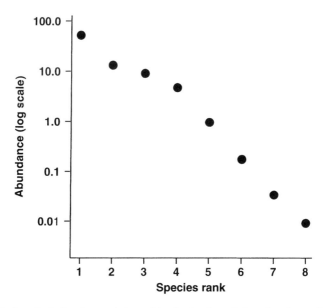

Fig. 5.13. A ranked-abundance plot, used to display the relative abundance distribution (RAD) (artificial data).

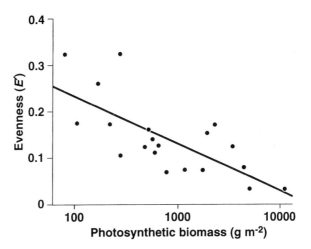

Fig. 5.14. The relation between evenness (measured with index E') and photosynthetic biomass, as seen in ten distinct communities in south-west New Zealand. (After Drobner *et al.*, 1998.)

However, some interesting trends in evenness can be seen. Wilson *et al.* (1996d) found that, in higher-phosphate plots of the Park Grass experiment, evenness was higher. They also found that, in two grasslands, evenness increased through succession, a topic on which there had been much speculation.

Shape of the relative abundance distribution (RAD)

The shape of the RAD contains information other than the slope. There are five simple shapes possible (Fig. 5.15). S-shaped curves (b) and approximately geometric slopes (c) seem common in the literature (e.g., Wilson, 1991a; Lee *et al.*, 1991; Watkins & Wilson, 1994). Concave-up slopes (a) are sometimes found. Reverse-S (d) and concave-down (e) shapes seem very uncommon. This is an assembly rule of a very general kind.

It has often been suggested that particular shapes of RAD will occur in particular conditions (see Wilson, 1991a). The usual way to examine this has been by comparison with models of RAD, for plants: Broken stick, Geometric, General Lognormal and Zipf–Mandelbrot (Wilson, 1991a).

Watkins and Wilson (1994) examined 12 communities, with replicate quadrats. They found highly significant differences in the RAD of different communities, but could find no rule to predict which sort of community would have what sort of RAD. Wilson *et al.* (1996d) compared a range of about 80 fertilizer treatments in the Park Grass experiment. There was no consistent relation between the RAD and any soil factor, except for a slight tendency for the Broken Stick model to fit less poorly where phosphate had been applied, and that apparently just because evenness was higher.

The discovery of consistent trends in more subtle aspects of the relative abundance pattern remains elusive. It cannot be predicted which shape of RAD will appear when, in spite of ancient claims in the literature (see Wilson, 1991a; Watkins & Wilson, 1994). It has to be suspected that the identity of individual species has too great an effect for general rules to be possible.

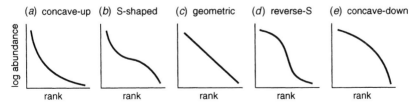

Fig. 5.15. Five basic shapes possible for relative abundance distributions (RADs).

Conclusions

The story so far

Assembly rules can be found in plant communities, but they are challenging to find. Rules can be seen for variance in richness, but interpretation of them is ambiguous. Some rules can be seen for texture convergence, limiting similarity and guild proportionality, and we can sometimes find intrinsic guilds. Currently, these give the best evidence. Evidence for biomass constancy can be found, with difficulty, but this is a trivial rule. Rules that are based on the abundance of the species and are more subtle than biomass constancy seem elusive. Some predictability can be seen in evenness, and in the general shapes of the RAD that are found, but predictivity of the shape of the RAD for a particular community is currently non-existent.

It is known that there is intense competition in natural communities. It is known that species differ widely in their morphology, and in their physiology. How can this not affect their ability to survive in the big wide world?

Probably the difficulty in finding assembly rules in plant communities is because:

(a) there is variation at all scales in environment and in patch history, and this represents noise in assembly-rule work,
(b) the true assembly rules are complex, e.g., because of:
    variation in niche width, and
    the existence of alternative plant strategies ('there is more than one way to kill a cat'), and
(c) the processes that produce assembly rules are subtle.

The way forward

The continued search for assembly rules requires large datasets, with high-quality data such as the 2810 point quadrats Wilson and Roxburgh (1994) or the 534 biomass quadrats of Wilson *et al.* (1996d). Data are required from relatively homogeneous environments, and/or sampling using a grid or a circular transect (Bycroft *et al.*, 1993) so that patch models and the like can be used.

More sophisticated methods of analysis are also required, especially to utilize abundance data. For example, methods have recently been developed for analyzing texture convergence that use the shape of the character distribution, not just the mean (B. Smith & J.B. Wilson, unpublished data), and also methods that can examine texture convergence even with complete overlap in the species present (J.B. Wilson & B. Smith, unpublished data).

Methods are needed that can better partition out environmental noise and cope with spatial autocorrelation (Watkins & Wilson, 1992; Palmer & van der Maarel, 1995).

Insight into the operation of assembly rules might also be obtained by examining the ecological and physiological mechanisms that are the basic cause of the rules, such as canopy and root interactions. For example, it would be interesting to know the species–species interactions that cause guild proportionality (e.g., in the lawn community of Wilson & Roxburgh, 1994). However, because the rules are usually only tendencies, it may be difficult to see the relevant mechanisms above the noise of other processes.

Where and when will assembly rules be found?

Generalizations should be sought on where and when community structure will be strong, giving assembly rules. Some might expect assembly rules to operate mainly in species-rich communities, since more species–species interactions are possible, and since species-rich communities may be closer to niche saturation. However, the best evidence so far has come from quite species-poor communities: lawn and salt marsh (Wilson & Roxburgh, 1994; Wilson & Whittaker, 1995).

Where assembly rules exist, when do they operate? Are they continually present? Or do they operate only in some years, for example during periods when resources are more limiting (Wilson *et al.*, 1995b; cf. Wiens 1977)?

Why assembly rules anyway?

If assembly rules are common, and are a significant community force, they may prove useful for predicting responses to environmental change and to biotic introductions, and for effective community restoration. Drake *et al.* (1996) propose, albeit on rather weak evidence, the 'Humpty Dumpty' effect: that it may not be possible to establish a community simply by bringing together the constituent species, but only by knowing the assembly sequence.

It would be premature to conclude, as some do, that assembly rules are absent. There is some evidence for them, and the search is worthwhile. Even if they are usually not there, it is important that is discovered.

Acknowledgments

I thank for comments on a draft W.S. Armbruster, W.G. Lee, E. Weiher and members of my research group.

References

Armbruster, W.S. (1986). Reproductive interactions between sympatric *Dalechampia* species: are natural assemblages 'random' or organized? *Ecology* 67: 522–533.

Armbruster, W.S. (1995). The origins and detection of plant community structure: reproductive versus vegetative processes. *Folia Geobotanica et Phytotaxonomica* 30: 483–497.

Armbruster, W.S., Edwards, M.E., & Debevec, E.M. (1994). Floral character displacement generates assemblage structure of Western Australian triggerplants (*Stylidium*). *Ecology* 75: 315–329.

Auerbach, M. & Shmidaa, A. (1993). Vegetation change along an altitudinal gradient on Mt Hermon, Israel – no evidence for discrete communities. *Journal of Ecology* 81: 25–33.

Austin, M.P., Nicholls, A.O., & Margules, C.R. (1990). Measurement of the realized qualitative niche: environmental niches of five *Eucalyptus* species. *Ecology* 60: 161–177.

Barkman, J.J. (1979). The investigation of vegetation texture and structure. In *The Study of Vegetation*, ed. M.J.A. Werger, pp. 123–160. The Hague, The Netherlands: Junk.

Bio, A.M.F., Alkemade, R., & Barendregt, A. (1998). Determining alternative models for vegetation response analysis: a non-parametric approach. *Journal of Vegetation Science* 9: 5–16.

Blondel, J., Vuilleumier, F., Marcus, L.F., & Terouanne, E. (1984). Is there ecomorphological convergence among Mediterranean bird communities of Chile, California, and France? *Evolutionary Biology* 18: 141–213.

Boutin, C. & Harper, J.L. (1991). A comparative study of the population dynamics of five species of *Veronica* in natural habitats. *Journal of Ecology* 79: 199–221.

Bycroft, C.M., Nicolaou, N., Smith, B., & Wilson, J.B. (1993). Community structure (niche limitation and guild proportionality) in relation to the effect of spatial scale, in a *Nothofagus* forest sampled with a circular transect. *New Zealand Journal of Ecology* 17: 95–101.

Caley, M.J. & Schluter, D. (1997). The relationship between local and regional diversity. *Ecology* 78: 70–80.

Cody, M.L. (1986). Structural niches in plant communities. In *Community Ecology* ed. J.M. Diamond & T.J. Case, pp. 381–405. NY: Harper & Row.

Cole, B.J. (1983). Assembly of mangrove an communities: patterns of geographical distribution. *Journal of Animal Ecology* 52: 339–347.

Colwell, R.K. & Winkler, C.W. (1984). A null model for null models in biogeography. In *Ecological Communities: Conceptual Issues and the Evidence*, ed. D.R. Strong, D. Simberloff, L.G. Abele, & A.B. Thistle, pp. 344–359. Princeton, NJ: Princeton University Press.

Connell, J.H. (1980). Diversity and the coevolution of competitors, or the ghost of competition past. *Oikos* 35: 131–138.

Connell, H.V. & Lawton, J.H. (1992). Species interactions, local and regional processes, and limits to the richness of ecological communities: a theoretical perspective. *Journal of Animal Ecology* 61: 1–12.

Connor, E.F. & Simberloff, D. (1979). The assembly of species communities: chance or competition? *Ecology* 60: 1132–1140.

Cresswell, J.E., Vidal-Martinez, V.M., & Crichton, N.J. (1995). The investigation of saturation in the species richness of communities: some comments on methodology. *Oikos* 72: 301–304.

Dale, M.R.T. (1984). The contiguity of upslope and downslope boundaries of species in a zoned community. *Oikos* 42: 92–96.

de Kroon, H. & Olff, H. (1995). On the use of the guild concept in plant ecology. *Folia Geobotanica et Phytotaxonomica* 30: 519–552.

Diamond, J.M. (1975). Assembly of species communities. In *Ecology and Evolution of Communities*, ed. M.L. Cody & J.M. Diamond, pp. 342–444. Cambridge, MA: Harvard University Press.

Diaz, S.; Acosta, A.; Cabido, M. (1994). Grazing and the phenology of flowering and fruiting in a montane grassland in Argentina – a niche approach. *Oikos*, 70: 287–295.

Dice, L.R. (1952). *Natural Communities*. Ann Arbor: University of Michigan Press.

Drake, J.A., Huxel, G.R., & Hewitt, C.L. (1996). Microcosms as models for generating and testing community theory. *Ecology* 77: 670–677.

Drobner, U., Bibby, J., Smith, B., & Wilson, J.B. (1998). The relation between community biomass and evenness: what does community theory predict, and can these predictions be tested? *Oikos* 82: 295–302.

Fleming, T.H. & Partridge, B.L. (1984). On the analysis of phenological overlap. *Oecologia (Berlin)* 62: 344–350.

Fox, B.J. & Brown, J.H. (1993). Assembly rules for functional groups in North American desert rodent communities. *Oikos* 67: 358–370.

Gaston, K.J., Warren, P.H., & Hammond, P.M. (1992). Predator: non-predator ratios in beetle assemblages. *Oecologia*, 90: 417–421.

Gilpin, M. (1994). Community-level competition: asymmetrical dominance. *Proceedings of the National Academy of Sciences, USA*, 91: 3252–3254.

Goldberg, D.E. (1995). Generating and testing predictions about community structure: which theory is relevant and can it be tested with observational data? *Folia Geobotanica et Phytotaxonomica* 30: 511–518.

Goodall, D.W. (1966). The nature of the mixed community. *Proceedings of the Ecological Society of Australia* 1: 84–96.

Graves, G.R. & Gotelli, N.J. (1993). Assembly of avian mixed-species flocks in Amazonia. *Proceedings of the National Academy of Sciences, USA*, 90: 1388–1391.

Grime, J.P. (1973). Competitive exclusion in herbaceous vegetation. *Nature* 242: 344–347.

Grime, J.P. (1979). *Plant Strategies and Vegetation Processes*. Chichester, UK: Wiley.

Hastings, A. (1987). Can competition be detected using species co-occurrence data? *Ecology* 68: 117–123.

Hunt, G.L. Jr (1991). Occurrence of polar seabirds at sea in relation to prey concentrations and oceanographic factors. *Polar Research* 10: 553–559.

Jeffries, M.J. & Lawton, J.H. (1985). Predator-prey ratios in communities of freshwater invertebrates: the role of enemy free space. *Freshwater Biology* 15: 105–112.

Keddy, P.A. (1992). Assembly and response rules – two goals for predictive community ecology. *Journal of Vegetation Science* 3: 157–164.

Keeley, J.E. (1992). A Californian's view of fynbos. In *The Ecology of Fynbos – Nutrients, Fire and Diversity*, ed. R.M. Cowling, pp. 372–388. Cape Town, South Africa: Oxford University Press.

Lawler, S.P. (1993). Direct and indirect effects in microcosm communities of protists. *Oecologia* 93: 184–190.

Lawton, J.H., Lewinsohn, T.M. & Compton, S.G. (1993). Patterns of diversity for the insect herbivores on bracken. In *Species Diversity in Ecological*

Communities, ed. R.E. Ricklefs & D. Schluter. pp. 178–184. Chicago, IL: University of Chicago Press.

Lee, W.G., Ward, C.M., & Wilson, J.B. (1991). Comparison of altitudinal sequences of vegetation on granite and on metasedimentary rocks in the Preservation Ecological District, southern Fiordland, New Zealand. *Journal of the Royal Society of New Zealand* 21: 261–276.

Leps, J. (1995). Variance deficit is not reliable evidence for niche limitation. *Folia Geobotanica et Phytotaxonomica* 30: 455–459.

Lockwood, J.L., Powell, R.D., Nott, M.P., & Pimm, S.L. (1997). Assembling ecological communities in time and space. *Oikos* 80: 549–553.

MacArthur, R.H. (1972). *Geographical Ecology: Patterns in the Distribution of Species*. NY: Harper & Row.

McIntosh, R.P. (1975). H.A. Gleason – 'Individualistic Ecologist' 1882–1975: his contribution to ecological theory. *Bulletin of the Torrey Botanical Club* 102: 253–273.

Manly B.F.J. (1991). *Randomization and Monte-Carlo Methods in Biology*. London: Chapman & Hall.

May, A.U. & MacArthur, R.H. (1972). Niche overlap as a function of environmental variability. *Proceedings of the National Academy of Sciences, USA* 69: 1109–1113.

Mohler, C.L. (1990). Co-occurrence of oak subgenera: implications for niche differentiation. *Bulletin of the Torrey Botanical Club* 117: 247–255.

Orians, G.H. & Paine, R.T. (1983). Convergent evolution at the community level. In *Coevolution*, ed. D.J. Futuyma & M. Slatkin, pp. 431–458. Sunderland, USA: Sinauer.

Pacala, S.W. & Tilman, D. (1994). Limiting similarity in mechanistic and spatial models of plant competition in heterogeneous environments. *American Naturalist* 143: 222–257.

Palmer, M.W. & van der Maarel, E. (1995). Variance in species richness, species association, and niche limitation. *Oikos* 73: 203–213.

Partel, M., Zobel, M., Zobel, K., & van der Maarel, E. (1996). The species pool and its relatio to species richness: evidence from Estonian plant communities. *Oikos* 75: 111–117.

Pianka, E.R. (1980). Guild structure in desert lizards. *Oikos* 35: 194–201.

Pianka, E.R. (1988). *Evolutionary Ecology*, 4th edn. NY: Harper and Row.

Pimm, S.L. (1991). *The Balance of Nature*? Chicago, IL: University of Chicago Press.

Pleasants, J.M. (1980). Competition for bumblebee pollinators in Rocky Mountain plant communities. *Ecology* 61: 1446–1459.

Pleasants, J.M. (1990). Null-model tests for competitive displacement: the fallacy of not focussing on the whole community. *Ecology* 71: 1078–1084.

Ranta, E., Teräs, I., & Lundberg, H. (1981). Phenological spread in flowering of bumblebee-pollinated plants. *Annales Botanici Fennici* 18: 229–236.

Ricklefs, R.E. (1987). Community diversity: relative roles of local and regional processes. *Science* 235: 167–171.

Roxburgh, S.H. & Chesson, P. (1998). The Random Patterns test: a new method for detecting species associations in the presence of spatial autocorrelation. *Ecology*, in press.

Schluter, D. (1986). Tests for similarity and convergence of finch communities. *Ecology* 67: 1073–1085.

Simberloff, D. (1982). The status of competition theory in ecology. *Annales Zoologici Fennici* 19: 241–253.

Smith, B. & Wilson, J.B. (1996). A consumer's guide to evenness indices. *Oikos* 76: 70–82.

Smith, B., Moore, S.H., Grove, P.B., Harris, N.S., Mann, S., & Wilson, J.B. (1994). Vegetation texture as an approach to community structure: community-level convergence in a New Zealand temperate rainforest. *New Zealand Journal of Ecology* 18: 41–50.

Stone, L., Dayan, T., & Simberloff, D. (1996). Community-wide assembly patterns unmasked: the importance of species' differing geographical ranges. *American Naturalist* 148: 997–1015.

Tokeshi, M. (1986). Resource utilization, overlap and temporal community dynamics: a null model analysis of an epiphytic chironomid community. *Journal of Animal Ecology* 55: 491–506

Warming, E. (1909). *Oecology of Plants: An Introduction to the Study of Plant-Communities*. Oxford, UK: Oxford University Press.

Watkins, A.J. & Wilson, J.B. (1992). Fine-scale community structure of lawns. *Journal of Ecology* 80: 15–24.

Watkins, A.J. & Wilson, J.B. (1994). Plant community structure, and its relation to the vertical complexity of communities: dominance/diversity and spatial rank consistency. *Oikos* 70: 91–98.

Weiher, E. & Keddy, P.A. (1995). Assembly rules, null models, and trait dispersion: new questions from old patterns. *Oikos* 74: 159–164.

Weiher, E., Clarke, G.D.P. & Keddy, P.A. (1998). Community assembly rules, morphological dispersion, and the coexistence of plant species. *Oikos* 81: 309–322.

Wiens, J.A. (1977). On competition and variable environments. *American Scientist*, 65: 590–597.

Wiens, J.A. (1991a). Ecological similarity of shrub-desert avifaunas of Australia and North America. *Ecology* 72: 479–495.

Wiens, J.A. (1991b). Ecomorphological comparisons of the shrub-desert avifaunas of Australia and North America. *Oikos* 60: 55–63.

Wilson, J.B. (1987). Methods for detecting non-randomness in species co-occurrences: a contribution. *Oecologia* 73: 579–582.

Wilson, J.B. (1989). A null model of guild proportionality, applied to stratification of a New Zealand temperate rain forest. *Oecologia* 80: 263–267.

Wilson, J.B. (1991a). Methods for fitting dominance / diversity curves. *Journal of Vegetation Science* 2: 35–46.

Wilson, J.B. (1991b). Does Vegetation Science exist? *Journal of Vegetation Science* 2: 289–290.

Wilson, J.B. (1994). Who makes the assembly rules? *Journal of Vegetation Science* 5: 275–278.

Wilson, J.B. (1995a). Testing for community structure: a Bayesian approach. *Folia Geobotanica et Phytotaxonomica* 30: 461–469.

Wilson, J.B. (1995b). Hypothesis testing, intrinsic guilds and assembly rules. *Folia Geobotanical et Phytotaxonomica* 30: 535–536.

Wilson, J.B. (1995c). Null models for assembly rules: the Jack Horner effect is more insidious than the Narcissus effect. *Oikos* 72: 139–143.

Wilson, J.B. (1995d). Variance in species richness, niche limitation, and vindication of patch models. *Oikos* 73: 277–279.

Wilson, J.B. (1996). The myth of constant predator: prey ratios. *Oecologia* 106: 272–276.

Wilson, J.B. & Lee, W.G. (1994). Niche overlap of congeners: a test using plant altitudinal distribution. *Oikos* 69: 469–475.

Wilson, J.B. & Roxburgh, S.H. (1994). A demonstration of guild-based assembly rules for a plant community, and determination of intrinsic guilds. *Oikos* 69: 267–276.

Wilson, J.B. & Watkins, A.J. (1994). Guilds and assembly rules in lawn communities. *Journal of Vegetation Science* 5: 591–600.

Wilson, J.B. & Gitay, H. (1995a). Limitations to species coexistence: evidence for competition from field observations, using a patch model. *Journal of Vegetation Science* 6: 369–376.

Wilson, J.B. & Gitay, H. (1995b). Community structure and assembly rules in a dune slack: variance in richness, guild proportionality, biomass constancy and dominance/diversity relations. *Vegetatio* 116: 93–106.

Wilson, J.B. & Whittaker, R.J. (1995). Assembly rules demonstrated in a saltmarsh community. *Journal of Ecology* 83: 801–807.

Wilson, J.B., Agnew, A.D.Q., & Gitay, H. (1987). Does niche limitation exist?. *Functional Ecology* 1: 391–397.

Wilson, J.B., Roxburgh, S.H., & Watkins, A.J. (1992). Limitation to plant species coexistence at a point: a study in a New Zealand lawn. *Journal of Vegetation Science* 3: 711–714.

Wilson, J.B., Agnew, A.D.Q., & Partridge, T.R. (1994). Carr texture in Britain and New Zealand: convergence compared with a null model. *Journal of Vegetation Science* 5: 109–116.

Wilson, J.B., Allen, R.B., & Lee, W.G. (1995a). An assembly rule in the ground and herbaceous strata of a New Zealand rainforest. *Functional Ecology* 9: 61–64.

Wilson, J.B., Sykes, M.T., & Peet, R.K. (1995b). Time and space in the community structure of a species-rich grassland. *Journal of Vegetation Science* 6: 729–740.

Wilson, J.B., Steel, J.B., Newman, J.E., & Tangney, R.S. (1995c). Are bryophyte communities different? *Journal of Bryology* 18: 689–705.

Wilson, J.B., Crawley, M.J., Dodd, M.E., & Silvertown, J. (1996a). Evidence for constraint on species coexistence in vegetation of the Park Grass experiment. *Vegetatio* 124: 183–190.

Wilson, J.B., Lines, C.E.M., & Silvertown, J. (1996b). Grassland community structure under different grazing regimes, with a method for examining species associations when local richness is constrained. *Folia Geobotanica et Phytotaxonomica* 31: 197–206.

Wilson, J.B., Ullmann, I., & Bannister, P. (1996c). Do species assemblages ever recur? *Journal of Ecology* 84: 471–474.

Wilson, J.B., Wells, T.C.E. Trueman, I.C., Jones, G., Atkinson, M.D. Crawley, M.J., Dodd, M.E., & Silvertown, J. (1996d). Are there assembly rules for plant species abundance: An investigation in relation to soil resources and successional trends. *Journal of Ecology* 84: 527–538.

Zobel, K., Zobel, M., & Peet, R.K. (1993). Changes in pattern diversity during secondary succession in Estonian forests. *Journal of Vegetation Science* 4: 489–498.

6

Assembly rules at different scales in plant and bird communities

Martin L. Cody

Introduction: assembly rules, scale, purpose and application

One of the basic premises in ecology is that communities are composed of collections of species that are subsets of a larger pool of available species (e.g., Cody & Diamond, 1975; Diamond & Case, 1986), and that the composition of these subsets is governed, potentially at least, by 'assembly rules' (Diamond, 1985). In particular, and as originally envisaged by Diamond (1985), if community size varies with, e.g., island size, vegetation structure, or some other extrinsic factor, then such rules will also govern community amplification. Ideally, assembly rules should cover both species composition and relative abundance, the two more basic variables or descriptors of the community.

Conceptually, assembly rules are of two basic categories. One type, which will be called 'type A', governs community membership or species composition as community size increases or decreases. Thus, type A rules address which species are added as community size increases, e.g., with island size (MacArthur & Wilson, 1967), or with the stature or complexity of vegetation (MacArthur *et al.*, 1966; Cody, 1973), or decreases in smaller or in more isolated habitat fragments with poor recolonization potential (Cody, 1973; Bolger *et al.*, 1991), or decreases with time as former landbridge islands equilibrate, either real islands (Diamond & Mayr, 1976; Diamond *et al.*, 1976) or habitat islands (Brown, 1971).

The extent to which there are identifiable differences between random samples of species and individuals drawn from a species-individuals pool and observed community replicates is the extent to which the second 'type B' category of assembly rules operate. Here, the reference is not to communities that vary in size in response to some extrinsic factor, but rather to those which are more or less constant in size yet differ in species composition. Assuming a fixed resource base, there may be alternative suites of species that comprise

165

stable or equilibrial consumer sets (e.g., Case & Casten, 1979). Assembly rules in this case evaluate ecological compatibility; some species are compatible by dint of some form or extent of different resource utilization and thus can coexist as community members, whereas others are precluded in combination(s) because of their ecological similarities and resultant competition.

Assembly rules of either type, A or B, may each address different components of diversity (see, e.g., Cody, 1975, 1986, 1993). At one level, assembly rules might control the number of locally coexisting species, or α-diversity, where the equilibrial species number may reflect a balance between colonization and extinction rates, as on islands or in fragmented habitat patches, or simply a matching of some subset of a wide array of potential consumers to the resources present. At a second level, assembly rules might relate community amplification and/or a changing species composition, i.e., species turnover, to a changing resource base, as on islands of increasing size or in habitats of increasingly complex structure, and thus address β-diversity. Third, such rules might regulate species turnover and alternative community composition within different parts of a habitat's range, and identify the qualifications of ecological counterparts that contribute to γ-diversity.

The determinants of community composition can be examined at several scales, and it is an examination of assembly rules at these various scales that is the focus of this chapter. Briefly, these are described as follows.

(a) The 'classical' case is that of species sets on islands, where islands of comparable size and isolation support comparable species numbers, and numbers are enhanced with increased island size and reduced distance from a source biota. Then, for a given island size and isolation, one can ask the questions: Is species number predictable? Is species composition predictable? Are species substitutable (ecological equivalents or counterparts)? Are species added in predictable sequence with increased size or decreased isolation? Here, such questions are asked of plant species on islands in Barkley Sound, Vancouver Island, British Columbia, using long-term census data that have been collected over more than a decade.

(b) Within a single, broadly distributed habitat type, questions of habitat patch size or isolation are less prominent. In this case, assembly rules are sought that define and govern the extent to which communities are replicated across the geographical range of the habitat. The relevant questions are: Are the species' totals, or α-diversities, constant across the habitat range? If they are not, do they vary predictably with, e.g., latitude, habitat structure, or some other extrinsic variable? Is species composition predictable and constant? If not, are ecological substitutes or counterparts identifiable, and are rules discernible that predict when the substitutions occur?

Reference is made here to several studies of this type that bear on these questions, but data collected from two habitat types are discussed and emphasized in the chapter: mulga bushland across Australia, and oak woodland in western North America, from Baja California north through Alta California to Washington.

(c) Within a site and within a given habitat type, vegetation structure typically varies, and with it the community may change at a local or patch scale. Between years, abiotic aspects of habitat may vary (e.g., wet years vs. dry years), and so also may the availability of candidates for community membership, for example with variable overwintering success in the birds that constitute the breeding community. Then questions can be asked such as: Is the assembly of the local (patch) subcommunity governed by assembly rules that are based on the vegetation or some other surrogate of resource availability, or is it haphazard? Are the densities of species within the habitat related to habitat suitability and its variation throughout the site, and are species distributed patchily within habitats in accordance with habitat suitability? Is the use of habitat patches by community subsets predictable, with knowledge of the species' habitat requirements and preferences? Is patch use affected by the presence or density of other resource consumers? Are ecological substitutes between patches readily identifiable, and what rules lead to predicting which of alternative species or species combinations might occupy a patch? These questions are addressed here in two habitat types in Grand Teton National Park for which long-term data are available on species occupancy and territory disposition. These are Grass-sage and Wet willows habitats, with emphasis on the latter.

The theme of assembly rules at each of these different scales has been addressed to some extent in previously published or related work, but to different degrees. Each of the following sections begins with a perspective on related or published work, introduces and discusses the data sets used in the respective sections, and follows with detailed analyzes of assembly rules in different systems at different scales.

Island systems

Background

Island distributions and nestedness

The theme of species distributions on islands has been the subject of extensive research over the last several decades since its re-invigoration by MacArthur and Wilson (1967), and the assembly rules approach to community structure was formulated in this context (Diamond, 1975). A good deal of the recent literature has addressed the notion of nestedness, or the degree to which the floras of smaller islands are subsets of those on larger islands. Various tests have been proposed for statistical significance of the nestedness criterion (Cody, 1992; Patterson & Atmar, 1986; Ryti & Gilpin, 1987; Simberloff & Levin, 1985). The phenomenon of nestedness has relevance with respect to whether the island biota is extinction or dispersal driven (Blake, 1991; Bolger et al., 1991; Cutler, 1991; Kadmon, 1995), and to possible conservation strategies (Patterson 1987; Simberloff & Levin 1985; Simberloff & Martin, 1991). Nestedness in island biotas signals type A assembly rules, but the rules are often not transparent, especially when island area is closely controlled.

Plant on islands in Barkley Sound

Plant distributions on islands were discussed in Cody (1992) from the point of view of nestedness, with reference to distributions on islands in Barkley Sound, on the outer coast of Vancouver Island, British Columbia. The climax vegetation is Coastal Coniferous Forest. The data set is part of a long-term study involving some 220 islands covering an island size range of over six orders of magnitude, with isolation distances from 'mainland' Vancouver Island between 0.001 to 15 km, and recording a flora of around 330 species. The Islands have been censused repeatedly between 1981 and 1996, at intervals of 1–4 y, providing information on species numbers and identities, immigration and extinction rates (Cody, unpublished data), and the evolution of reduced dispersal in island populations of anemochorous species (Cody & Overton, 1996).

The mid-sized islands

To remove the most obvious source of variation in island species lists, I consider only a set of mid-sized islands, of areas 1100–6300 m^2. There are 36 such islands in the data set, on which 132 plant species have been found; on the most diverse islands about half the species total occurs. In this larger picture, species do not show significant nestedness with respect to island occupied, although islands are significantly nested in terms of the species they support (Cody, 1992).

Here, plant distributions are discussed for three habitat-specific subsets of the island floras: forest species, shoreline species, and 'edge' species (a category of largely weedy plants occupying the island perimeters between the interior forest and the shoreline). Nestedness and assembly rules are investigated in each of the three species subsets, with differences that appear to be the consequence of the stability of the habitats occupied by the plants, and the species turnover (colonization/extinction dynamics) within them.

Forest species

After subdividing the island flora by habitats, some 19 species (Fig. 6.1) constitute a forest component, and by a number of alternative criteria these species display significant nestedness (NB four of the 36 islands support none of these forest plants; the size range of the mid-sized islands is intermediate between smaller islands with essentially no forest plants and larger islands with many.)

Assembly rules are not obvious from a display and positive test of nestedness, in that inspection of the data does not reveal what they might be; however, species likely are not accumulated on islands randomly, and positive nestedness indicate that such rules may exist. A reorganization of the species-by-sites matrix (SSM) by plant guilds might be productive, but here it is not (Fig. 6.1(b)). Separating plants into guilds by growth form (trees, shrubs, ferns, and forbs – mostly monocot geophytes) does not suggest by what means some similarly sized islands have different plants than others. Note that if there were alternative forest species or ecological vicariants on the islands, they should show up (with disjunct distributions) in a guild-level reorganization of the SSM.

Eliminating the five species that occur on only one island and eight islands with zero or just one forest species leaves 14 species and 28 islands. A reorganization of this trimmed SSM is more useful (Fig. 6.1(c)), as the islands now sort readily into three quite distinct groups:

(a) those with Pacific Yew (*Taxus brevifolius*): Class 1;

(b) those with either or both Douglas Fir (*Pseudotsuga menziesii*) and Lodgepole (Shore) Pine (*Pinus contorta*): Class 2;

(c) and those with none of these three trees: Class 3; scrubbed islands, with 0–1 forest species, are called Class 4.

The 11 Class 1 islands with *Taxus* have, in addition, several shrub species (mostly ericads) that are lacking in the other islands. Six Class 2 islands, lacking *Taxus and* supporting *Pseudotsuga/Pinus* (PP), have the ferns *Polypodium vulgare* (Pv) and *Polystichum munitum* (Pm) with lower and higher probabilities, respectively. Lastly, the 11 Class 3 islands without the three trees have the lily *Maianthemum dilatatum* with incidence comparable to that on all other mid-sized islands, the fern *Pm* with similar probability to *PP* islands, and reduced incidences of fern *Pv* and *Boschniakia* (a parasite on ericad shrubs).

The islands in these three groups are clearly distinct in species composition, but what accounts for the distinction? The term 'intrinsic guilds' is suitable

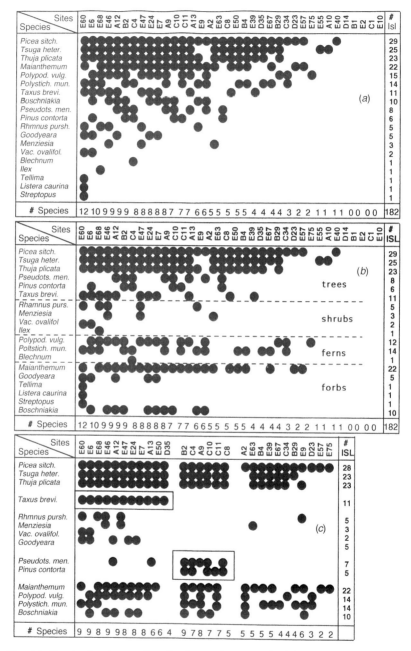

Fig. 6.1. Species by site matrices for forest plants on 36 mid-sized islands in Barkley Sound, British Columbia. (*a*) Islands ranked by species numbers and species ranked by island occurrences, with significant nestedness apparent; (*b*) species grouped by growth form; (*c*) islands grouped into classes, I. to r., 1: *Taxus* islands, 2: dry conifer (*Pinus/Pseudotsuga)* islands, and 3: others. Class 4: few forest plants, omitted.

for species groups that co-occur but without any obvious a priori reason for their co-occurrences (cf. Wilson & Roxburgh, 1994). Note that we have already run a strong island area filter (such that there is no longer a significant correlation between log(area) and log(species number); $r = 0.424$, $P > 0.05$). In fact, the *Taxus* islands are wetter with taller trees (see below), and deeper soils and shade (Cody, unpublished data); in contrast, *P. menziesii* and *P. contorta* are both indicators of dry conditions, which are found on steeper, rockier islands on which soils do not readily accumulate, runoff is rapid, and tree stature is in general lower; the correlation between islands class and vegetation height, $r = -0.78$, is significant). Both of the *PP* species are absent from the nearby mainland, but they recur commonly on the interior, drier side of Vancouver Island.

Sorting islands by larger ($>1.3 \times 10^3$ m^2) vs. smaller, taller (island summit >7 m above the *Fucus* line) vs. lower, and exposed (island groups B and C, towards the outer reaches or mouth of the sound) versus sheltered (groups A, D and E – islands interior or close to the mainland) illuminates some of the conditions that favor one island class over another, with their particular plant assemblages (Fig. 6.2). Class 4 islands tend to be well represented among the small and low islands, class 3 islands are tall and rocky, class 2 islands are those with greater exposure, and class 1 islands are generally larger and more sheltered. ANOVA shows that there are significant effects of all of these factors on island class ($P < 0.01$).

Shoreline species

There are 29 species that are characteristically shoreline plants; they are listed in Fig. 6.3(*a*) which, as with forest plants, displays significant nestedness (Wilcoxon rank sum test, $P < .05$; correlation between species numbers and island size = 0.31, NS). The most widespread shoreline plants occur on nearly all islands (5 species on $\geq 30/36$ islands), but only two islands support $>50\%$ of the plant list. In Fig. 6.3(*b*) the matrix is rearranged by rows to reflect plant affinity with three different shoreline substrates, respectively, muddy/gravelly beaches, sandy beaches, and rocky shores. All islands have rocky habitat, and the incidence of the plants of rocky shorelines does not differ among the three island categories (0.58, 0.58, 0.57, respectively, left to right Fig. 6.3(*b*)). But some islands have, in addition, sandy beaches and others muddy beaches as well. Those islands with muddy beaches tend to be larger (1455 m^2 vs. 1326 m^2, with more precipitation runoff reaching the shoreline) but not significantly so. Such islands, however, do support significantly more shoreline species (12.22 vs. 7.81, $P < 0.01$ by t-test). Islands

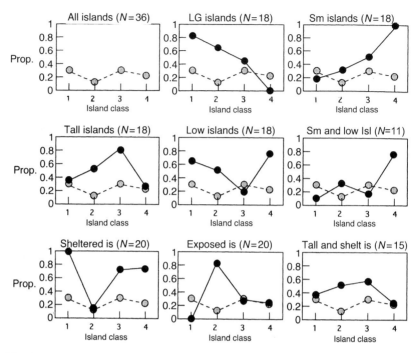

Fig. 6.2. Island classes (1–4 according to forest plant associations) are differentially represented among large vs. small, tall vs. low, and sheltered vs. exposed islands. The proportional data (from all islands lumped together) is repeated to aid in comparisons.

with sandy beaches (left-hand two groups) show similar incidences of plants typical of this habitat (0.36 and 0.38, respectively), but sandy beaches do not significantly increase the total count of shoreline species on these islands.

Edge species

The third component of the island flora are referred to as 'edge species', those plants that occur between the interior forest and the shoreline; they form the most diverse category of island plants, totaling 84 species. Many of the edge species are weedy, opportunistic and widely distributed, many have conspicuous dispersal abilities (e.g., anemochorous, parachuted achenes, as is many Asteraceae and Onagraceae, minute sticky seeds as in Brassicaceae, bird-vectored fruits as in most of the shrubs in the families Ericaceae, Caprifoliaceae, Rhamnaceae, and Rosaceae). Unlike the forest and shoreline floras, the edge flora is not significantly nested (by casual inspection or by the usual statistical tests), nor is any subset of it. Very few of the edge species are widely distributed over the mid-sized islands, and most islands have but a small proportion of the total edge flora; the mid-island average of 6.6 (± 4.2) species constitutes just 12% of the species total in this category.

In Fig. 6.4(*a*) the herbaceous component of the edge flora is listed, some 54 species that exclude the woody taxa. The two largest families are Poaceae and Asteraceae (around a dozen species each). Rearranging the matrix by familial or growth form affinity does not generate nestedness, and an arrangement of islands following that of the forest plant classes adds nothing to pattern or conciseness of the edge species matrix. Fig. 6.4(*b*) reflects this for the two largest families; the rank order of both islands and taxa is preserved from Fig. 6.4(*a*). There is no obvious nestedness to the species, by appearance or test, nor any suggestion of a more subtle form of assembly to these weedy communities.

The incidences of the plant species are given in the right-most column of the figure, and sum to the number of taxa present on the average island (e.g., 1.81 for Asteraceae, 1.75 for Poaceae). These incidences can also be used to generate the variance in expected species richness on the islands under the hypothesis of random distribution and contra either nestedness or disjunction of species over islands. The expected variance is $\Sigma_i(J_i)(1-J_i)$, *fide* D. Ylvisaker (pers. comm.); for the Asteraceae, observed/expected variance is 1.42/0.82, and by the *F*-test ($F = 1.73$, df = 35, 35, $P > 0.05$) the random hypothesis is upheld. For Poaceae, expected variance in species/island is 1.23, significantly less than the observed 3.29, indicating that there are more islands than expected with many grasses and no grasses. Some degree of nestedness in islands is therefore indicated for species of Poaceae, although rules for their abundance on some islands and dearth on others are still obscure.

Whereas species numbers of forest plants respond significantly to island characteristics such as rock height and area, the edge species do not (Fig. 6.5); island isolation or exposure do not contribute to an explanation of variance in species numbers. Thus edge species remain quite unpredictable in number and kind, and if there are assembly rules for them, they remain undiscovered.

Community replication in wide-ranging habitats
Background
Birds in scrub, woodland, and forest

An alternative approach to rules for community structure is by seeking type B assembly rules via replicated censuses or repeated samples from within a single, broad habitat type. For example, the bird communities of the scrubby

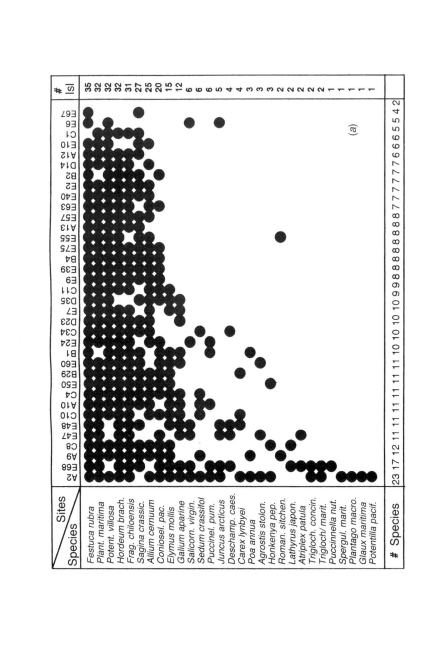

(a)

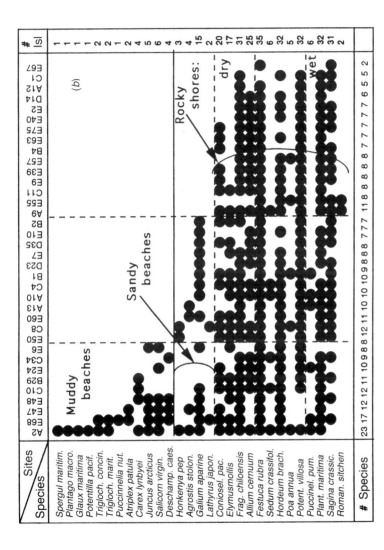

Fig. 6.3. (a) Species-by-site matrix for shoreline plants on mid-sized islands, with islands and species ranked and showing significant nestedness. (b) species are grouped by shoreline habitats, which are added on islands from right to left.

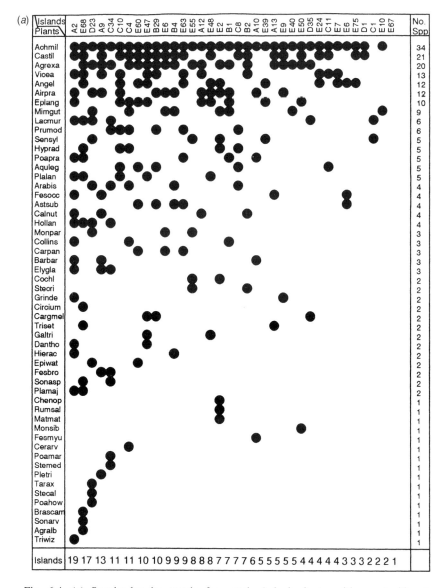

Fig. 6.4. (*a*) Species-by-site matrix for weedy 'edge' plants, with no significant nestedness. (*b*) Distributions of species in the two largest families of edge plants, Asteraceae and Poaceae. The weedy composites are randomly distributed over islands, but the grasses do show a degree of nestedness, although no patterns of species co-occurrence are apparent in either family.

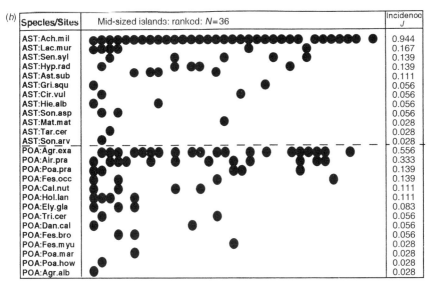

Fig. 6.4. (continued)

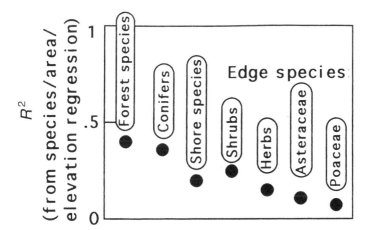

Fig. 6.5. The predictability of species numbers (R^2, ordinate) as a function of island area and elevation varies among different components of the island floras. It is most predictable in forest species and least in weedy, edge species.

sclerophyllous vegetation typical of Mediterranean-type climates world wide, and variously called chaparral, macchia, fynbos, heath or maquis, have been extensively studied (Cody, 1973, 1974; Cody & Mooney, 1978; Cody, 1983b; Cody, 1986; Cody, 1994a), with conclusions on the regulation of community size (species numbers, α-diversity), composition (species' relative abundances and ecological characteristics), and membership attributes (sizes and relative morphology, foraging behavior, etc.). When tests for assembly rules are made across continents, phylogenetic constraints are often weak or absent (as quite unrelated species may occupy parallel or convergently similar ecological niches in different biogeographic regions); any case for community replication is therefore more compelling. Analyses of bird communities within continents, both in chaparral or its equivalent and in other habitats in the same climatic region, e.g., renosterveld and kloof woodland in southern Africa (Cody, 1983a), or over broader geographical areas, e.g., in *Eucalyptus* woodlands, forest, and rainforest in Australia (Cody, 1993), have also been made.

'Habitat islands' is a phrase that refers to isolated habitat fragments separated from each other by different vegetation; habitat islands constitute an intermediate situation between true islands and continuously distributed vegetation. Assembly rules for species' retention in attenuating bird communities in increasing smaller and more isolated patches of Afromontane woodland were described in detail by Cody (1983b).

Mulga bushland and healthland in Australia

Analyses of replicated bird censuses across two widely distributed habitat types in Australia, namely mulga bushland and protead heathland, were recently published (Cody, 1994a, b), with strongly contrasting results. The Australian heathland is a fire-prone scrub vegetation dominated by evergreen sclerophylls, especially Proteaceae and in particular *Banksia*. This habitat occurs on the poorest soils (laterite and sand-over-laterite), more extensively in winter-rainfall areas of the south and southwest, but also beyond this area, around the periphery of the continent and especially in isolated patches near the eastern and northern coasts. Mulga is an open bushland vegetation dominated completely by the phyllodinious *Acacia aneura*, sometimes with a couple of smaller and scarcer acacias and Myoporaceae shrubs. It burns infrequently, occurs on the more productive red earths across interior Australia, mostly in areas of both summer and winter rainfall with 180–220 mm annual precipitation, and is nearly contiguous from the west coast to central Queensland. In many characteristics the bird communities of these two habitats are at opposite extremes.

Analysis of the bird communities of protead heathlands compared census results from 17 sites continent-wide, which produced a total species list of 96 taxa. These communities varied widely in size, from 5 to 22 species, and also in species composition from site to site, with a component of species turnover between sites related to both differences in vegetation structure (β-diversity) and distance apart of the sites (γ-diversity). Only two bird species occur in >50% of the heathland censuses, and thereby qualify as 'core' species (i.e., occur more often than would be the case if communities were assembled by a random draw of species from the total species pool). Because α-diversity averages about 12 species (and shows no significant variation across the continent), the contribution of core species to the local α-diversity is minor (about 1/6). Overall, community composition is unpredictable, and conforms closely to expectations from a random selection of species according to binomial null models (Cody, 1986, 1992). Thus most birds in the protead heathlands are rather erratic and unpredictable, have low incidence in the censuses (>75% of the recorded species occur in <25% of the censuses), and their identities change substantially even between adjacent sites.

However, despite wide variations in community composition, some general assembly rules do apply:

(a) Species richness is modest; mean α-diversity of 11.6 ± 1.9 s.d. does not vary regionally, but 96% of its variation depends on inter-site differences in vegetation structure, many of which are in turn related to temporal stages in post-fire regrowth;

(b) Heathland bird density is low, averaging around four individuals per hectare (l/H) – the lowest of all Australian shrub habitats censused;

(c) Insectivores are poorly represented, by species that are both relatively constant (e.g., malurid wrens) and relatively rare (*ca.* 2 l/H);

(d) Nectarivorous birds (Meliphagidae, and especially *Phyllidonyris* spp.) are dominant in both species richness and bird density; meliphagids comprise the majority of heathland endemics and nearly 50% of the total species list, add two to six species to local α-diversity, and generally total at least 50% of total bird biomass;

(e) Overall, species numbers and identities are sensitive to vegetation structure, which is largely a function of time since the last fire. Heathland bird resources are successional stage specific, and their consumers are wide-ranging resource specialists in pursuit of suitable resources (Cody, 1994a).

The breeding birds of the mulga bushlands are generally quite different from those of protead heathlands; mulga species richness averages nearly twice that of heath, and in contrast to heathlands, the communities vary in composition scarcely at all across thousands of kilometers of the habitat's range. With predictable community membership, species turnover between sites is low, even cross-continentally, and unrelated to both differences in vegetation structure (β-diversity) and distance apart of the sites (γ-diversity); peripheral species in the mulga communities are more a product of the proximity and type of adjacent habitats (Cody, 1994b). The following assembly criteria, rules to descriptive generalities, have been formulated:

(a) Mulga α-diversity is narrowly constrained (mean 21.5 ± 0.8 species) across the mulga (n = 20 census sites), with little systematic variation attributable to geographical position or latitude, and none attributable to variations in the physiognomic structure of the vegetation;

(b) Bird densities in mulga are around 12 indiv. ha^{-1}, varying little among mulga sites and generally about three times those in heathland;

(c) Core species are prominent in the community make-up, comprising 2/3 of the local α-diversity (14/21 species) and *ca.* 3/4 of the total bird density, even though they comprise only 26/81 species encountered in the censuses;

(d) Ecological replacements occur in core niches, with different (often congeneric) species substituting in different part of the habitat range;

(e) Few mulga birds are habitat endemics (perhaps 3–4 species), but the top nine core species all reach higher densities in mulga than in other habitats;

(f) Insectivores, both ground- and foliage-foraging, are the largest component of the community, and constitute two-thirds of the total bird density;

(g) Nectarivores are very rare (*ca.* 0.2 indiv. ha^{-1}), mostly utilizing mulga mistletoes), and the 1–2 meliphagid species present are omnivorous or frugivorous;

(h) Some assembly rules for core niches are simple (e.g., one or another crow *Corvus* depending on region, one or another butcherbird *Cracticus*, species interchangeable, identity unspecified);

(i) Other assembly rules for core niches are more involved, e.g., one midsized omnivore species (babblers *Pomatostomus* spp.), with specific identity favoring the wide-ranging *P. superciliosus*, but reverting to *P. temporalis* in mulga sites adjacent to woodland, or to *P. hallii* or *P. ruficeps* within the more limited southeastern ranges of these latter two species; or two perch-and-pounce insectivores (robins), one of which is always *P. goodenovii*, the second either a *Microeca* or a *Melanodryas* species depending on geographic region;

(j) Yet other rules, e.g., for the composition of the guild of small low-foraging insectivores (mostly thornbills), are complex. Eight candidate species occur in the mulga sites (6 *Acanthiza*, 1 *Sericornis* and 1 *Aphelocephala*), and some five of these have ranges that include the average mulga site. Exclusively summer-rainfall mulga (Northern Territory) supports two-species guilds, summer plus winter rainfall (central-western

Queensland) three-species guilds, and winter rainfall sites (southwest Australia) four to five species guilds. Guild composition varies regionally, with Chestnut-rumped thornbill (*A. uropygialis*) and Yellow-rumped thornbill (*A. chrysorrhoa*) regular in most sites, Broad-tailed thornbill (*A. apicalis*) more casual countrywide, Redthroat (*Sericornis brunneus*), Slate-backed thornbill (*Acanthiza robustirostris*) and southern whiteface (*Aphelocephala leucopsis*) in SW Australia, and Buff-rumped thornbill (*A. reguloides*) regular, Yellow thornbill (*A. nana*) casual, in Queensland.

Assembly rules in oak woodland birds

Oak woodlands in Western North America

In this section, data from censuses of oak woodland habitats are used to illustrate a further example of the approach. Woodlands dominated by oaks (*Quercus* spp.) are widespread in western North America, and occur from the Cape region of Baja California north through Alta California, Oregon, and Washington to southern British Columbia, covering over 20° of latitude and including a dozen oak species. Elsewhere, oak-dominated habitats occur also from south-central Colorado south through the Rocky Mountains, from Ontario south to Florida, and thence south in México into Central America, but here only the westernmost oak woodlands along the Pacific coast will be discussed.

Some 70 breeding bird censuses are in hand from the Pacific series of oak woodlands, from 40 different sites, some censused more than once, in different years). Of these, 30 censuses are the author's, conducted mostly in May–June 1994, on standardized 5 ha sites using standard census techniques; the remaining censuses are the published results of others and producing numbers and identities of breeding species, plus estimates of breeding bird densities.

Overall species composition and diversity

The geographical distribution of 40 bird census sites in oak woodland along the Pacific coast spans about 25° of latitude, extending from the Cape region of Baja California north to Washington (Fig. 6.6). At these sites, α-diversity averages 28 species, and there is no detectable trend with latitude (*P* = 0.34). However, variation in α-diversity is related to variation in vegetation structure; profile area (area under a plot of vegetation density versus height, a measure of total vegetation biomass at the site) accounts for a modest 15% (p = 0.04) of this variation (Fig. 6.7(*a*)). Total bird density also responds to the same vegetation variable (r^2 = 0.23, *P* =.008; Fig. 6.7(*b*)).

West Coast Oak Woodlands

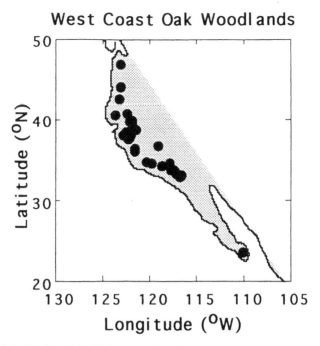

Fig. 6.6. Distribution of 40 bird census sites in oak woodlands along the west coast of North America, covering about 25° of latitude.

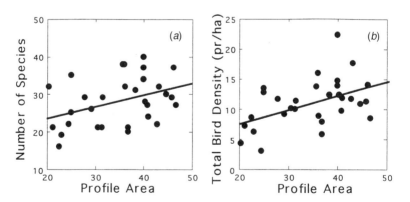

Fig. 6.7. Numbers of bird species in oak woodland sites show modest increases with total vegetation density (profile area, abscissa; $r^2 = 0.15$, $P = 0.04$), and total bird density also increases with the same woodland variable ($r^2 = 0.23$, $P = 0.008$).

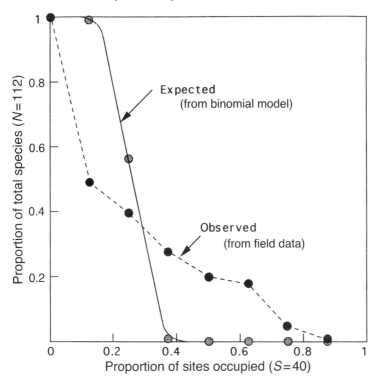

Fig. 6.8. Bird community composition among 40 oak woodland census sites differs significantly from a collection of random subsamples of the avifauna. A null binomial model P(k; n,p) is constructed with probability P = (average α-diversity)/(total species list in all sites) = 28/112, n = 40 census sites, and k = 'successes', or number of sites at which species are recorded. Cumulative proportion of total species (ordinate) is drawn against decreasing proportion of sites occupied (abscissa) for observed and expected functions. Over one-third of the oak woodland birds, 40/112 species, qualify as core species in being more predictable at census sites than chance would predict.

In 70 censuses at 40 sites, 112 bird species were recorded. A null binomial model can be used to compare species incidence over sites with that predicted from a random species assortment (see Cody. 1994b). Here probability of k 'successes' (species' records at a site) with success probability p (= 28/112 = 0.25) in n (= 40) replicates) is computed. Any species that occurs with incidence $J > 0.3$ is significantly more constant in the censuses than is predicted from a random draw; expected and observed distributions are shown in Fig. 6.8. Some 40 species fall into this category, and are termed the 'core species'. These 40 species are shown schematically in Fig. 6.9; although they account for only 36% of the species list, they are the major component of the oak

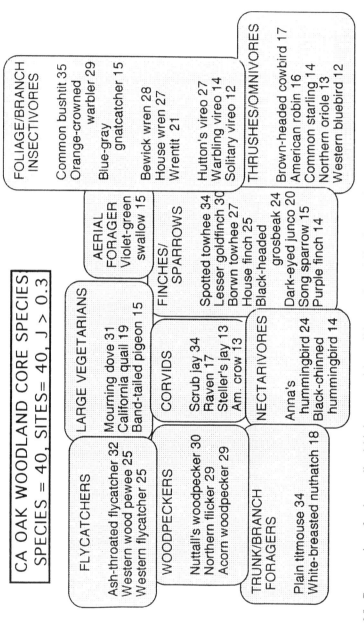

CA OAK WOODLAND CORE SPECIES
SPECIES = 40, SITES= 40, J > 0.3

FLYCATCHERS

Ash-throated flycatcher 32
Western wood pewee 25
Western flycatcher 25

WOODPECKERS

Nuttall's woodpecker 30
Northern flicker 29
Acorn woodpecker 29

TRUNK/BRANCH FORAGERS

Plain titmouse 34
White-breasted nuthatch 18

LARGE VEGETARIANS

Mourning dove 31
California quail 19
Band-tailed pigeon 15

CORVIDS

Scrub jay 34
Raven 17
Steller's jay 13
Am. crow 13

NECTARIVORES

Anna's hummingbird 24
Black-chinned hummingbird 14

AERIAL FORAGER

Violet-green swallow 15

FINCHES/ SPARROWS

Spotted towhee 34
Lesser goldfinch 30
Borwn towhee 27
House finch 25
Black-headed grosbeak 24
Dark-eyed junco 20
Song sparrow 15
Purple finch 14

FOLIAGE/BRANCH INSECTIVORES

Common bushtit 35
Orange-crowned warbler 29
Blue-gray gnatcatcher 15

Bewick wren 28
House wren 27
Wrentit 21

Hutton's vireo 27
Warbling vireo 14
Solitary vireo 12

THRUSHES/OMNIVORES

Brown-headed cowbird 17
American robin 16
Common starling 14
Northern oriole 13
Western bluebird 12

Fig. 6.9. Core species of oak woodland birds, grouped by guilds; numbers following species names are the number of sites at which the species was censused.

woodland bird community, comprising an average of 80% of the α-diversity over sites, and 82% of the total bird density.

As in the mulga, the identification of core species in oak woodlands is compromised by the occurrence of species that are ecological replacements in essentially the same core role or niche. For example, an oak woodland generally supports one or two wren species, an oriole, a chickadee or titmouse, but overall there are more candidate species for these core niches than are supported by local site α-diversity. Here, the supernumerary species competing for limited niche representation are five wrens (House, Bewick's, Cactus, Canyon, Rock), two orioles (Northern, Hooded), and four parids (Mountain, Chestnut-backed, and Black-capped chickadee, Plain titmouse). These 'extra' species inflate the total list and obscure the identification of core species unless their roles as equivalent contenders for core niches are recognized.

Oak woodland wrens

The assembly of species in particular guilds within the oak woodlands is based on various criteria. In wrens, for example, both House and Bewick wren are common in the woodlands overall (incidence $J = 0.27, 0.28$, respectively), and there is tendency to favor House wren over Bewick wren at lower latitudes in taller woodland with reduced understory. But where they are both common ($n = 25$ sites with combined densities > 0.5 pr/ha) each species is the predominant factor in the other's density, and densities of the two species vary inversely ($r = -0.39$, $p = 0.05$; Fig. 6.10).

Oak woodland woodpeckers

Some guilds are readily understood by reference to body size. Oak woodlands have three core species of woodpeckers, the large Northern flicker (142 g), the mid-sized Acorn woodpecker (83 g), and the small Nuttall's woodpecker (38 g). About 3/4 of the oak woodland sites, particularly those in central and southern California, support this three-species combination (Fig. 6.11). However, other combinations of woodpecker species occur in other oak woodlands; in the more northerly sites a similar three-species combination substitutes Hairy woodpecker (70 g) for Acorn woodpecker and Downy woodpecker (27 g) for Nuttall's woodpecker, and in oak woodland sites near desert habitats (Baja California, SW California) Gila woodpecker (70 g) fills the mid-size woodpecker slot and Ladder-backed woodpecker (30 g) the small size slot. Whereas the names of the woodpecker species change, their segregation by body size differences of around 2× does not, and the pattern is maintained throughout the woodlands (Fig. 6.11).

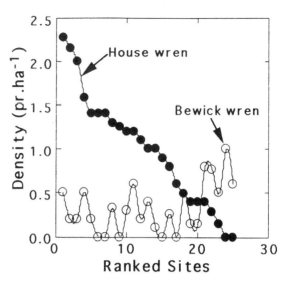

Fig. 6.10. The two common wrens of oak woodlands, Bewick and House wren, tend to vary in density inversely ($r = -0.39$, $P = 0.05$) at 25 sites (ranked along abscissa) where both species are common (combined density > 0.50 pairs ha^{-1}). Either or both wrens are present in most sites, and these species appear to be more or less substitutable in the oak woodland community.

Oak woodland vireos

Vireos are slow-searching foliage insectivores; there are three species of vireos in the oak woodlands, and all qualify as core species. They differ somewhat in body size (by an average 20%), but not as conspicuously as the woodpeckers. The woodland sites average 1.33 vireo species; Hutton's vireo (11.6 g) is common ($J = 0.90$), Warbling vireo (14.8 g) and Solitary vireo (16.6 g) less so ($J = 0.45$, 0.3, respectively). Warbling vireo incidence is rather constant over latitude, but that of Hutton's vireo declines sharply to the north whereas Solitary vireo is more common at higher latitudes (Fig. 6.12). Based on each species' incidence, no particular two-species combination is favored over another, and all combinations (of one to three species) are no different in incidence than is predicted from the incidences of individual species.

Oak woodland flycatchers

Sallying flycatchers are a conspicuous component of the oak woodland bird community, and there are three common core species: the large Ash-throated flycatcher (symbol L: 27 g; $J = 0.80$), the intermediate sized Western wood pewee (M: 13 g; $J = 0.63$) and the small Western (or Pacific slope) flycatcher

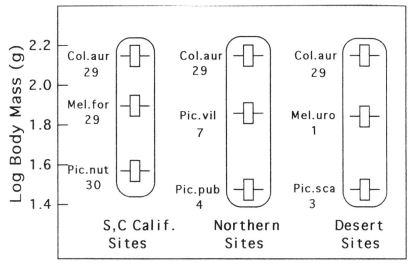

Fig. 6.11. Almost all oak woodland sites support three woodpecker species that vary by body size by a factor of *ca.* 2.5, small to medium, or 1.75, medium to large. However, the usual trio (left; site incidences are shown below species names) is replaced by a similarly sized trio in northern sites (center) and again in desert-edge sites (right). 'Col.aur.' = *Colaptes auratus*, 'Mel.for.' = *Melanerpes formicivorus*, 'Mel.uro.' = *M. uropygialis*, 'Pic.nut.' = *Picoides nuttallii*, 'Pic.vil.' = *P. villosus*, 'Pic.pub.' = *P. pubescens*.

(S: 10 g; J = 0.63). Four other flycatchers are rare in the censuses (as indicated in Fig. 6.13). The three core species constitute the prevalent three-species combination (14/40 sites, indicated in Fig. 13 by 'LMS'). Only two other species combinations are common, a two-species combination of LS (at 6/40 sites) and a single species L (at 7/40 sites; see Fig. 6.13).

The incidence of particular species' combinations is related chiefly to vegetation structure in the oak woodlands (Fig. 6.14). The lowest and least dense woodlands support only L (Ash-throated flycatcher). Most oak woodlands of intermediate height and density have the three species LMS, but the medium-sized species M tends to drop out in very dense woodlands which are typically LS (Fig. 6.14). Taller woodlands, with the addition of the largest Olive-sided flycatcher (X: 32 g; J = 0.10), may have four species (XLMS), or a trimmed two-species combination (MS) that occurs in 25 m oak woodlands at the highest latitudes. Thus assembly rules in the flycatchers are based on body sizes as in woodpeckers, but with preferred combinations varying largely with the structural aspects of the woodland vegetation.

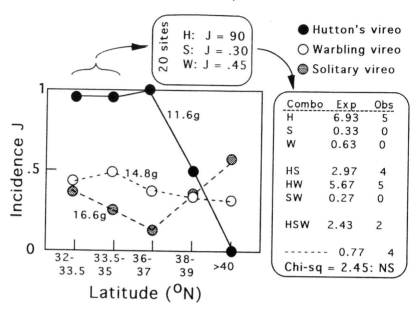

Fig. 6.12. The incidence of three vireo species in oak woodland varies over latitude, while the mean species number (1.33) remains about constant. In 20 southerly sites, the three vireo species occur in various 0–3 species combinations (see Obs) that are in the same proportions as predicted (by Chi-sq., $P > 0.05$) from random assembly (see Exp), using the species incidences J_i shown at top to generate predicted frequencies.

SPECIES:	# SITES:	COMBINATIONS:	
X: Olive-sided flycatcher (32g)	4	4 Spp: 2 sites	LMS: 14 XLM: 2 XMS: 1
L: Ash-throated flycatcher (27g)	32	3 Spp: 17 sites	
M: Western wood pewee (13g)	25	2 Spp: 12 sites	LS: 6 MS: 3 LM: 3
S: Western flycatcher (10g)	25	1 Sp: 9 sites	
Other Empidonax	2		L: 7 M: 1
Black phoebe	5		
Western kingbird	2		S: 1

Fig. 6.13. The three common flycatchers of the oak woodlands differ in size, and are represented as large, medium and small (L, M, S; left-side of figure). Three particular species combinations are relatively common: LMS, LS, and L, with the combination LMS is especially prevalent; other combinations are observed rarely (see right-hand side) or never (unlisted).

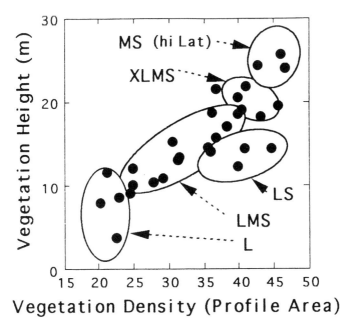

Fig. 6.14. The occurrence of various flycatcher assemblages is related to the woodland vegetation, with assemblage size increasing with vegetation height and profile
area (vegetation density), the medium-sized species squeezed out in very dense
vegetation, and high-latitude sites supporting just medium and small species.

Assembly rules and community composition within sites

Background

Insectivorous bird communities

Within- and between-site habitat variation, and concomitant variation in the
composition of bird assemblages across the changing vegetation, has been
extensively studied in Old World warblers Sylviinae, and previously reported
(Cody & Walter, 1976; Cody, 1978, 1983a, 1983b, 1984). The approach used
is to plot the territories of the breeding species within a heterogeneous site,
and then relate use of the habitat by the various species present to (a) the
vegetation within the study site and within species' territories, (b) the habitat
requirements of the species, in terms of vegetation heights, densities, and
foraging substrates, and (c) the presence of other species with similar but
variously divergent habitat requirements.

One example of this approach was discussed by Cody and Walter (1976), a
study of the five *Sylvia* warblers of Mediterranean-type scrub and woodland

in SW Sardinia. Ranked from lowest to tallest in terms of preferred vegetation height, these are *Sylvia sarda* −S, *S. undata* −U, *S. melanocephala* −M, *S. cantillans* −C, and *S. atricapilla* −A, with letters indicating species abbreviations. In low, meter-high vegetation (*Cistus-* dominated scrub at Narcao), three species occurred (S, U, and M) with little interspecific difference in territories classified by vegetation height and density. Yet the species pairs S-U and U-M were spatially segregated, with a demonstrated avoidance of interspecific overlap. In somewhat taller macchia (*Erica, Arbutus, Phyllarea*) at Bau Pressiu, a fourth species (C) is added. Again, strong behavioral interactions serve to segregate territories spatially between species U and M, while species pairs S–U and U–C show weaker interactions. In yet taller macchia at Terrubia, S drops out and A is added; here the interactions between U and M and U and C that occur in lower vegetation disappear, while a strong interaction characterizes species C–A. Overall, species interactions are predictable knowing the preferred vegetation and foraging heights of each species, and the extent to which the vegetation within a site allows the various species to forage at different heights (in which case their territories overlap) or constrains them to forage at similar heights (in which case they exhibit interspecific territoriality).

A second example is taken from the sylviine warblers in scrub (*Rosa, Juniperus*) to low woodland (*Quercus, Betula*) habitats in southern Sweden (Cody, 1978, 1985). Seven sylviine warblers breed in the area (with additional species present in marshlands); these are five *Sylvia* species (*S. communis* −U, *S. nisoria* −N, *S. borin* −B, *S. curruca* −R, *S. atricapilla* −A), a *Hippolais (icterina* −H) and a *Phylloscopus (trochilus* −P), with letters again indicating species abbreviations. The species are listed in order of ascending foraging height and preferred vegetation heights within territories, but there are considerable overlaps among species. Within sites, the occupancy of the habitat by each warbler species was evaluated at the scale of 15 × 15 m quadrats, in each of which the vegetation structure (density over height) was also measured.

As vegetation height within quadrats increases, the number of coexisting warbler species using the patch increases from one to two to three and up to four species. At Bejershamn, the transitions are U, to U or N, to N, N or B, B or C, BC, BHP or ACH, and the combination ACPH is rather uncommon in the tallest woodland. The species pairs U–N and N–B are strongly interspecifically territorial in habitat intermediate between their specific preferences, and the pair B–C is weakly interactive. The assembly of increasing large suites of warblers in taller vegetation is dependent on the match between their combined foraging height distributions and the vegetation profile (plotting density as a function of height). Thus the stable species combinations are products of the conformity between species' foraging requirements and the foraging opportunities provided by the vegetation, and given the wide overlaps in habitat preferences and foraging height distributions, the combinations are adjusted or refined by direct interspecific interactions.

Birds in grass-sagebrush, Grand Teton National Park

Emberizid finches have been the subject of a long-term study in a grass-sagebrush habitat by Cody (1974, 1996). Four species breed in a 4.7 ha site in Grand Teton National Park, Wyoming: White-crowned sparrow *Zonotrichia leucophrys* −W, Brewer's sparrow *Spizella breweri* −B, Vesper sparrow *Pooecetes gramineus* −V, and Savannah sparrow *Passerculus sandwichensis* −S, with a fifth, Chipping sparrow *Spizella passerina*, of sporadic occurrence. At a scale of 15 × 15 m quadrats, each of 208 quadrats was classified as being core, secondary, or marginal with respect to the habitat preferences of the emberizids. Within the site, W is restricted to areas of the tallest sage and S to the grassier parts of the site, and their territories are located in similar positions among years. The habitat requirements, core plus secondary, of both B and V are satisfied over about 50% of the study area, and these two species are the commoner of the four. But for both species their preferred habitat is interspersed with less suitable vegetation, and the disposition of their territories changes somewhat from year-to-year.

The densities of the emberizid finches varies among years, most obviously in response to wetter or drier conditions at the site. Savannah sparrow, for example, is rarer is drier years, when its occupancy of quadrats drops to 11% (vs. up to 36% in wet years). But its habitat occupancy depends also on the densities of, and habitat use by, the other species present; these sparrows are less tolerant of shared habitat (with B and V) in dry years than in wet years. Although habitat choice is similar in Brewer's and Vesper sparrows, the two species tend to oscillate in density asynchronously, with V commoner in drier years and B in the wetter. The two species demonstrate an aversion to territorial overlap in years when the overall sparrow density at the site is high, but a significant tendency to shared use of their joint core quadrats when densities are low; in years of intermediate density, there is neither positive nor negative association of the two species within the site. In this study, the assembly of the finches at the quadrat scale is understood by a hierarchical set of factors: (a) the extrinsic conditions that affect species' densities in the area (perhaps a function of overwintering success); (b) the extrinsic conditions within the site, likely a function of late spring–summer weather conditions there; (c) distinct habitat preferences in some species (W, S) and similar

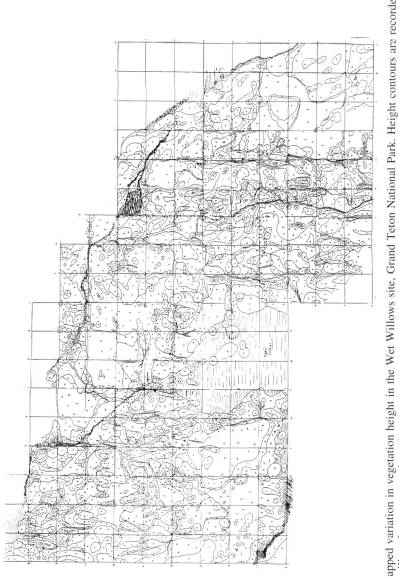

Fig. 6.15. Mapped variation in vegetation height in the Wet Willows site, Grand Teton National Park. Height contours are recorded in feet, with grassy willow-free areas, pools and streams, and 15 m × 15 m quadrats as indicated.

requirements in others (B, V), all potentially satisfied at the site; (d) behavioral interactions that moderate interspecific territorial overlap, interactions that vary between years with local conditions and with local sparrow densities (Cody, 1996).

The willows site, Grand Teton National Park

The habitat, birds, and methodology

Close by the grass-sagebrush site just discussed is a very different habitat that similarly has been studied over a long time period, a marshy site dominated by various species of willows 1–3 m high and interspersed with small streams, ponds, grassy and sedgy glades (Cody, 1974, 1996). The bird community of the 3.3 ha willows site has been recorded intermittently since 1966 (and was earlier studied by Salt, 1957a, b). The site supports two main bird guilds: five species of paruline warblers (Yellow warbler *Dendroica petechia.* Common yellowthroat *Geothlypis trichas.* Wilson's warbler *Wilsonia pusilla,* MacGillivray's warbler *Oporornis tolmei* and Northern waterthrush *Seiurus novaboracensis,* with the first three common and the last two rare; and six emberizid finches (Song sparrow *Melospiza melodia,* Lincoln's sparrow *Melospiza lincolnii,* Fox sparrow *Passerella iliaca,* White-crowned sparrow, Clay-colored sparrow *Spizella pallida,* and Savannah sparrow), with the first four common and the last two rare; the willows community is rounded out with a flycatcher, magpie, wren, thrush, and a hummingbird.

The site is mapped by vegetation height (Fig. 6.15), and the frequency distribution of willow heights within, and its variation among, 15×15 m quadrats forms the basis of this study. Thus the organization of the willows bird community and its assembly guides are studied at this local 'patch' resolution, at which scale habitat use is assessed and territory locations plotted.

The emberizid sparrows

Six emberizid sparrow species have bred within the site over the last 30 years (Song −S, 20 g, Fox −F, 32 g, Lincoln's −L, 17.4 g, White-crowned −W, 29 g, Savannah −V, 19 g and Clay-colored Sparrow −C, 12 g), with the first four species common and consistently present, the penultimate sporadic, and the last common before the early 1970s but not present since. All species feed mostly on the ground except the Clay-colored and White-crowned sparrows, which often forage somewhat higher in the vegetation. The common species show extensive similarities in quadrat use (see plots of species over principal components of vegetation, Fig. 6.16(a)), although discriminant function analysis (DFA) reveals differences in the mean habitat preferences of

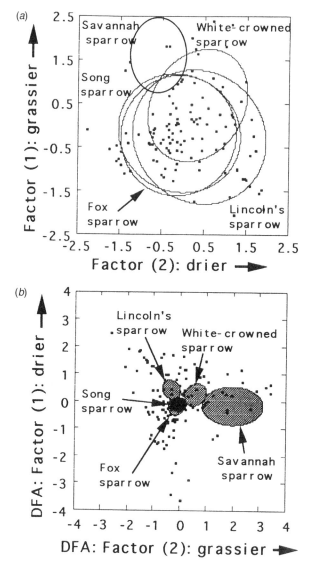

Fig. 6.16. (*a*) Factor analysis (principal components) of vegetation characteristics for emberizid sparrows in the willows site, showing distinct habitat preference in Savannah sparrow for grassy areas (high F{1}scores), a less distinct preference for grassy openings in White-crowned sparrow, and similar habitat choice in the remaining three species. (*b*) Discriminant function analysis of the same data show that mean habitat preferences differ among species (95% centroids) in all except Song and Fox sparrows.

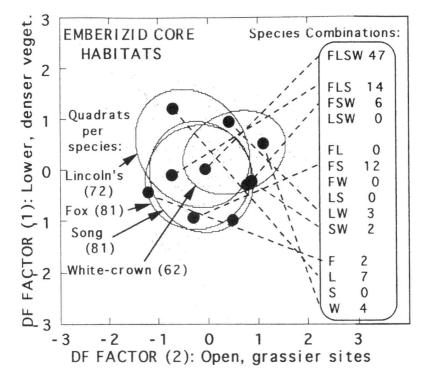

Fig. 6.17. Core habitat (50% confidence ellipses) is shown for the four common emberizid sparrows at the willows site. Numbers of occupied quadrats Q, by species, are shown at left, while the numbers of quadrats potentially supporting various 1–4 species combinations are shown at the right. See text for further discussion of actual species combinations found.

all species except Song and Fox sparrows (see habitat centroids, Fig. 6.16(b)). Given that habitat preferences overlap extensively among species, predicting the identity of a quadrat's occupant based on vegetational characteristics is successful in just 38% (90/237) of the cases. 50% confidence ellipses were used on 1994 data to define core habitat (Fig. 6.17) for the four common emberizids (F,L,S,W), which amounts to 97 quadrats or two-thirds of the site. Of this combined core, half is core habitat for all four species (47/97), while most of the remainder is core for either the trio FLS (14 quadrats) or the duo FS (12 quadrats); see Fig. 6.17 for the complete breakdown.

Given that core habitat is identified and the numbers of quadrats that fall within different categories of core habitat known, the next question was how each species occupies its core habitat, and specifically whether its occupancy is affected by (a) whether its core quadrats are co-occupiable by other sparrows and (b) whether they are actually occupied by other species. The results

Table 6.1. *Analysis of habitat occupancy by emberizid sparrows in the Tetons wet willows habitat*

(a) Lincoln's sparrow core habitat: Q = 71; 26/71 occupied (prop. 0.366)

Quadrats occupied

Core type	# Q	# Occ	Prop.	#Exp	by +	L	L'	#Exp
FSLW	47	14	.298	17.20	FSW	2	6	2.93
FLS	14	6	.429	5.12	∪(FS,FW,SW)	5	17	8.05
LW	3	1	.333	1.10	∪(F,S,W)	8	6	9.88
L	7	5	.714	2.56	None	8	6	5.12

Chi-sq. = 1.24, df = 3, NS   Chi-sq. = 3.2, df = 3, NS

	S	7	29	13.12 (i)
(i) Chi-sq. = 4.49, 4.71, df = 1, p < .05	S'	19	16	12.81 (i)

	F	11	29	14.64 (ii)
(ii) Chi-sq. = 1.43, 1.85, NS	F'	15	16	11.35 (ii)

(iii) Chi-sq. = 2.57, NS	SF	5	19	8.78 (iii)
(iv) Chi-sq. = 20.79, p <<.05;	S'F'	12	0	4.29 (iv)
(v) Chi-sq. = 1.79, NS	S or F	9	26	12.8 (v)

(b) Song sparrow core habitat: Q = 81; 47/81 occupied (prop. 0.580)

Quadrats occupied

Core type	# Q	# Occ	Prop.	#Exp	by +	S	S'	#Exp
FSLW	47	26	.553	27.26	FLW	1	2	1.74
FLS	14	8	.571	8.12	∪(FL,FW,LW)	12	14	15.08
FSW	6	5	.833	3.48	∪(F,L,W)	21	8	16.82
FS	12	7	.583	6.96	None	13	8	12.18
SW	2	1	.500	1.16				

Chi-sq. = 0.71, df = 4, NS   Chi-sq. = 2.04, df = 3, NS

	L	8	16	13.92
Chi-sq. = 8.70, df = 3, p < .05	L'	39	18	33.06

	F	11	15	14.64
Chi-sq. = 3.10, DF = 3, NS	F'	15	16	11.35

	FL	5	5	8.78
Chi-sq. = 0.26, 0.01, 0.01, NS	F' L'	13	9	4.29
	ForL	28	21	12.8

(c) Fox sparrow core habitat: Q = 81; 46/81 occupied (prop. 0.568)

Quadrats occupied

Core type	# Q	# Occ	Prop.	#Exp	by +	F	F'	#Exp
FSLW	47	25	.532	26.70	LSW	1	1	1.14
FLS	14	11	.786	7.95	∪(LS,LW,SW)	14	7	11.93
FSW	6	3	.500	3.41	∪(L,S,W)	25	19	24.99
FS	12	5	.417	6.82	None	6	8	7.95
F	2	2	1.00	1.14				

Chi-sq. = 2.46, df = 4, NS   Chi-sq. = 0.85, df = 3, NS

Table 6.1. (*continued*)

(d) White-crowned sparrow core habitat: Q = 62; 21/62 occupied (prop. 0.339)

Core type	# Q	# Occ	Prop.	#Exp	Quadrats occupied by	+	W	W′	#Exp
FSLW	47	13	.277	15.93	FLS	1	1	4	1.70
FSW	6	3	.500	2.03	∪(FL,FS,LS)	9	9	11	6.78
LW	3	1	.333	1.02	∪(F,L,S)	11	11	21	10.85
SW	2	0000		0.68	None	0	0	5	1.70
W	4	4	1.00	1.36					

Chi-sq. = 2.20, df = 1, NS   Chi-sq. = 2.42, df = 2, NS

Q = # of quadrats, #Occ = # of quadrats occupied, #Exp = expected number of quadrats occupied. Species abbreviations are L: Lincoln's sparrow, F: Fox sparrow, S: Song sparrow, W: White-crowned sparrow. The ∪ is the set theory symbol for union.

of this analysis are shown in Table 6.1. For each species, the occupancy rate (quadrats occupied/quadrats available) of its core habitat is not affected by whether or not its core is also core for the other three species, for various two-species combinations, or for other species individually (Table 6.1, left side). Further, in no species does occupancy of core habitat deviate from random expectations depending on whether its core quadrats, grouped by numbers of co-occupant species, are actually occupied by the trio, various duos or various other single species (Table 6.1, upper right for each species). However, two species, Lincoln's and Song sparrow, do show a significant dependence on each others' distributions and tending to avoid quadrat co-occupancy with the other (Table 6.1, lower right for the two species). Lincoln's sparrow has a significantly higher occupancy rate in quadrats without Song sparrows (S′) and especially those without Song and Fox sparrow (S′F′), and a significantly lower occupancy rate where Song sparrow is present (+S). The corresponding analysis for Song sparrow occupancy shows a significantly reduced quadrat use where Lincoln's sparrow is present (+L); Fig. 6.18 graphically summarizes these results.

Overall, the assembly of emberizid species' combinations in various habitat quadrats is understood by (a) species specific habitat preferences, which are distinct among species for the most part, but broadly overlapping in the four commoner species, especially in Song and Fox sparrows (S, F), less similar in Lincoln's sparrow (L); (b) Fox sparrow is a much larger species, which may explain why it appears not to interact with S or L; (c) Song and Lincoln's sparrows are similarly sized, and tend to avoid shared use of their common core habitat. (d) Further evidence supports a behavioral interaction

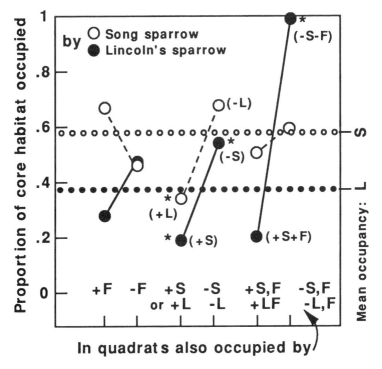

Fig. 6.18. Two emberizid species, Song and Lincoln's sparrow, avoid interspecific co-occupancy of quadrats via behavioral interactions. The mean quadrat occupancy rate for each species is shown on ordinate at the right, and each species' occupancy rate is shown in the figure (open dots for Song sparrow, closed for Lincoln's sparrow) as a function of quadrat occupancy by the other species (at bottom, abscissa). Note that Lincoln's sparrow occupies significantly more quadrats in its core habitat when Song (–S) or both Song and Fox sparrows (–S,F) are absent, significantly fewer when Song sparrow is present (+S), and is unaffected by Fox sparrow presence. Song sparrow shows a corresponding but weaker avoidance of Lincoln's sparrow, occupying significant fewer of its core quadrats when Lincoln's sparrow is absent (–L). * indicates a significant deviation from the expected occupancy rates.

between S and L, as the two species occasionally display interspecific aggression, with song duels and chasing between allospecific territorial neighbors. Lastly, (e) in dry years Lincoln's sparrow is more common and Song sparrow densities are lower, while in wet years the opposite is true; in any one year the two tend to occupy the site in a complementary fashion.

The paruline warblers

The suite of warblers breeding in the wet willows site is comprised of three 10 g species: Yellowthroat (H), Yellow (Y), and MacGillivray's warbler (G), the slightly smaller Wilson's warbler (W, 8 g), and the larger (and rare) Northern waterthrush (N, 18 g). In rank density, Y and H are abundant, W is usually common but varies in density among years, and G and N are both scarce at the site, the former present most years and the latter only in wet years (Cody, 1996). While the waterthrush forages on or near the ground in the more waterlogged quadrats, the other four species are foliage insectivores within the willows vegetation. Wilson's warbler does a little aerial flycatching, but apart from that all four of the species engage in similar foraging behavior (but employed at distinctly different heights above; see below).

Like the emberizids, the parulines also show interspecific differences in habitat preference, but with much overlap. A factor analysis of vegetation characteristics shows the distributions occupied quadrats for each species (Fig. 6.19(*a*)). The waterthrush is most distinct, with a preference for watery sites in which only the tallest willows survive. MacGillivray's warbler prefers taller willows with less open water, and Yellowthroat preferentially occupies the shorter willows. The results of DFA on the warbler habitat preferences are shown in Fig. 6.19(*b*). The centroids depicted are 95% confidence ellipses around the mean; while the means are fairly distinct among species, there is overall extensive interspecific overlap, and just 95/243 quadrats (39%) can be correctly classified to specific occupant.

An analysis of the core habitats for the three more common species (Y, H, W) was undertaken as for the emberizids (above). It shows there are no interactions among species over habitat; a quadrat is occupied or not independently of whether another one or two species also include the quadrat within their territories.

The assembly of warblers in habitat patches as represented by different quadrats is determined largely by habitat structure. The basis of the assembly rules is provided by the foraging height distributions of the different species, which rank G-W-Y-H from higher to lower in foraging activity within the vegetation; Fig. 6.20 (left-hand side; from Cody, 1996) illustrates the differences among species in this predominant aspect of foraging ecology. The warblers select habitat patches within the site that provide vegetation of heights corresponding to their foraging activity. Thus the representation of tall vegetation in quadrats occupied by high-foraging MacGillivray's warbler significantly exceeds the site average, that of low vegetation in Common yellowthroat quadrats, a low-foraging species, is higher than the site average, and Yellow warbler, which is nearly ubiquitous within the plot, is least distinct in quadrats occupied and forages at intermediate heights that are well represented throughout the site (Fig. 6.20, right-hand side).

In summary, different quadrats support combinations of warbler species in

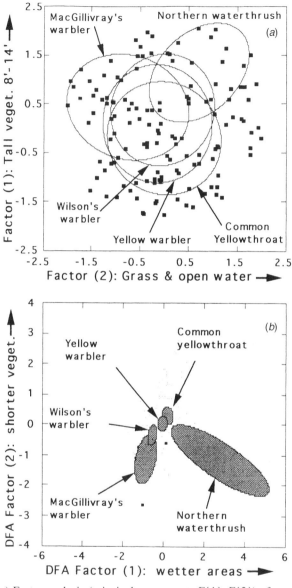

Fig. 6.19. (*a*) Factor analysis (principal components F{1}, F{2}) of vegetation characteristics at the willows site, and the distributions of paruline warblers over these
factors. Northern waterthrush shows a distinct habitat preference for wet habitat (high
F{2} scores) and for taller vegetation (high F{1} scores), whereas MacGillivray's
warbler occupies tall vegetation but in dry quadrats. Rather similar habitat preferences
characterize the remaining three species. (*b*) Discriminant function analysis of the
same data shows that mean habitat preferences tend to be distinct among all species
(95% centroids shown), except that MacGillivray's and Wilson's warblers exhibit a
considerable overlap.

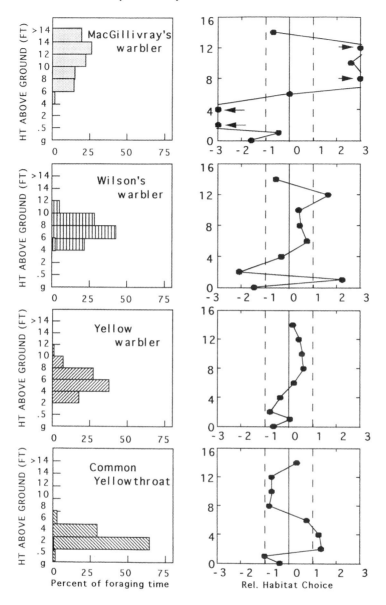

Fig. 6.20. Foraging height distributions differ significantly (by Chi-squared tests) in the four common foliage insectivorous warblers (left-hand side). MacGillivray's warbler forages highest in the vegetation, Common yellowthroat lowest, with Wilson's and Yellow warblers at intermediate heights. Foraging heights correspond to the predominant vegetation heights in the quadrats occupied by each species (right-hand side); 'relative habitat choice' is measured here by the deviation of vegetation in occupied quadrats from the site as a whole, in standard deviation units. Off-scale points labeled with arrows.

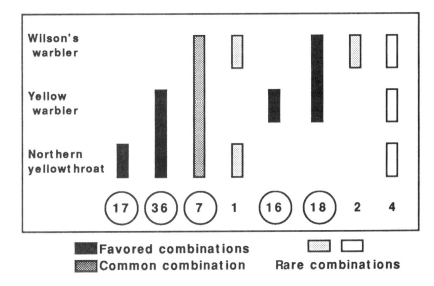

Fig. 6.21. Quadrats support various wabler combinations in accordance to the distribution of vegetation height within them. The common combinations are circled, indicating the numbers of quadrats in which these combinations occur.

accordance with the availability of willows in the required height classes. Favored combinations are those supported by willows that are low (with H only), low to intermediate (H + Y), of intermediate height (Y only) or are intermediate to tall (Y + W; see Fig. 6.21.) The three species combination (H+Y+G) is also common. Other combinations are rare, and the low incidence of MacGillivray's warbler at the site is readily explained by the dearth of willows over 2m in height (viz. Fig. 6.15).

Conclusions

In this chapter the application of assembly rules to communities has been discussed in three general classes of comparison, at three different levels of resolution. Data from a variety of studies were reviewed for each of the three classes, which were illustrated further and discussed in detail using additional studies and examples: (a) amongst islands with discrete plant species assemblages, (b) in oak woodland bird communities replicated across a considerable range of the habitat type, and (c) within a single heterogeneous willows site, in which the composition of emberizid sparrow and paruline warbler guilds was assessed at the level of individual quadrats. Clearly, assembly rules can be sought at each of these different levels of resolution.

Patterns of species co-occurrence on islands were used to identify a pos-teriori island characteristics that support one or another assemblage of forest plants. While α-diversity is not closely constrained on the islands, differences in island size, elevation, and sheltered vs. exposed positions contribute to species turnover and hence to β-diversity. In other components of the flora, the weedy edge species and especially the Asteraceae allow for no predictable species assemblages, and species replacements among islands contribute not to β- but to γ-diversity.

Oak woodland bird communities, on the other hand, appear tightly regulated in α-diversity, although the community composition does change throughout the range of the oaks. Some species replacements appear unpre-dictable, with the substitution of ecologically related species unrelated to shifts in vegetation structure, latitude, or other geographical aspects of site position (e.g., wrens, vireo guilds in the southern sites). In some cases the composition of favored assemblages is related to variation in the structure of the wood-land (flycatchers) or the geographical position of the site (woodpeckers). Thus changes among sites in community composition contribute variously to both β- and γ-diversity.

Within sites, ecologically related species may coexist by dint of morpho-logical or behavioral differences, and their coexistence moderated by the char-acteristics of the habitat at the quadrat scale. Species that are particularly close, ecologically, may coexist only in certain years (e.g., high productivity years) or in certain sites or that permit their divergence, but otherwise they segregate by habitat use. Thus favored combinations will include one or the other but not both species together. Old World sylvine warblers in various pair-wise combinations, Brewer's and Vesper sparrows, and Song and Lincoln's sparrows all provided examples of this phenomenon, and further illuminated conditions under which their coexistence might in fact be permitted.

References

Blake, J.G. (1991). Nested subsets and the distribution of birds on isolated woodlots. *Conservation Biology* 5: 58–66.

Bolger, D.T., Alberts, A.C., & Soulé M.E. (1991). Occurrence patters of bird species in habitat fragments: sampling, extinction, and nested species subsets. *American Naturalist*, 137: 155–166.

Brown, J.H. (1971). Mammals on moutaintops: nonequilibrium insular biogeography. *American Naturalist* 105: 467–478.

Case, T.J. & Casten, R. (1979). Global stability and multiple domains of attraction in ecological systems. *American Naturalist* 113: 705–714.

Cody, M.L. (1973). Parallel evolution and bird niches. In *Mediterranean-Type Ecosystems: Analysis and Synthesis*, ed. F. diCastri & H.A. Mooney Ecological Studies 7, Chapter 3, pp. 307–338. Wien: Springer-Verlag.

Cody, M.L. (1974). *Competition and the Structure of Bird Communities.* Monographs in Population Biology 7, Princeton NJ: Princeton University Press.

Cody, M.L. (1975). Towards a theory of continental species diversities. In *Ecology and Evolution of Communities*, ed. M.L. Cody & J. Diamond Chapter 10, pp. 214–257.

Cody, M.L. (1978). Habitat selection and interspecific interactions among the sylviid warblers of England and Sweden. *Ecology Monographs* 351–396.

Cody, M.L. (1983a). Bird species diversity and density in Afromontane woodlands. *Oecologia* 59: 210–215.

Cody, M.L. (1983). Continental diversity patterns and convergent evolution in bird communities. In *Ecological Studies 43*, ed. F. Kruger D.T. Mitchell & J.U.M. Jarvis Chapter 20, pp. 347–402. Wien: Springer-Verlag.

Cody, M.L. (1985). Habitat selection in the sylviine warblers of western Europe and north Africa. In *Habitat Selection in Birds*, ed. M.L. Cody, Chapter 3, pp. 85–129. Orlando, San Diego: Academic Press.

Cody, M.L. (1986). Diversity and rarity in Mediterranean ecosystems. Ch. 7 In *Conservation Biology: The Science of Scarcity and Diversity,* ed. M. Soulé, pp. 122–152. Sunderland, MA: Sinauer Assoc.

Cody, M.L. (1992). Some theoretical and empirical aspects of habitat fragmentation. *Proceedings of the Society of California Academy of Science Symposium, Occidental* College, CA.

Cody, M.L. (1993). Bird diversity patterns and components across Australia. In *Species Diversity in Ecological Communities: Historical and Geographical Perspectives*, ed. R.E. Ricklefs & D. Schluter, Chapter 13, pp. 147–158. Chicago, IL: University of Chicago Press.

Cody, M.L. (1994a). Bird diversity components in heathlands: comparisons SW and NE Australia. *Tasks in Vegetation Science*, ed. R. Groves & M. Farragitaki, vol. XX, pp. 47–61. Holland: Kluwer.

Cody, M.L. (1994b). Mulga bird communities. I. Species composition and predictability across Australia. *Australian Journal of Ecology* 19(6): 206–219.

Cody, M.L. (1994c). Bird communities of Australian *Eucalyptus* and north-temperate *Quercus* woodlands. *Journal für Ornithologie* 135: 468.

Cody, M.L. (1996). Birds of the central Rocky Mountains. In *Long-Term Studies of Vertebrate Communities.* ed. M.L. Cody, & J. Smallwood Chapter 11, pp. 291–342. Orlando, San Diego: Academic Press.

Cody, M.L. & Diamond, J.M. (1975). *Ecology and Evolution of Communities.* Cambridge, MA: Belknap Press of Harvard University Press.

Cody, M.L. & Walter, H. (1976). Habitat selection and interspecific territoriality among Sardinian *Sylvia* warblers. *Oikos* 27: 210–223.

Cody, M.L. & Mooney, H.E. (1978). Convergent and divergent evolution in chaparral, plant and bird communities on four continents. *Annual Review of Ecology and Systematics* 9: 265–321.

Cody, M.L. & Overton, J.McL. (1996). Evolution in achene morphology of composite plants on small islands. *Journal of Ecology* 84(1): 53–61.

Cutler, A. (1991). Nested faunas and extinction in fragmented habitats. *Conservation Biology* 5: 496–505.

Diamond, J.M. (1975). Assembly of species communities. In Chapter 14, *Ecology and Evolution of Communities*, ed. M.L. Cody, & J.M. Diamond, pp. 342–444. Cambridge, MA: Belknap Press of Harvard University Press.

Diamond, J.M. & Mayr, E. (1976). Species–area relation for birds of the Solomon

Archipelago. *Proceedings of the National Academy of Sciences, USA* 73: 262–266.

Diamond, J.M., Gilpin M.E. & Mayr, E. (1976). Species–distance relation for birds of the Solomon Archipelago: the paradox of the great speciators. *Proceedings of the National Academy of Sciences USA*, 73: 2160–2164.

Diamond, J. & Case, T.J. (1986). *Community Ecology*. NY: Harper & Row.

Kadmon, R. (1995). Nested species subsets and geographic isolation: a case study. *Ecology* 76: 458–465.

MacArthur, R.H., Recher, H. & Cody, M.L. (1996). On the relation between habitat selection and bird species diversity. *American Naturalist* 100: 319–332.

MacArthur, R.H. & Wilson, E.O. (1967). *The Theory of Island Biogeography*. Monographs Population Biol. 1. Princeton, NJ: Princeton University Press.

Patterson, B.D. (1987). The principle of nested subsets and its implications for biological conservation. *Conservation Biology* 1: 323–334.

Patterson, B.D. & Atmar, W. (1986). Nested subsets and the structure of insular mammalian faunas and archipelagoes. *Biological Journal of the Linnaean Society* 28: 65–82.

Ryti, R.T., & Gilpin, M.E. (1987). The comparative analysis of species occurrence patterns on archipelagos. *Oecologia* 73: 282–287.

Salt, G.W. (1957a). An analysis of avifaunas in the Teton Mountains and Jackson Hole, Wyoming. *Condor* 59: 373–393.

Salt, G. W. (1957b). Song, Lincoln's and fox sparrows in a Tetons willow thicket. *Auk* 74: 258.

Simberloff, D. & Levin, B. (1985). Predictable sequences of species loss with decreasing island area – and birds in two archipelagoes. *New Zealand Journal of Zoology* 8: 11–20.

Simberloff, D. & Martin, J-L. (1991). Nestedness of insular avifaunas; simple summary statistics masking complex species patterns. *Ornis Fennica* 68: 178–192.

Wilson, B. & Roxburgh, X. (1994). *Oikos* 69: 267–276.

Wilson, J.B. & Roxburgh, S.H. (1994). A demonstration of guild-based assembly rules for a plant community, and determination of intrinsic guilds. *Oikos* 69: 267–276.

7

Impact of language, history, and choice of system on the study of assembly rules

Barbara D. Booth and Douglas W. Larson

Introduction

Assembly rules provide a useful framework that allows ecologists to make predictions about community change and development over time. However, three features of the discussion about ecological assembly rules lessen its potential impact on ecologists and those who apply ecology to management. First, the current language of the idea has become unnecessarily complex. It is not always clear what researchers mean when they refer to assembly rules. Some argue that assembly rules only apply to biotic interactions (Wilson & Gitay, 1995) whereas others consider all constraints on community development (Keddy, 1992). Some workers look for rules that apply at the small scale, or to specific species distributions, whereas others look for larger, more general patterns (Drake *et al.*, 1993).

Second, the current discussion appears to have little regard for the history or origin of work on the constraints on community development. Diamond (1975) is usually cited as having coined the phrase 'assembly rule', but in reality, many early ecologists such as Clements (1916, 1936) and Gleason (1917, 1926) addressed many of the questions currently being considered. By ignoring their work, we run the risk of reinventing ideas already well established in the literature.

Finally, the primary evidence used to develop and test ideas on assembly rules is often derived from complex ecosystems that do not necessarily follow simple trajectories (Drake, 1991). Thus, community ecologists may be starting with the most complex systems available. An alternative strategy may be to start with simple natural systems where assembly rules are easier to find. After being able to observe them, it may be possible to expand the horizon to more complex systems. It is suggested that communities in unproductive, stressful environments are more likely to be predictable and therefore are good systems to use for preliminary investigations of assembly rules.

In this chapter, it is argued that, if workers persist in ignoring these problems, little progress will be made in the study of assembly rules. Each of these three points will be addressed below. In each case the objective is to simplify the debate, not to complicate it. It is important to remember that if the history of science and the way humans participate in it are ignored the mistakes of the past are very likely to be repeated.

Language considerations

Since the publication of Diamond's book chapter (Diamond, 1975), there has been a proliferation of papers trying to promote, refute, or test the idea that there are sets of constraints (rules) on community formation and maintenance (assembly). Some argue that particular combinations of species are the result of the non-random consequences of competition (Diamond & Gilpin, 1982; Fox, 1987). Others suggest that the observed patterns could also be explained by chance (Connor & Simberloff, 1979) or historical factors (Drake, 1990a). Others take the view that assembly rules should be discussed within a grammatical context that includes many other equally vague but potentially useful terms such as 'organism', 'species', or 'ecosystem' (Haefner, 1978, 1981). Some have argued that the assembly rules can include any constraint on the species pool (Keddy, 1992), while others have a more narrow view that assembly rules can only include constraints placed upon species by other species (Lawton, 1987; Wilson & Gitay, 1995; Wilson & Whittaker, 1995). Clearly, there is no general concensus as to what assembly rules are and how they should be defined, although usage of the term falls under two general schools of thought: those who believe assembly rules should describe the mechanisms of biotic interactions only, and those who recognize that a community is formed from a species pool and that biotic and abiotic mechanisms will act to determine which of these species will form the community (Table 7.1). Even within these groups, there is some variation in the definition. Here we take the latter view.

Many researchers have tried to articulate assembly rules for various ecological and artificial communities. In more controlled environments the rules generated are general in nature and can be used to establish principles of community assembly. For example, Drake (1991) showed, in laboratory experiments, that larger systems with greater total productivity are more prone to historical events than small systems with low productivity. He found that species invasion sequence was more likely to have an effect on the species

Table 7.1. *Some definitions and statements used to define or describe assembly rules*

Biotic interactions

Diamond (1975). 'Do properties of each species considered individually tell us most of what we need to know in order to predict how communities are assembled from the total species pool.' p. 385

'It seems likely that competition for resources is a major factor underlying assembly rules.' p. 423

Lawton (1987). 'I take the view that assembly rules arise from species interactions.'

Wilson & Watkins (1994). '... generalized constraints on species interactions'

Wilson & Whittaker (1995). '... generalized restrictions on species presence or abundance that are based on the presence or abundance of one or several other species or types of species (not simply the response of individual species to the environment).'

Biotic and abiotic interactions

Haefner (1981). describing the 'ecosystem assembly problem'. 'Construct an algorithm such that, given an arbitrary collection of environmental factors, the output of the algorithm is a list of species associated with the environment.'

'The model must be general by being applicable to any given environment and species pool ... [and] produce specific predictions of the spatial distributions of particular species.'

Keddy (1989). 'Given (a) a species pool and (b) an environment, can we predict the abundance of the organisms actually found in that environment?' p. 156

Roughgarden (1989). 'A community's membership is structured by the transport processes that bring species to it, and it is structured by the population dynamics, including species interactions of its members; and furthermore, it is rarely possible to focus on only one of these sides. That is to say, a community reflects both its applicant pool and its admission policies.' p. 218

Drake (1990b). 'Given a species pool and environment, what rules influence community structure?'

Drake *et al.* (1993). '... how mechanisms [ecological processes and environmental variation] function to produce ecological patterns.'

Grover (1994). 'When local communities are assembled from a regional species pool, assembly rules state which of the species from this pool can coexist.'

---

composition of more productive systems. Similarly, Law and Morton (1993) used computer simulations to show that the predictability of community composition decreased as the number of species increased, and that historical factors influencing the order of invasion were extremely important to the outcome. These generalities can now be tested in suitable ecological communities.

Some workers have described constraints in field situations; however, these tend to be limited in scope as they are species or trait specific and rarely describe the entire community. Fox and Brown (1993), for example, found in rodent communities in southwestern deserts of North America that each new species entering a community will tend to come from a different functional group. When all functional groups are filled, the rule repeats. They were able to apply this rule to different groups of desert rodents and at different spatial scales; however, it applies only to rodents and focuses just on competitive interactions of rodents. Similarly, van der Valk (1981) showed that the invasion of wetlands involves the primary assembly rule that plants must be able to tolerate flooding during germination. This rule is useful as it focuses on the mechanisms important to establishment and allows us to predict which species may become part of the community; however, is limited because it does not take competition into consideration nor does it predict final species composition and abundance. One further difficulty with empirical field studies is determining whether the identified mechanisms are important to community development or whether they are simply important in maintaining the current community. This has been discussed extensively by Drake (1991).

In other settings we find that: '[The assembly rules] demonstrated are not strong ones. Either there are many other processes at work, obscuring the assembly rules, or more powerful methods are required to find the rules' (Wilson & Watkins 1994). Other assembly rules are defined and then concluded to be ephemeral or contextual in their appearance (Wilson *et al.*, 1995b), or are defined in the introductions of papers and then never mentioned in the Discussion sections (Wilson *et al.*, 1996).

What seems to have been missed is that the original chapter that offered the term assembly rule simply took advantage of an often used technique of ecological pedagogy: when faced with a truth that more or less applies at least some of the time, try to present that truth in the form of a 'law' or, when even greater uncertainty is involved, as a 'rule'. Hence Bergmann's rule predicts that homeotherms in polar regions should be relatively larger than similar organisms near the equator, Allen's rule claiming that appendages should be shorter in polar regions, and Gloger's rule stating that animals from hot dry regions ought to be less pigmented than ones from wet cold habitats (Pianka, 1983). Few practising ecologists expect such rules to do more than describe very general trends that might reflect some underlying physiological or evolutionary principle. That is their purpose. But from time to time, there are those who try to test empirically some of the predictions from these rules, only to find that the truth is obscured by the variance (Peters, 1988).

By trying to make assembly rules that apply all of the time, we are preventing ourselves from establishing general principles that apply most of the time.

Some of the difficulty in finding general rules that apply most of the time is due simply to the complexity of ecological systems, and the different scales at which they are examined. Further, ecologists should not be embarrassed to derive rules or laws to describe patterns or processes that apply only part of the time, or for part of the world. As Diamond and Gilpin (1982) also tried to show, ecologists should not permit a single philosophical view to be used to dissect the structure of nature. The evidence from the assembly rules debate suggests that ecologists are trying to search for greater clarity in the wording of various assembly rules based on the available data. However, the debate continues over how best to do this. Diamond and Gilpin (1982), and later Keddy (1989), correctly pointed out that, even the decision about what null model to test patterns against requires more information than we have about the structure and function of ecosystems. Wilson (1994) argues that actual organizational features of communities should be tested against null (or alternate) models. He does not address the issue that few statistical tests would be able to discriminate among any of the individual species distribution patterns he plotted. Thus, the choice of a particular null model could entirely explain the finding of an assembly rule in a New Zealand rain forest (small plants and the herb guild in the ground flora will be relatively constant whereas large, less frequent plants will not) (Wilson *et al.*, 1995a).

So what scientific progress has been made by the use of new terminology in this area – progress that would not have been made if simple and clear terminology had been used? The debate has certainly resulted in the generation of more primary data, and despite Keddy's assertion that we have enough primary data (Keddy, 1989), many believe that we do not. Other than this vastly greater amount of information on the composition of specific plant and animal communities world-wide, what new and consistently verifiable predictions have been made because of the new language? Few. What has actually been accomplished is that workers previously exhausted with 70 years of debate over the complex controls of community composition in succession have been re-energized to study essentially the same problem dressed in new clothes.

One of the questions that seems important here is this: what new questions does the idea of assembly rules permit that were not possible with the original language of succession or even the language that was reviewed by Drury and Nisbet (1973)? It can be argued that if the word constraint is used to substitute for the word rule and the word development is used to substitute for the word assembly, then the current discussion of assembly rules is reduced to a discussion of developmental constraints on community structure: an idea fully explored by Clements, Gleason and many other ecologists. Wilson and Whittaker's (1995) discussion of guild proportionality would likely not surprise any of the past workers in the area of plant or animal communities undergoing succession because the idea of there being developmental constraints on the form and number of plants and animals was widely accepted at the turn of the century.

The language of the assembly rules debate has hidden the initial intention of the term, and has added to the confusion that ecologists, conservationists and land managers have over whether the idea encapsulates new and important concepts and applications. In the next section, the development of the ideas of constraints on community structure is discussed in an attempt to remind ourselves of the already established principles.

Historical considerations

The idea of ecological or community assembly rules is not new, nor is it fair to attribute the idea to Diamond (1975) as many have done (Connor & Simberloff, 1979; Fox, 1987; Keddy, 1989; Wilson, 1991, 1994; Drake *et al.*, 1993; Wilson *et al.*, 1994, 1995a,b). What is new is the emergence of the term 'assembly rules', used initially within the context of animal ecologists trying to explain the constraints on species abundance and composition on tropical islands. The attempt to explain these island biogeographic patterns itself was not new and is dated at least to the summary work by MacArthur and Wilson (1967). As is often the case in biology, these complicated species composition patterns – both their genesis and maintenance – were being explored by one group of ecologists, animal ecologists, without explicit discussion of the fact that the problem had already been examined at length by other groups, most notably the plant ecologists. For example, in 1916 Clements wrote:

Reasons why plants appear at certain stages – Migrules are carried into an area more or less continually during the course of its development. This is doubtless true of permobile seeds, such as those of aspen. As a rule, however, species reach the area concerned at different times, the time of appearance depending chiefly upon mobility and distance. As a consequence, migration determines in some degree when certain stages will appear. The real control, however, is exerted by the factors of the habitat, since these govern ecesis[1] and hence the degree of occupation. The habitat determines the

[1]'Ecesis is the adjustment of the plant to a new home. It consists of three essential processes, germination, growth and reproduction. It is the normal consequence of migration ...' Clements (1916) (p. 68).

character of the initial stage by its selective action in the ecesis of migrules.... After the initial stage the development of succeeding ones is predominantly, if not wholly, a matter of reaction, more or less affected by competition. In addition, some stages owe their presence to the fact that certain species develop more rapidly and become characteristic or dominant, while others which entered at the same time are growing slowly. (p. 102)

In this text, taken from the classic work 'Succession', there is a concern for the very problem considered by Diamond (1975) and other island biogeographers before him: the question of what constraints determine if, and when, a species will be successful at a certain site.

Very shortly after Clements wrote the paragraph quoted above, H.A. Gleason wrote in 1917 and reiterated the following in 1926:

It has sometimes been assumed that the various stages in a successional series follow each other in a regular and fixed sequence, but that frequently is not the case. The next vegetation will depend entirely on the nature of the immigration which takes place in the particular period when environmental change reaches the critical stage. Who can predict the future for any one of the little ponds considered above? In one, as the bottom silts up, the chance migration of willow seeds will produce a willow thicket, in a second a thicket of *Cephalanthus* may develop, while a third, which happens to get no shrubby immigrants, may be converted into a miniature meadow of *Calamagrostis canadensis*. A glance at the diagram of observed successions in the Beach Area, Illinois, as published by Gates, will show at once how extraordinarily complicated the matter may become, and how far vegetation may fail to follow simple, pre-supposed successional series. (p. 21).

Here is the idea of historical artifact or randomness being important to species invasion patterns. In addition, what is really interesting about these quotations from Clements and Gleason, and the vast literature that grew out of the debate between the 'organismic' and the 'individualistic' view of succession, is that the primary evidence used to support the two warring schools of thought was identical, or nearly so.

The fight continued through the 1960s when Odum (1969) attributed some measure of Clements' superorganismic thought to the process of succession, as revealed by the new buzzwords: 'The strategy of ecosystem development' (Odum, 1969). Odum believed that succession was 'community controlled' – that there were some rules governing community development. This idea was discussed by many, including Grime (1977, 1979) who used a relatively simple triangular model to describe the rate of biomass accumulation and the sequence of life-forms along successional pathways. Drury and Nisbet's brave 1973 attempt to tow ecologists out of the gumbo of rhetoric about the actual patterns and mechanisms behind ecological succession was applauded by some (Golley, 1977), but ignored by most. Drury and Nisbet promoted the idea that patterns resulted from differential growth, colonizing ability, and longevity of constituent species, rather than from autogenic processes where earlier plants ameliorated conditions for later colonizing species as Odum had suggested. Connell and Slatyer (1977) argued that competitive interactions (between sessile organisms and with herbivores, predators and pathogens) were an important force and outlined three models of mechanisms (facilitation, tolerance, inhibition) that produce successional change. It was at this time that the new idea of ecological communities being controlled by assembly rules was first offered by animal ecologists concerned with the control of species mixtures.

In the history of ideas about how communities are assembled and maintained, we see that many of the same problems identified at the beginning of ecology are still with us, unsolved and (perhaps) forever clouded by complex details that are the reality of nature. The core of the current discussion repeats similar debates over the last century.

The choice of system used to search for assembly rules

The last feature of the discussion of assembly rules is the part that attracted us to this debate and to a symposium on the subject in the first place. Regardless of the set of language used, why is it that specific sets of constraints on community composition are so hard to consistently find in real plant communities? Again, a return to the historical literature may help: Under the chapter dealing with ecesis causes, Clements (1916) noted:

Properly speaking, competition exists only when plants are more or less equal. The relation between ... dominant tree and secondary herbs on the forest floor [is not competition]. The latter has adapted itself to the conditions made by the trees, and is in no sense a competitor of the latter. Indeed, as in many shade plants, it may be beneficiary. The case is different, however, when the seedlings of the tree find themselves alongside the herbs and drawing upon the same supply of water and light. They meet upon more or less equal terms, and the process is essentially similar to the competition between seedlings alone, on the one hand, or herbs on the other. The immediate outcome will be determined by the nature of their roots and shoots, and not by the dominance of the species. Naturally, it is not at all rare that the seedling tree succumbs. When it persists, it gains an increasing advantage each succeeding year, and the time comes when competition between tree and herb is replaced by dominance and subordination. This is the course in every bare area and in each stage of the sere which develops upon it. The distinction between competition and dominance is best seen in the development of a layered forest in a secondary area, such as a burn. All the individuals compete with each other at first in so far as they form intimate groups. Within the growth of shrubs, the latter become dominant over the herbs and are in turn dominated by the trees. Herbs still compete with herbs, and shrubs with shrubs, as well as with younger individuals of the next higher layer. Within the dominant tree-layer, individuals compete with individuals and species with species. Each layer exemplifies the rule that plants similar in demands compete when in the same area, while those with dissimilar demands show the relation of dominance and subordination. (page 72)

This text sets out the idea that community development will have a significant degree of organization by life form, but that organization will be strongly scale dependent at the species level. This has been confirmed by Wilson *et al.* (1995b). That life-form itself can be predicted is, to a certain degree, a morphological tautology. Küppers (1985) showed that the invasion sequence and final structure of European hedgerows was controlled almost completely by the architecture of the plants shading the ground. Physiological differences among the plants had no role in regulating the final configuration of these systems. This text of Clements also implies that complex systems will be strongly influenced by the sequence of species introductions, as has been shown by several ecologists (Abrams *et al.*, 1985; Robinson & Dickerson, 1987; Drake, 1991; Drake *et al.*, 1993). McCune and Allen (1985) showed that only 10% of canopy tree composition in very productive west-coast coniferous forest could be predicted by site factors, whereas the other 90% were attributable to historical (and other unknown) factors. In other words, the details of the species composition, including the dominant species, are impossible to predict because the number, timing, and consequences of competitive interactions and other historical events are too great to compute.

In contrast to the complex and unpredictable species composition that applies to productive forest or grassland, Clements notes that for unproductive communities:

The selective action of bare areas upon the germules brought into them is exerted by ecesis.... The two extremes, water and rock, are the extremes for ecesis, the one impossible for plants whose leaves live in the air and the light, the other for those whose roots must reach water. The plants which can ecize in such extremes are necessarily restricted in number and specialized in character, but they are of the widest distribution, since the habitats which produced them are universal. From the standpoint of ecesis, succession is a process which brings the habitat nearer the optimum for germination and growth, and thus permits the invasion of an increasingly larger population. The fundamental reason why primary succession is long in comparison with secondary, lies in the fact that the physical conditions are for a long time too severe for the vast majority of migrants, as well as too severe for the rapid increase of the pioneers. (page 71)

This selection of text suggests that a small array of species that share an ability to tolerate environmental extremes will occur in habitats that are 'bare'. Without stating it explicitly, Clements suggested that bare areas are more likely to have communities of organisms that share many features in common, and that these features permit the slow and relentless increase in biomass over long periods of time during which invasions of the habitat by other organisms are not possible. In bare areas, interactions among taxa, and especially between each taxon and the physical environment, ought to be more easily predicted. These same interactions that take place in productive habitats where there are many more species and therefore more interactions, will not be as easy to predict.

The predictability of succession in unproductive areas was also discussed by Grime (1979). He wrote:

> Where the productivity of the habitat is low, the role of ruderal plants and competitors in secondary succession is much contracted, and stress-tolerant herbs, shrubs, and trees become relatively important at an earlier stage. The growth form and identity of the climax species varies according to the nature and intensity of the stresses occurring in the habitat.... Under more severe stress, such as that occurring in arctic and alpine habitats, ruderal and competitive species may be totally excluded and here both primary and secondary successions may simply involve colonization by lichens, certain bryophytes, small herbs, and dwarf shrubs. (page 150)

Grime observed that the successional trajectories for (in his terms) highly stressed communities appear to be very flat, or perhaps even non-existent. Plant biomass remains low throughout and moves from ruderals to stress-tolerant species, bypassing the middle phase of intense competition. This idea was also thoroughly explored and supported by Svoboda and Henry (1987), who showed that many arctic tundras behave as systems that undergo non-directional and non-replacement succession. Communities of plants and animals that develop in such sites engage in more or less permanent low-level interactions with each other. As a result, they show little historical contingency, and little ability to exploit the full array of developmental, successional or assembly options that more productive systems show. Schulman (1954) and Bond (1989) also presented the idea that many conifer woodlands contain communities of slow-growing organisms that are poor competitors with more aggressive higher plants. These non-competitive communities also have enormous capacity for displaying individual plant and community longevity. Baker (1992) shows that timescales of nearly 2000 years are needed if one wishes to detect pulses of recruitment in Bristlecone Pine forests. This is also true for other conifers (Barden, 1988; Larson, 1990) and even some plant communities in tropical latitudes (Aplet & Vitousek, 1994).

Constraints on community structure of the Niagara Escarpment

It appears to us that an unproductive, stressed community (one that is 'bare') might be ideal to examine constraints on community structure. These systems are likely to be deterministic, stable, and not prone to control by historical events. This led us to our work on the Niagara Escarpment. Cliffs of the Niagara Escarpment, southern Ontario, Canada, support one of the oldest and least disturbed forests in the world (Larson & Kelly, 1991; Kelly *et al.*, 1994) (Fig. 7.1(*a*),(*b*)). Uneven-aged stands of *Thuja occidentalis* (eastern white cedar) occur on the cliffs with individual tree ages extending to 1890 years. Older generations of cedar are represented in the coarse woody debris which accumulates and persists for over 3500 years in the talus at the base of the cliff face. In addition to the tree component of the community that has been precisely dated, community composition is also surprisingly predictable over many spatial scales (Larson *et al.*, 1989; Cox & Larson, 1993; Gerrath *et al.*, 1995) and timescales (Ursic *et al.*, 1997) (Fig. 7.2). This is true for animals as well (Matheson, 1995). In fact, for almost all of the plant groups so far examined, the species richness, species composition, and plant/animal abundance on cliff faces are predictable from site to site and over time. This constancy of the form of the community is present along the entire Niagara Escarpment, even though the cliff face community itself is surrounded by lush deciduous forest in the south near Niagara Falls, and near-Boreal forest in the north near Sault Ste. Marie, Michigan, USA. At the life-form level, and sometimes even at the level of genus, this predictability may apply to cliffs globally. The cliffs therefore appear to support an extremely persistent and permanent ecosystem.

The aspect of community composition on cliffs of the Niagara Escarpment that is the most surprising is that only a single species of conifer occurs as the dominant canopy-forming tree, namely *T. occidentalis*. Recent work by Young (1996) and Walker (1987) shows that such dominance by *Thuja* (along with its predictable array of herbs, lichens and mosses) on limestone cliffs has been a feature of this community for perhaps the entire interglacial and post-glacial period (*ca.* 30 000 years). This community of plants still persists throughout some southern US states (North Carolina, Virginia, Ohio, Tennessee, Kentucky, Pennsylvania) when the limestone cliffs face due north, but even in settings where cliff aspects deviate from this, the dominant cliff tree simply switches to *Juniperus virginiana*. Most of the other plant species are retained.

The question therefore, becomes, 'why does such a predictable community of plants emerge from a landscape with so many sources of "migrules"?' If

(*a*)

(*b*)

Fig. 7.1. (*a*) Aerial and (*b*) ground view of the Niagara Escarpment, Ontario, Canada. (Courtesy of Uta Matthes-Sears.)

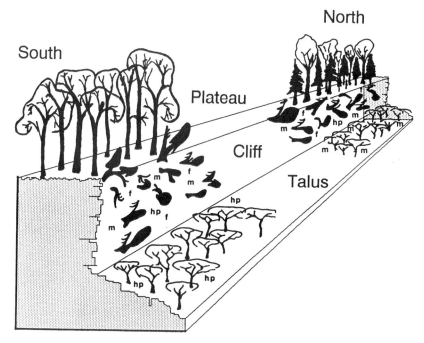

Fig. 7.2. Illustration of the predictable nature of the cliff face community and the less predictable nature of plateau and talus communities along the Niagara Escarpment. The cliff face community supports the same species of higher plants (hp), ferns (f) and mosses (m) from North to South and the same microclimate conditions. Conversely, plateau forests species such as red oak (*Quercus rubra*), white ash (*Fraxinus americana*), and sugar maple (*Acer saccharum*) are common in the South, while aspen poplar (*Populus tremuloides*) and white spruce (*Picea glauca*) are common in the North. The talus slope supports a greater diversity of higher plants in the South and more mosses in the North.

we consider the approach suggested by Keddy (1992), the question can be answered following three steps. First, by assembling a list of potential colonists (the species pool). Second, by collecting morphological, physiological and ecological information about each to form a trait matrix. Finally, the resultant community can be found by determining what traits, and therefore what species, will be deleted from the species pool. In our case we decided to narrow the search by focusing experimental work on just the tree species that anchors the entire system. Thus a new question emerges: namely, how does the cliff environment filter all trees, except *Thuja* from the pool of possible colonists?

Methods

To answer this question, two cliffs typical of the Niagara Escarpment were selected for study (Milton and Dufferin) and transects were set up running from the plateau, down the clifff face and into the talus. Samples were collected from the three communities. This design allowed comparisons of assembly characteristics of the adjacent, yet highly distinct communities. The seed rain (seeds arriving at a site) was sampled by placing 'sticky traps' (modeled on Werner, 1975 and Rabinowitz and Rapp, 1980) in the plateau and talus, and at meter intervals down the cliff face in five transects at both sites over two years. Similarly, soil samples were collected to examine the seed bank (seeds being deposited in soil and litter) from the three positions in five transects at both sites in the spring and fall over a three-year period. The low rate of natural seedling recruitment had already been determined in demographic work (Larson & Kelly, 1991). To examine seedling establishment we planted newly germinated seeds of seven tree species at a third cliff site. Forty randomized locations were used for each species. Individual plants were followed over time, and the rate and timing of mortality was compared among taxa. After three years, surviving seedlings were harvested and biomass and root: shoot ratios calculated. These experiments allowed us to establish a species pool for the cliff face and to determine at what stages species were removed from the pool.

Results

A variety of tree seeds arrived on the cliff in the seed rain (Table 7.2) and became incorporated into the seed bank (Table 7.3). Species lists of the three communities are almost identical. Seed rain and seed bank density was higher in the talus than on the plateau or cliff face (Figs. 7.3 and 7.4). In addition, although many species were present, seed density of tree species other than *T. occidentalis* and *B. papyrifera* was very low (Figs. 7.3 and 7.4). Therefore, the seed rain and seed bank acted only as partial constraints on community composition at these earliest of stages in plant development. Early establishment patterns of the various forest trees also differed among species. A variety of tree species became established on the cliff face and survived into their second year, however, the most interesting results from the seedling establishment experiments came when the plants were entering their third year (Fig. 7.5). By this time a slow filtering of the non-viable taxa was evident. Some seedlings of most species survived until this time; however, while mortality of cedars decreased, the mortality of other species continued. Some

Table 7.2. *List of tree species found in the plateau, cliff face and talus seed rain at the Milton (M) and Dufferin (D) sites in 1993 and 1994*

Latin name	Common name	Plateau	Cliff face	Talus
Acer saccharum	Sugar Maple	M D	M D	M D
Betula papyrifera	White Birch	M D	M D	M D
Betula alleghaniensis	Yellow Birch	D	M D	M D
Ostrya virginiana	Ironwood	M D	M D	M D
Quercus rubra	Red Oak	M	M D	
Thuja occidentalis	Eastern White Cedar	M D	M D	M D

Table 7.3. *List of tree species found in the plateau, cliff face and talus seed bank at the Milton (M) and Dufferin (D) sites from 1993 to 1995*

Latin name	Common name	Plateau	Cliff face	Talus
Acer saccharum	Sugar Maple	M	D	
Betula papyrifera	White Birch	M D	M D	M D
Betula alleghaniensis	Yellow Birch	M D	M D	M D
Fagus grandifolia	Beech	M	M D	
Fraxinus americana	White Ash	D	M D	D
Ostrya virginiana	Ironwood	M D	M D	M
Quercus rubra	Red Oak	M	D	
Thuja occidentalis	Eastern White Cedar	M D	M D	M D

of these patterns were hinted at by the end of the first year, but it was not until the third year that the similarity between the field manipulations and the natural Niagara Escarpment vegetation became clear. Therefore, a variety of tree species can arrive on the cliff face in the seed rain, become incorporated into the seed bank and become established as seedlings on the cliff face. However, at each of these stages there is a filtering or removal of individuals from the community that contributes to final species composition.

The next step in the investigation was to determine what traits of cedar allow it to be successful on cliff faces, and what traits of other species result in them being filtered out of the community. Results to date are preliminary. At the end of the experiment, in the summer of 1996, all the trees remaining alive from the seedling experiments were harvested. Seedling characteristics and the biomass of roots and shoots was measured. Most of the surviving seedlings were exceptionally small plants. All species except white pine and tamarack were less than 3 cm tall and were less than approximately three grams in weight after two years growth (Table 7.4). White birch was 1.5 cm

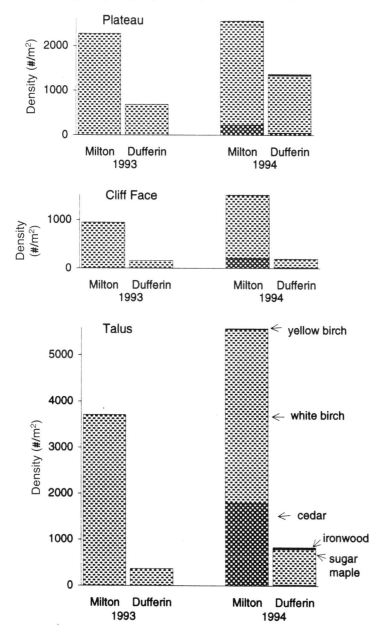

Fig. 7.3. Seed rain density (number of seeds per square metre) of eastern white cedar (*Thuja occidentalis*), white birch (*Betula papyrifera*), yellow birch (*Betula alleghaniensis*), sugar maple (*Acer saccharum*), and ironwood (*Ostrya virginiana*) on the plateau, cliff face and talus at two sites (Milton and Dufferin) in 1993 and 1994.

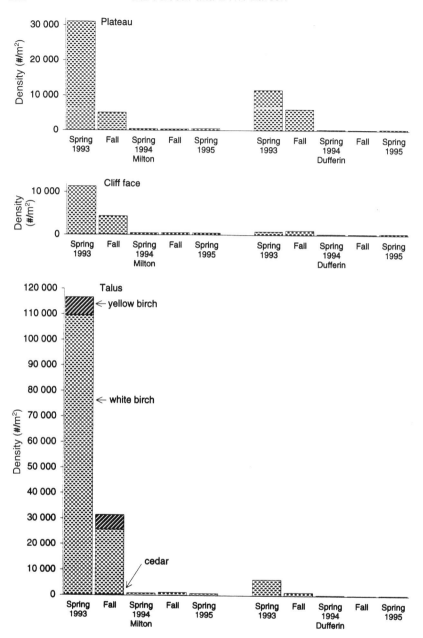

Fig. 7.4. Seed bank density (number of seeds per square metre) of eastern white cedar (*Thuja occidentalis*), white birch (*Betula papyrifera*), yellow birch (*Betula alleghaniensis*), sugar maple (*Acer saccharum*), and ironwood (*Ostrya virginiana*) on the plateau, cliff face and talus at two sites (Milton and Dufferin). Seed bank samples were collected in the spring and fall from 1993 to 1995.

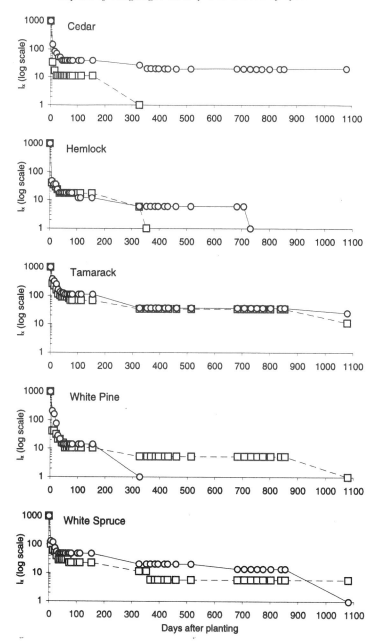

Fig. 7.5. Seedling survivorship (lx) of eastern white cedar (*Thuja occidentalis*), hemlock (*Tsuga canadensis*), tamarack (*Larix laricina*), white pine (*Pinus strobus*) and white spruce (*Picea glauca*) on natural cliff faces over a three-year period. Seeds were planted in May 1993 on crevices (squares) and ledges (circles).

Table 7.4. *Table of seed weight, second and third year survivorship, and character-istics of second year seedlings planted on the cliff face. Species planted were yellow birch (Betula alleghaniensis), white birch (Betula papyrifera), tamarack (Larix laricina), white spruce (Picea glauca), white pine (Pinus strobus), eastern white cedar (Thuja occidentalis) and hemlock (Tsuga canadensis)*

Species	Seed mass (mg)	Final survivorship year 2	Final survivorship year 3	Final height (cm)	Longest root (cm)	Final biomass (g)	Root: shoot
Betula papyrifera	0.3	1%	–	1.5	11.0	0.030	1:1.7
Betula alleghaniensis	1.0	0%	–	–	–	–	–
Thuja occidentalis	1.3	2%	2%	2.8	6.2	0.032	1:2.7
Larix laricina	1.4	3%	2%	4.7	5.3	0.047	1:1.8
Picea glauca	2.0	1%	0.3%	3.0	7.3	0.029	1:1.6
Tsuga canadensis	2.4	4%	0%	–	–	–	–
Pinus strobus	17.2	0.3%	0%	5.9	9.1	0.147	1:2.0

tall (Table 7.4). In addition, most surviving cedar trees showed much smaller root systems compared to shoots than was expected. Root–shoot ratios in healthy surviving cedars were about 1:2.7 while values for other species were less than 1:2 (Table 7.4). These experimental cedars has similar root: shoots ratios to naturally occuring cedars (Matthes-Sears *et al.*, 1995). White pine seedlings, which have the largest seeds and seedlings, were completely excluded by the end of the experiment. Many of these seedlings appeared healthy, but pushed themselves out of their small crevices. In addition, the large-seeded species also suffered heavy seed predation. Hemlock and yellow birch were also excluded by the end of the experiment.

Significance

The experiments were set up to explore the species pool on cliffs, and then to explore the constraints against the retention of species in the woody plant component of the community. The results show that the small array of trees that occur on cliffs is not totally restricted at the stage of seed rain or seed bank. Starting at the germination stage, however, the relationship between the size and potential productivity of the plant comes into play. Large-seeded species, those with high growth rates, or those with limited capacity to grow large shoots with small roots were selected against. By the end of the experiment, the woody plant composition was similar to that shown during the late stages of primary succession on the limestone cliff faces of abandoned limestone quarries (Ursic *et al.*, 1997). This suggests that the deletion constraints or assembly rules that operate on natural cliff faces are the same as those operating on artificial quarry cliff faces.

Crawley (1987) has shown that cliffs are habitats with almost zero invasibility. In fact, this appears to be not the case. Instead, it appears that most organisms on cliffs have low survival, and that only certain physiological, anatomical, morphological, and developmental characteristics permit the colonization of cliffs. If conditions on the cliff change, then the characteristics that permit species to colonize cliffs will change. If productivity increases, then cliff species may grow faster; however, other species may then become established and outcompete them.

One might ask, what rules can be derived from these results? We offer the following:

(a) Any component of a local biota has access to cliffs.
(b) Propagules of the local biota may persist on cliffs in a dormant phase.
(c) The size of propagule, and the size, growth rate, and competitive ability of a seedling are all negatively related to survival on cliffs.
(d) Any agent that increases the potential growth rate of cliff species increases the likelihood of new invasion by other components of the species pool. This may result in a faster growth rate of species, but in a less predictable community.
(e) Any agent that decreases the potential growth rate of cliff species decreases the likelihood of new invasions from the species pool. A slower growing, but more predictable community is the result.
(f) Cliff communities are permanent and non-survivable for most taxa.

Unlike many of the other ecological settings within which assembly rules have been sought, cliffs and other low productivity systems (such as caves or arctic tundra) offer a set of advantages that ecologists can exploit. First, the processes that occur on cliffs do so very slowly, lessening the likelihood that key events in the establishment of the community will be missed. Second, soil is nearly absent, so the historical contingency imposed by earlier site conditions of that soil do not apply. Third, because the environments start off 'without things', it is relatively easy to conduct resource addition experiments (as already done Matthes-Sears *et al.*, 1995). Thus, if one wished to experimentally examine how assembly constraints vary with productivity, the experiments would be easy to carry out.

Conclusions

It seems compelling to consider what low-productivity environments offer to ecologists by way of overall conclusions and recommendations for future work. If community ecology, at least in the area of assembly rules, has suffered from anything, it has been (a) the failure to keep the language simple and unpretentious, and (b) the failure to be aware of history. But, these two problems probably cannot be resolved even by drawing attention to them. A possible solution is to simplify the way assembly theory is tested. By using familiar forest, grassland, or wetland systems as test sites, ecologists are almost universally examining systems with high and similar amounts of productivity. As Drake (1991) and others have very clearly shown, such systems will have enormous amounts of unknown (and perhaps unknowable) historical contingency, therefore these systems cannot ever permit a detailed testing of the validity of the concept. Of all the factors cited by Kodric-Brown and Brown (1993) as being important to explain the remarkably predictable species composition of fish communities in Australian springs, the lack of historical differences in their development was the most important.

We conclude that by examining systems at (or very near) the limits of survival for most organisms, the processes can be viewed more clearly without historical contingency. Thus, low productivity ecosystems that start off with their 'backs against the wall' may actually be the best places to use as test systems. Ecosystems that have so far been marginalized in the literature may actually offer the opportunity to study the dynamic aspects of community ecology. They are sufficiently simple and predictable that the organizational 'rules' that apply to them may be easy to find, and it may be possible to test them empirically. Clements (1916) was aware of this opportunity 80 years ago.

Acknowledgments

The authors would like to thank Drs Richard Reader, Tom Nudds and Brian Husband for many discussions that led up to this text, and Evan Weiher and Paul Keddy for being brave enough to let us say some of these things. The research work was supported by the Natural Sciences and Engineering Research Council of Canada.

References

Abrams, M.D., Sprugel, D.G., & Dickmann, D.I. (1985). Multiple successional pathways on recently disturbed jack pine sites in Michigan. *Forest Ecology Management* 10: 31–48.

Aplet, G.H. & Vitousek, P.M. (1994). An age-altitude matrix analysis of Hawaiian rainforest succession. *Journal of Ecology* 82: 137–148.

Baker, W.L. (1992). Structure, disturbance, and change in the bristlecone pine forests of Colorado, USA. *Arctic and Alpine Research* 24: 17–26.

Barden, L.S. (1988). Drought and survival in a self-perpetuating Pinus pungens population: equilibrium or nonequilibrium? *American Midland Naturalist* 119: 253–257.

Bond, J.W. (1989). The tortoise and the hare: ecology of angiosperm dominance and gymnosperm persistence. *Biological Journal of Linnaean Society London* 36: 227–249.

Clements, F.E. (1916). *Plant Succession.* Publication Number 242. Washington: The Carnegie Institution.

Clements, F.E. (1936). The nature and structure of the climax. *Journal of Ecology* 22: 9–68.

Connell, J.H. & Slatyer, R.O. (1977). Mechanisms of succession in natural communities and their role in community stability and organization. *American Naturalist* 111: 1119–1144.

Connor, E.R. & Simberloff, D. (1979). The assembly of species communities: chance or competition? *Ecology* 60: 1132–1140.

Cox, J.E. & Larson, D.W. (1993). Spatial heterogeneity of vegetation and environmental factors on talus slopes of the Niagara Escarpment. *Canadian Journal Botany* 71: 323–332.

Crawley, M.J. (1987). What makes a community invasible? In *Colonization, Succession and Stability*, ed. A.J. Gray, M.J. Crawley, P.J. Edwards, pp. 429–453. Oxford: Blackwell Scientific Publications.

Diamond, J.M. (1975). Assembly of species communities. In *Ecology and Evolution of Communities*, ed. M.L. Cody & J.M. Diamond, pp. 342–444. Cambridge (USA): Belknap Press, Harvard University Press.

Diamond, J & Gilpin, M.E. (1982). Examination of the 'null' model of Connor and Simberloff for species co-occurrences on islands. *Oecologia* 52: 64–74.

Drake, J.A. (1990a). Communities as assembled structures: do rules govern pattern? *Trends in Ecology and Evolution* 5: 159–163.

Drake, J.A. (1990b). The mechanisms of community assembly and succession. *Journal of Theoretical Biology* 147: 213–233.

Drake, J.A. (1991). Community-assembly mechanics and the structure of an experimental species ensemble. *American Naturalist* 137: 1–26.

Drake, J.A., Flum, T.E., Witteman, G.J., Voskuil, T., Hoylman, A.M., Creson, C., Kenny, D.A., Huxel, G.R., Larue, C.S., & Duncan, J.R. (1993). The construction and assembly of an ecological landscape. *Journal of Animal Ecology* 62: 117–130.

Drury, W.H. & Nisbet, I.C.T. (1973). Succession. The *Arnold Arboretum Journal* 54: 331–368. In *Ecological Succession.* Benchmark papers in Ecology /5, ed. F.B. Golley, (1977), pp. 287–324. Stroudburg: Dowden, Hutchinson & Ross.

Fox, B.J. (1987). Species assembly and the evolution of community structure. *Evolutionary Ecology* 1: 210–213.

Fox, B.J. & Brown, J.H. (1993). Assembly rules for functional groups of North American desert rodent communities. *Oikos* 67: 358–370.

Gerrath, J.F., Gerrath, J.A., & Larson, D.W. (1995). A preliminary account of endolithic algae of limestone cliffs of the Niagara Escarpment. *Canadian Journal of Botany* 73: 788–793.

Gleason, H.A. (1917). The structure and development of the plant association. *Bulletin of the Torrey Botany Club* 44: 463–481.

Gleason, H.A. (1926). The individualistic concept of the plant association. *Bulletin of the Torrey Botany Club* 53: 7–26.

Golley, F.B. (1977). *Ecological Succession.* Benchmark papers in Ecology /5. Stroudburg: Dowden, Hutchinson & Ross.

Grime, J.P. (1977). Evidence for the existence of three primary strategies in plants and its relevance to ecological and evolutionary theory. *American Naturalist* 111: 1169–1194.

Grime, J.P. (1979). *Plant Strategies and Vegetation Processes*. Chichester: John Wiley.

Grover, J.P. (1994). Assembly rules for communities of nutrient-limited plants and specialist herbivores. *American Naturalist* 143: 258–282.

Haefner, J.W. (1978). Ecosystem assembly grammares: generative capacity and emprical adequacy. *Journal of Theoretical Biology* 73: 293–318.

Haefner, J.W. (1981). Avian community assembly rules: the foliage-gleaning guild. *Oecologia* 50: 131–142.

Keddy, P.A. (1989). *Competition*. London: Chapman & Hall.

Keddy, P.A. (1992). Assembly and response rules: two goals for predictive community ecology. *Journal Vegetation Science* 3: 157–164.

Kelly, P.E., Cook, E.R., & Larson, D.W. (1994). A 1397-year tree-ring chronology of *Thuja occidentalis* from cliffs of the Niagara Escarpment, southern Ontario, Canada. *Canadian Journal of Forestry Research* 24: 1049–1057.

Kodric-Brown, A., & Brown, J.H. (1993). Highly structured fish communities in Australian desert springs. *Ecology* 74: 1847–1855.

Küppers, M. (1985). Carbon relations and competition between woody species in a central European hedgerow. *Oecologia* 66: 343–352.

Larson, D.W. (1990). Effects of disturbance on old-growth *Thuja occidentalis* at cliff edges. *Canadian Journal of Botany* 68: 1147–1155.

Larson, D.W. & Kelly, P.E. (1991). The extent of old-growth *Thuja occidentalis* on cliffs of the Niagara Escarpment. *Canadian Journal of Botany* 69: 1628–1636.

Larson, D.W., Spring, S.H., Matthes-Sears, U., & Bartlett, R.M. (1989). Organization of the Niagara Escarpment cliff community. *Canadian Journal of Botany* 67: 2731–2742.

Law, R. & Morton, R.D. (1993). Alternative permanent states of ecological communities. *Ecology* 74: 1347–1361.

Lawton, J.H. (1987). Are there assembly rules for successional communities? In ed. A.J. Gray, M.J. Crawley, & P.J. Edwards, *Colonization, Succession and Stability*, pp. 225–244. Oxford: Blackwell Scientific Publications.

MacArthur, R.H. & Wilson, E.O. (1967). *The Theory of Island Biogeography*. Monographs in population biology. Princeton, NJ: Princeton University Press.

McCune, B. & Allen, T.F.H. (1985). Will similar forests develop on similar sites. *Canadian Journal of Botany* 63: 367–376.

Matheson, J.D. (1995). Organization of forest bird and small mammal communities of the Niagara Escarpment, Canada. MSc. thesis, University of Guelph.

Matthes-Sears, U., Nash, C.H., & Larson, D.W. (1995). Constrained growth of trees in a hostile environment: the role of water and nutrient availability for *Thuja occidentalis* on cliff faces. International Journal of *Plant Science*, 156: 311–319.

Odum, E.P. (1969). The strategy of ecosystem development. *Science* 64: 262–270.

Peters, R.H. (1988). The relevance of allometric comparisons to growth, reproduction and nutrition in primates and man. In *Symposium on Comparative Nutrition*, ed. K. Blaxter & I. MacDonald, pp. 1–19. London: Libbey.

Pianka, E.R. (1983). *Evolutionary Ecology*. NY: Harper and Row.

Rabinowitz, D. & Rapp, J.K. (1980). Seed rain in a North American tall grass prairie. *Journal of Applied Ecology* 17: 793–802.

Robinson, J.V & Dickerson, Jr., J.E. (1987). Does invasion sequence affect community structure. *Ecology* 68: 587–595.

Roughgarden, J. (1989). The structure and assembly of communities. In: *Perspectives in Ecological Theory.* ed. J. Roughgarden, R.M. May, & S.A. Levin, pp. 203–226. NJ: Princeton University Press.

Schulman, E.R. (1954). Longevity under adversity in conifers. *Science* 119: 396–399.

Svoboda, J. & Henry, G.H.R. (1987). Succession in marginal arctic envrironments. *Arctic and Alpine Research* 19: 373–384.

Ursic, K., Kenkel, N.C., & Larson, D.W. (1997). Revegetation dynamics of cliff faces in abandoned limestone quarries. *Journal of Applied Ecology* 34: 289–303.

van der Valk, A.G. (1981). Succession in wetlands: a Gleasonian approach. *Ecology* 62: 688–696.

Walker, G.L. (1987). Ecology and population biology of Thuja occidentalis L. in its southern disjunct range. PhD thesis, University of Tennessee, Knoxville.

Werner, P.A. (1975). A seed trap for determining pattern of seed deposition in terrestrial plants. *Canadian Journal of Botany* 53: 810–813.

Wilson, J.B. (1991). Does vegetation science exist? *Journal of Vegetation Science* 2: 189–190.

Wilson, J.B. (1994). Who makes the assembly rules? *Journal of Vegetation Science* 5: 275–278.

Wilson, J.B. & Watkins (1994). Guilds and the assembly rules in lawn communities. *Journal of Vegetation Science* 5: 590–600.

Wilson, J.B. & Gitay, H. (1995). Limitations to species coexistence: evidence for competition from field experiments using patch model. *Journal of Vegetation Science* 6: 369–376.

Wilson, J.B. & Whittaker, R.J. (1995). Assembly rules demonstrated in a saltmarsh community. *Journal of Ecology* 83: 801–808.

Wilson, J.B., Agnew, A.D.Q., & Partridge, T.R. (1994). Carr texture in Britain and New Zealand: community convergence compared with a null model. *Journal of Vegetation Science* 5: 109–116.

Wilson, J.B., Allen, R.B., & Lee, W.G. (1995a). An assembly rule in the ground and herbaceous strata of a New Zealand rain forest. *Functional Ecology* 9: 61–64.

Wilson, J.B., Sykes, M.T., & Peet, R.K. (1995b). Time and space in the community structure of a species-rich limestone grassland. *Journal of Vegetation Science* 6: 729–740.

Wilson, J.B., Wells, T.C.E, Trueman, I.C., Jones, G., Atkinson, M.D., Crawley, M.J., Dodd, M.E., & Silvertown, J. (1996). Are there assembly rules for plant species adundance? An investigation in relation to soil resources and successional trends. *Journal of Ecology* 84: 527–538.

Young, J.M. (1996). The cliff ecology and genetic structure of northern white cedar (*Thuja occidentalis* L.) in its southern disjunt range. MSc thesis, University of Tennessee, Knoxville.

Part II:

Other perspectives on community assembly

8

On the nature of the assembly trajectory

James A. Drake, Craig R. Zimmermann, Tom Purucker, and Carmen Rojo

Introduction

Assembly is a basic device of nature where a broad range of entities, spanning many levels of scale, interact across time and space to produce oberserved pattern. Entities such as individuals, phenotypes, populations, guilds, and higher levels of organization like hierarchical structures are all subject to the processes of assembly. Because all biological systems are assembled in a dynamical sense, any generality in the process would prove valuable to our understanding of nature. Arguably, such an understanding is essential to fully appreciate the action of mechanisms as they are played out in evolutionary and ecological time.

Community assembly is ultimately driven by the invasion (e.g., speciation, immigration) and extinction of species played out against a complex background of environmental constraint. While the environment acts as a filter, eliminating some species and promoting others, it also provides spatio-temporal complexities which serve as a resource upon which ecological strategies can be built. Assembly processes and rules which operate within one environment may exhibit entirely different outcomes as a function of even minor environmental variation. Despite such obvious descriptions of the course of nature, the essence of the assembly trajectory remains little more than an elusive metaphor. Here, the character of the assembly trajectory is evaluated and a general framework offered which serves to interface the operation of ecological mechanisms and the mechanics of community assembly within the more general realm of complex systems.

An assembly perspective

In order to begin our exploration into the nature of the assembly trajectory, a perspective is first presented from which we base our arguments. It is suggested that a variety of phenomena serve to regulate the assembly processes in space and time. The outcome of an assembly process is the nature of the action of specific mechanisms. This perspective is developed in this chapter by outlining several of these phenomena and their respective action during community assembly.

Ecological communities are aggregations of entities (e.g., individuals, populations, guilds) variously meshed and integrated across many levels of scale. Hence, it is likely that the process of assembly is both manifested and regulated at many levels of scale. A direct consequence of this biologically driven complexity is that assembly processes, although regulating system organization, are often invisible or distorted at many levels of observation. For example, if indeed *forbidden* species combinations (*sensu* Diamond, 1975) exist, it is unlikely that an examination focused exclusively on the species directly involved can resolve the reasons for the observed patterns. Competition may, indeed, be operating, but to demonstrate the direct mechanism of exclusion is simply not enough. It must be known what factors lend competition its dynamical character, given the situation at hand. This dynamical character comes in part from assembly processes operating synergistically and anatgonistically at a multitude of scales replete with positive and negative feedbacks (Drake *et al.*, 1996). An understanding of the full dynamics comes only when the dynamical step, here the inclusion/exclusion of a species, is placed in the perspective of the entire system assembly.

This multiplicity of control is illustrated by presenting a simple scenario. Consider generic communities which are subject to colonizing populations of different species. The trajectory[1] produced by fitting together populations in time and space can derive its directionality from a variety of assembly processes operating at different levels of scale. This directionality may be simply observed as changes in species relative abundance or in the wholesale substitution of extant species. Temporal dynamics could range from random to quasiequilibrium behavior or even exhibit criticality (Bak, 1996) or chaos. As invasions occur at points along the trajectory wholsesale changes in the direction and nature of the trajectory can occur as the community exhibits sensitivity to a new set of initial conditions. Small differences in some parameter, say the fecundity of colonizing individuals, have been shown to produce substantive changes in some assembly trajectories but not others (Drake, 1991; Drake *et al.*, 1993).

At many points along the trajectory, however, assembly processes result

---

[1]At this point we define the trajectory as a sequence of community states. We acknowledge the powerful role played by environmental heterogeneity and offer assembly against that background. We modify this definition subsequently.

in robust trajectories. Here, variation in individual fecundity may not be an issue and, despite large variance in the fecundity of an invader, the system responds indifferently (Drake, 1991). In this case, the rules which govern assembly, if indeed rules are operating, do not operate at the level of variance in invader fecundity. Even a local population may be a trivial level of organization if the system is governed by metascale processes and the incorporation of the population into a community occurs at that scale. These are clearly essential aspects of assembly, but to date these processes have barely been explored. It is necessary to ask at what levels of scale is assembly operating and what is the nature of the interface between the component dynamics and the the context of whole system response.

Assembly as a general mechanic of nature

Assembly is not a mechanism but a mechanic. It is the process of fitting together the dynamically variable pieces which comprise a system, pieces operating at disparate levels of spatial and temporal scale. Variation in the pieces result from the action of ecological mechanisms. The mechanisms of ecology are familiar to all: species interactions like competition or predation and successional processes like inhibition or facilitation. Such mechanisms and processes form the nuts and bolts of contemporary ecological thought and their dynamical nature in ecological and evolutionary time form the prevailing ecological paradigm. Mechanisms operate against a background of dynamic constraint driven by assembly mechanics, however, and consequently, often appear idiosyncratic when viewed outside this context (Drake, 1991; Drake *et al.*, 1996).

Assembly mechanics are the regulatory agents and processes which define the suite of plausible system stages or transitions through which a system can proceed. As such, assembly comprises the historical events and processes which have acted to drive the particular system to its present state. Notably, assembly mechanics can define how the operation of mechanism becomes manifest. Variability in the operation of a mechanism can result from either:

- constraints which vary in time and space,
- indeterminism in the action of the mechanism which is a direct product of some assembly trajectories, or
- environmental and stochastic events which modify the action of the mechanism.

The outcome of the operation can also serve as a mechanic for future states. For example, a case where competition has resulted in some specific pattern of resource utilization can be easily conceived which may, in itself, drive subsequent assembly steps. An understanding of the operation of any mechanism therefore, must also include an understanding of the mechanics which variously permit and modify its operation.

The levels of assembly

At this point, we define an assembly rule as:

an operator which exists as a function or consequence of some force, dynamical necessity, or context which provides directionality to a trajectory. The nature of this direction includes movement toward a specific state, some subset of all possible states, or a dynamical realm of definable character. Operators are the mechanics of community assembly.

The dichotomy between mechanism and operator is an essential distinction. An assembly rule exists as an 'if-then-else' or 'yes-no-maybe' type of grammatical switch, while a mechanism is something like competition manifest in terms of the classic R* (Tilman, 1988). In ecological terms, assembly rules define reachable and unreachable community states, the community being some complete set of species (bacteria on up) exhibiting limited membership (see McIntosh, 1995).

So defined, assembly rules appear to operate on two fundamental levels. The first level is *categorical* – defined by the presence of functioning levels of organization (e.g., individuals, populations, guilds) operating at various levels of scale. Assembly at this level is defined by the character of the pieces and the functional level of organization at which such pieces operate within the context of assembly processes. Assembly rules at this level may include things like the result of competition thus far, although not competition itself. The manner in which the operation of competition cascaded through the community will define the suite of plausible subsequent states.

The second level of assembly is *topological* – defined as the dynamical fabric upon which the system operates. One can visualize this fabric as a complex changing landscape or manifold which maps all plausible assembly space. We do not view this fabric in static terms, that is there can be no fixed template against which ecology functions. A trajectory which becomes unstable will warp the realized manifold into a new topology of plausible states. As such, this fabric – the set of all possible states and transitions – reflects the complex, evolving nature of the assembly process.

The dynamics and rules of assembly at the topological level are of a decidedly different nature than the rules which govern the categorical level. The fabric may be composed of a handful of trajectories that possess a common solution or exhibit similar dynamics. Some regions may initiate self-organizing behavior, while others do not.

The fabric might include an edge between ordered and chaotic regimes – the point of maximum information and diversity – at which trajectories approach or reach a critical state (Bak, 1996; Hastings *et al.*, 1993; Jorgenson, 1995; Solé & Manrubia, 1995). These are all higher-ordered components of trajectory directionality which are not necessarily, and perhaps never, a direct product of interspecific interactions. Further, the domain may be characterized by behavior that is either:

- deterministic – yielding a specific community configuration,
- probabilistic – yielding a density function of plausible states,
- chaotic, or entirely random and unpredictable (Hastings *et al.*, 1993).

Exposing the relationship between rules at the topological and those at the categorical level is critical to understanding the structure and organization in assembling ecological systems.

Given this view of the nature of systems and assembly, where and when are rules of assembly likely to exist? At the categorical level, assembly rules may be manifest in patterns of species co-occurrence. For example, certain combinations of species are simply so unstable, due perhaps to competition, that it is unlikely such species combinations will be observed in nature (Diamond, 1975). Trajectories containing that species set as a solution exhibit instability when realized. Here, the system can either evolve a higher-ordered structure or collapse to a former configuration. Interestingly, the species which initiate system-wide collapse can also function as an assembly mechanic because the subsequent community state may very well be a state which cannot otherwise be reached (cf. 'humpty-dumpty effects', Lewin, 1992). Whatever the scenario, assembly rules take the form of the manner in which the manifestation of the mechanism directs subsequent assembly steps.

Rules may also be architectural in nature because some food web topologies are either observed frequently or are impossible (Higashi & Burns, 1991; Martinez, 1992; Sugihara, 1985). Here, rules of assembly enforce specific interaction patterns apparently without regard to the species involved. Clearly, assembly pervades every aspect of biological structure and organization, but the general nature of control within and among systems remains elusive. For example, the way in which assembly leads to top-down or bottom-up control might offer a solution to controversies over the nature of short-term ecological contol. Imagine the uility of being able to switch a system from top-down to bottom-up control.

Self-organization and the assembly trajectory

Ecological systems are non-equilibrium, nonlinear, spatially extended, dissipative systems at all but the most trivial levels of scale. Yes, the local forest looks much the same from day to day but such apparent constancy belies a world of massive change. It must be accepted that there can never be a 'balance of nature' at least in terms of equilibrium (Drake *et al.*, 1994; Grover & Lawton, 1994). Indeed, the fact that evolution and extinction are not rare events, but commonplace, quashes arguments of ecological stability and constancy. This realization demands that alternative approaches and explanations be turned to questions of organization and structure in ecological communities (Brown, 1994, 1995). The characteristics listed above would seem to qualify ecological systems as one of the very best candidates for a so-called complex systems approach. The proof is in the pudding and while an arsenal of complexity-oriented tools is available, compelling demonstrations of the utility of such an approach are needed.

Ecological consequences of non-linearity

Dissipative systems, unlike conservative systems, give rise to time-irreversible processes which invite a powerful role for history. In an open energy environment, these processes can result in the formation of *dissipative structures* – structures which are created and maintained at a thermodynamically far-from-equilibrium state by the continuous uptake and transformation of energy. Entropy, as a byproduct of internal metabolic processes, is exported out of the system by the exchange of matter and energy across system boundaries (Brooks & Wiley, 1988; Johnson, 1995; Kauffman, 1993; Nicolis & Prigogine, 1989; Prigogine, 1980). When the emergent structure varies, energy flow varies, and when energy input changes so does the emergent structure. Dissipative structures do not exist without reference to the entities contained in the system, so clearly species substitutions matter.

   An essential characteristic of many nonequilibrium, dissipative systems is the operation of self-organization: the emergence of integrated structures through nonlinear interaction and feedback mechanisms between system components (Goodwin, 1987; Haken, 1988). Such structures are characterized by spatial and temporal regularities which could not be predicted with even a precise knowledge of initial conditions. Common themes generated by self-organization form attractors, which provide the coarse weave of the assembly fabric. These attractors define the possible states that this structure can attain even before mechanisms like competition and predation are played out[2].

[2]A reassessment of the null model may be in order.

Phase transitions between attractors occur when a critical value in one or more system parameters make the current system configuration unstable. The original attractor dissipates and a new attractor, stable with respect to the fluctuation, restructures the system under a new set of dynamic constraints. Attractors, therefore, are ephemeral entities which form, fluctuate, and dissipate as the self-organizing system evolves. This does not mean that attractors form and dissipate willy-nilly although such a situation is concievable. Rather specific attractors represent recurrent themes in ecological assembly. Coupled with the dynamical realities of competition, predation, and other mechanisms and processes, these attractors represent solutions to the operators or rules of assembly.

Self-organizing dynamics are only partially deterministic. Chance plays an essential role at all scales – a phenomenon well known to ecologists. Indeed, it has been argued that self-organizing processes have a absolute necessity for random fluctuations at some point in their histories (Pattee, 1987; Yates, 1987). Regardless of its ultimate source, biologists have long contended that stochasticity plays a role in biological systems (Chesson, 1986; Chesson & Case, 1986; Gilpin & Soulé, 1986; Goodman, 1987; Sale, 1977, 1979). It now appears likely that chance plays a fundamental role in the emergence of order in all dissipative systems.

It is asserted that the assembly trajectory must be approached as a self-organizing process built from a variety of components (e.g., individuals, populations, guilds) which are themselves historically derived and self-organized. It is recognized, however, that not all community assemblages have been shaped by self-organization. Communities likely exist on a continuum from what are essentially random assemblages up to highly integrated self-organized structures. Two factors which may be fundamental to whether a given system exhibits self-organization are environmental constancy and component diversity.

First, some degree of environmental constancy may be required. Highly volatile systems as a result of frequent natural or anthropogenic disturbance may not self-organize. Variable disturbance frequency and magnitude can strongly affect which community state is attained (Connell, 1978; Paine & Levin, 1981). Where the rate of disturbance is low compared to rate of community development, disturbance may act to reset the system back over one or more attractors. Forest gap dynamics demonstrate such a repeating cycle of community states. Cyclical disturbance can also act to maintain a trajectory within the basin of a given attractor. Fire-climax communities, for example, involve evolutionary adaptions by the constituent community members such that the presence of periodic fires has been incorporated as a necessary factor in the persistence of the community state. Disturbance, in this case, could be constructively considered as the cessation or suppression of fires. Finally, very frequent, intense or highly variable disturbances at any point during assembly process may prevent a self-organized state from emerging at all but the longest timescales, spans of time meaningless to the system.

Transient species and the ensuing variations in system dynamics may also adversely affect self-organization. Communities which experience high rates of disrupting invasions may never settle onto a self-organized trajectory. Huxel (1995), for example, found that communities subjected to simultaneous multiple invasions never attained a stable, invasion-resistant state. This is in stark contrast to the structural patterning which occurs when species invade over a period of time allowing community dynamics to be fully or largely expressed before subsequent invasion (Drake, 1991, 1993).

Secondly, some minimal species diversity and abundance may be necessary for self-organization to occur. If so, species-poor systems or systems in the very early stages of development would be expected to exhibit assemblages that are less structured in nature. In other cases, the available species pools might simply have 'holes' wherein species that would otherwise have a key role in the assembly process are missing. How efficiently self-organization can proceed in lieu of these missing species may depend on the strength and resilience of the attractors involved.

These exceptions notwithstanding, a self-organization approach provides a useful framework for understanding the formation of complex ecological systems. Ecological succession, for example, may be a self-organizing assembly process under the influence of one or more attractors. These attractors not only influence the assembly process within each seral stage but also govern transitions between stages. Transitions can initiated in two ways. First, transitions can be biotically initiated by the addition of new hierarchical levels or by key additions, substitutions, or extinctions of species within existing levels. Restructuring of the system, at this point, can occur at all scales and levels. The effect of adding a predator, for example, on restructuring food web topologies is well known. Predators can alter the interactions between competing consumers, such that consumer coexistence is either favored or prohibited (Caswell, 1978; Connell, 1971; Paine, 1966). The alteration of interaction at the consumer level can cascade downward altering interactions at lower levels (e.g., producers and abiotic interactions). Circular feedback, ultimately, can also affect the predator-level itself. A predator which promotes multiple species coexistence, for example, provides the opportunity for other predators to become established.

Phase transitions can also be abiotically initiated as autogenic processes of the assembling community alter the environment in such a way that fosters the entry of more competitive species. Connell and Slayter (1977) termed this process facilitation. Such transitions could be accompanied by relatively discrete transition phases wherein large species turnover occurs as food-web hierarchies crumble and are rebuilt. The system may approach a self-organized critical state at the phase transition. At this point the system is either driven to a more complex, higher-order structure or it is returned to a simpler state. The region around the critical phase transition may not be a favorable place to stay. Kaufmann (1993), for example, found that his model systems resided in the ordered regime just away from the order-chaos boundary.

Of course, not all assembly patterns show discrete transitions and varying degrees of discontinuity in species membership are often exhibited. Some patterns show discrete turnover of a core of dominant species only, while others display a gradual turnover of all species throughout the process. In the first case, a phase transition between two attractors can be imagined that only affects the community core in the short term. Replacement of peripheral species then gradually occurs as the trajectory moves deeper into the basin of the governing attractor. In the second case, the system may be under the influence of an attractor with a shallow, but wide basin of attraction. Critical parameter changes do not occur and a phase transition is not incurred. Rather, such a trajectory might slowly spiral into the attracting basin producing the observed gradual species replacement. Finally, the relationship between time and space mentioned earlier begs a comparison between the temporal behavior discussed here and similar spatial transitions (Milne *et al.*, 1996). Exploration of this relationship is left to the so inclined reader.

Evidence for self-organization

While the process of self-organization presents a theoretically compelling framework, the extent and manner in which these processes are responsible for the patterns of nature has yet to be fully appreciated. Indeed, the relationship between self-organization, natural selection, and the mechanisms and assembly operators of ecology are simply unknown despite a growing theoretical effort (e.g., Depew & Weber, 1995; Kauffman, 1993). Demonstrations of self-organization at higher-levels of organization generally rely on a statistical similarity between the system at hand and the behavior of a model (e.g., cellular automata) which exhibits self-organization. While such attempts represent bold explorations of the nature of the biological world, they remain correlative. Experimentation is needed.

These cautions offered, a variety of studies have implicated self-organization as a fundamental process in the production of pattern in many biological and physical systems. In fact, a database search using the term self-organization reveals an absolute explosion of research spanning a wide variety of physical, biological, and social systems. Solé and Manrubia (1995), for example, observed that the fractal-like patterns of rainforest canopy gap distribution were patterns readily recreated with a simple cellular automata model. Their model exhibited self-organization and an approach to a critical state[3], resulting in patterns which bore a striking similarity to those observed in nature. Similarly, Keitt and Marquet (1996) suggest that the patterns of species extinction in the Hawaiian avifauna can be explained by nonequilibrium dynamics leading to a such a critical state. Plotnick and McKinney (1993) have suggested that patterns of species extinction observed in the fossil record are also consistent with models of species turnover which exhibit self-organization and approaches to a self-organized critical state.

Experimental explorations of the process of self-organization have been conducted in a wide variety of physical and biochemical systems where the process has taken on an almost matter-of-fact air. Experimental verification is clearly a more daunting task at higher-levels of organization where the elements of the system (e.g., species, populations, guilds) are variable and process function on very long timescales. Regardless of the logistical difficulties of verification, the assembly trajectory exhibits properties and processes which are consistent with a self-organized structure. It is suggested that laboratory and microcosm analyses are the best candidates for initial demonstrations of self-organization in ecological systems.

If, indeed, self-organization withstands further scrutiny, the implications for ecology and biology in general are staggering. How much of the pattern and process is a result of self-organization and the operators of the assembly trajectory? In what follows we offer some thoughts and ideas, couched in terms of a general theory of assembly.

Probabilistic elements of assembly

With each point of an assembling trajectory, we associate an *assembly operator* in terms of a probability function – perhaps a logic switching function – which defines the suite of plausible directions the assembly trajectory can take. The assembly function at each point is a unique expression of the relative roles of constraint and chance as defined by:

---

[3]A critical state is an attractor which exhibits correlations at all length and spatial scales (Bak *et al.*, 1988). In terms of species extinctions or turnover the frequency and magnitude of collapse follows a power-law distribution of the form $D(s) \approx s^{-\alpha}$, where s is the size of collapse, $D(s)$ is the frequency of occurrence of collapse size s, and α is the spectral exponent.

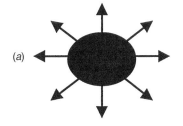
(a)

- No attractors influencing behavior
- Behavior fully stochastic
- All future states equally probable

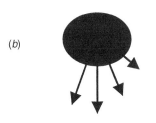
(b)

- Under influence of attractor(s)
- Behavior partially indeterministic
- Future states constrained

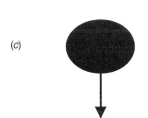
(c)

- Fully governed by attractor
- Behavior fully deterministic
- Single future state

Fig. 8.1. Graphical representation of three assembly functions, which display behavior ranging from fully stochastic to fully deterministic.

- of available species, their ecologies, and relative abundance;
- the basin portrait as defined by the system's parameter space;
- the historical inertia of the system; and
- vagaries of the environment.

Three assembly functions are offered ranging from fully random to fully deterministic. The first state (Fig. 8.1(*a*)) demonstrates fully random behavior. Here the state is not under the influence of any attractor and all plausible future states are equally probable. A successive string of such states would, thus, constitute a random walk. The second state (Fig. 8.1(*b*)) may be under the constraining influence of one or more attractors. While behavior, at this point, can become somewhat more predictable, varying degrees of sensitivity to initial conditions can still maintain a highly probable nature toward future states. Chaotic attractors, for example, demonstrate high sensitivity to initial conditions. Here, two arbitrarily close points can diverge at an exponential rate. The last state (Fig. 8.1(*c*)) shows a totally constrained, fully deterministic assembly function. This trajectory is locked deep within the basin of an attractor. Such an attractor could be considered a climax state as long as system parameter changes are insufficient to perturb the system out of its basin. Clearly, system behavior runs the gamut from that which is completely random to that which is fully deterministic. However, there is every reason to believe that most communities exhibit a composite of these behaviors. We would expect some degree of constraint and stochasticity to play a role at all steps in the assembly process.

Beginning at some initial point in state space, that initial state can be coupled with each of the plausible subsequent states as defined by the assembly probability function at that point. Repeating this process at progressive time steps will produce a Markovian-like assembly tree of branching alternative assembly paths. For a given elapsed time interval, a path probability can be determined for each alternative trajectory by simply taking the product of the function-defined probabilities at each time step within that interval. The probability that an assembling community will reach some defined point at the end of some time interval is the sum of the path probabilities for all trajectories which reach that point. Given some degree of stochasticity at each step, it is easy to imagine that, for any given assembling community, there could exist a very large number of alternative assembly paths. While some could be markedly different in nature, many others would simply be random variations on a theme. Path probabilities, consequently, are likely to be quite small and prediction of any given trajectory would be largely impossible.

For any given elapsed time, we can determine the instantaneous probability density of the state space for the assembling community based on the sum of all alternative path probabilities for each point in state space. If early stages of assembly are largely random in nature, the probability density at any point will be low as alternative trajectories are spread out over relatively large regions of state space. If the assembly process is self-organizing, however, these trajectories will begin to coalesce into smaller regions as they fall under the influence of various basins of attraction. Therefore, increasing the amount elapsed time would produce a probability density displaying regions of higher and higher probability. Consequently, predictions could be made as to where in state space an assembling community is likely to be found at any given time. Unfortunately, the best hope is for rather fuzzy predictions.

Elements of a general theory of assembly

A conceptual view of the nature of the assembly trajectory has been offered by evaluating constraint and context at levels of scale where controlling factors might emerge. Themes common across systems point to elements of a general theory of assembly. Such a theory must accommodate the existence and operation of a wide range of dynamics from pure determinism to sheer chance as well as higher-order processes such as self-organization and approaches to critical states. Clearly, the specifics of each ecological system, from the biology of constituent populations to ensembles of particular species which resist invasion, are system dependent and change with time. Competition here is not competition there. Top-down control works here but not there. Diversity is thusly maintained here but not there. Vagaries aside, however, biological systems display commonalties which point to the essential elements of structure. Here, a set of informal propositions, corollaries and comments are offered which we believed will have broad application.

Proposition 1: *All biological systems are assembled, historically contingent structures.*
> Corollary 1.1: The structure and organization of the present state cannot be fully understood without reference to the past.
> Corollary 1.2: The configuration of subsequent community states is strongly affected by the nature and character of the presently observed state.
> Corollary 1.3: Myriad factors such as disturbance are capable of erasing or rewriting history.

Proposition 2: *Biological systems are generally non-linear, dissipative, spatially extended, dynamic systems which are far from equilibrium over most of the plausible parameter and state space.*
> Corollary 2.1: Dissipative systems are time irreversible. Therefore, an understanding of a dissipative system requires an understanding of its evolution and history.
> Corollary 2.2: Non-linear systems can exhibit sensitivity to initial conditions. Such systems can quickly amplify small differences in system parameter values such that assembly trajectories diverge at an exponential rate. This insures that anything less than a perfect measurement of some state can lead to a rapid deterioration in predictive ability (See Corollary 2.5).
> Corollary 2.3: Feedbacks are highly significant.
> Corollary 2.4: Dynamics can range from pure chance to stable equilibrium to deterministic chaos.
> Corollary 2.5: Predictability decays exponentially where sensitive dependence on initial conditions exists. Thus, at some levels of scale prediction beyond the short term is difficult, if not impossible.

Proposition 3: *Directionality in the assembly trajectory is governed by critical dependence on the interplay between events, processes and structures which emerge in space and time.*

Corollary 3.1: Ecological systems (e.g., food webs) are dynamic across scales. Short term studies necessarily capture but a glimpse of reality.

Corollary 3.2: When the system is spatially extended (e.g., metapopulation processes) the trajectory is also spatially extended. Visualize a complex landscape which itself operates as a point on a trajectory within another landscape.

Corollary 3.3: Disturbance frequency and magnitude interplay with assembly in a complex fashion.

Corollary 4.2: Stochasticity plays a role at many points along the trajectory. The role played by stochasticity on future states reflects the nature of the assembly fabric at that point. At points, stochastic fluctuations are immaterial while, at other points, they dominate the system

Corollary 3.4: Some assembly scenarios can uniquely expose dynamical realms which are not possible in other scenarios even given the same species and environments.

Comment: The extent to which such a system or a variable of that system exhibits sensitive dependence is a function of properties which emerge during assembly. Specific properties of the system can either initiate/amplify or eliminate/dampen sensitive dependence. Such control develops as a function of system context. For example, during the assembly of an experimental community we have observed trajectories where variation in the fecundity of an invader led to large community impacts, while such variation in other trajectories was largely immaterial (Drake, 1991; Drake *et al.*, 1993). Such dynamics are suggestive of the operation of an assembly operator.

Proposition 4: *The action of mechanism is under the control of the assembly operator. This operator is context sensitive which can lead to variation in the manifestation of the mechanism.*

Proposition 5: *Self-organization provides selectable structures.* Assuming that self-organizing is occurring as is competition among a set of species along some assembly trajectory, what relationship is there between self-organization (the nature, direction and product of the SO trajectory) and the occurrence of competition?

Proposition 6: *No single type, nor cadre of types, of dynamical behavior is (are) capable of characterizing ecological reality. Rather, nature is a complex blend of dynamics spanning and combining a wide range of elements each operating with different constraints at a variety of spatio-temporal scales.*

Corollary 6.1: Structure can be a product of past deterministic events. At other times, structure is simply a product of chance. Unfortunately, the structures which result from both extremes can be indistinguishable.

Corollary 6.2: Just as the nature of control varies widely across space even within the confines of a given ecological system (e.g. a lake or grassland), control also varies with time.

Prospectus

One can either approach ecological systems for what they are or take solace in the notion that pieces equal the whole. Contemporary views are largely based on information obtained from highly reductionistic analyses. As with most complex systems, the general analytical approach has been one of a reduction of components, properties and processes to a logistically manageable level. Unfortunately, nature proceeds without regard to the logistic limitations of the observer. Faster hardware and increasingly sophisticated measurement devices generally serve to temporarily rescue the reductionistic approach until the limits of that level of resolution have again been reached.

A far more serious problem faced by the reductionistic approach, however, is that the focus of reduction – the system's pieces and mechanisms – frequently do not themselves possess many of the critical 'system-level' properties which influence structure and process. Understanding mechanism is undeniably an essential ingredient in understanding system properties, but an understanding of mechanism alone is not sufficient to explain the dynamics of ecological systems. Biological systems are more than simply an additive function of the system's components and even a perfect knowledge of mechanism cannot illuminate the nature of higher ordered phenomena. Such properties are emergent and their detection hinges on analytical scales capable of exposing those properties. Emergent properties form the essence of the system and exert a strong controlling effect on both the nature of the system and its component parts.

We offer that the manner in which assembly mechanics interface with, and modify, the expression of a mechanism is perhaps even more essential to advancing our understanding of nature than is the further refinement of pure mechanism. This is not a call to abandon the mechanistic approach, but rather to enhance that effort by conducting experiments within a developing theoretical context of both assembly and the general nature of complex systems. Assembly rules are the interface between the topological level of assembly space and the categorical level of functioning components. Assembly rules are the quantifiable expression of the assembly mechanic and their detection can only come by first understanding the interface between these levels – the interface is where ecology occurs.

Several aspects of assembly behavior which we have focused on suggest that there are higher-ordered processes, in excess of basic ecological mechanism, which structure community states. At this point, however, it is difficult to specify precisely what the states are which comprise the assembly trajectory. There is undoubtedly a hierarchy of significant and insignificant levels of organization which are expressed along the trajectory. Yet, what levels of organization play the most dominant role in defining structure changes along the course of the trajectory? Are there recognizable phase transitions during the course of assembly that provide information about the course of system development or, perhaps, even allow prediction of the final state? While it has been argued that ecological systems are indeed complex non-linear systems, systems inherently difficult to predict, tractable questions do exist if the system is approached with an adequate epistemological framework. While some tentative elements of such a framework have been offered here, the answers are not all here. Ecology is far too complex and hard-fought ground remains. Further progress will hinge on the development of a general and robust theory of complex systems, self-organization, and the creation of emergent order.

References

Bak, P. (1996). *How Nature Works: The Science of Self-Organized Criticality.* Copernicus.

Brooks, D.R. & Wiley, E.O. (1988). *Evolution as Entropy: Toward a Unified Theory of Biology.* Chicago: University of Chicago Press.

Brown, J.H. (1994). Complex ecological systems. In *Complexity: Metaphors, Models, and Reality,* ed. G. Cowan, D. Pines, & D. Meltzer, SFI Studies in the Science of Complexity, Proc. Vol. XIX, Addison-Wesley.

Brown, J.H. (1995). *Macroecology.* Chicago, IL: University of Chicago Press.

Cambel, A.B. (1993). *Applied Chaos Theory: A Paradigm for Complexity.* Boston, MA: Academic Press.

Caswell, H. (1978). Predator-mediated coexistence: a non-equilibrium mode. *American Naturalist,* 112: 127–154.

Chesson, P.L. (1986). Environmental variation and the coexistence of species. In: *Community Ecology,* ed. J. Diamond & T.J. Case. NY: Harper & Row.

Chesson P.L. & Case, T.J. (1986). Overview: non-equilibrium community theories, chance, variability, history, and coexistence. In *Community Ecology,* ed. J. Diamond & T.J. Case, NY: Harper & Row.

Connell, J.H. (1971). On the role of natural enemies in preventing competitive exclusion in some marine animals and in rain forest trees. In *Dynamics of Populations,* ed. P.J. den Boer & G. Gradwell. Wageningen, The Netherlands: Center for Agricultural Publishing and Documentation.

Connell, J.H. (1978). Diversity in tropical rainforests and coral reefs. *Science,* 199: 1302–1310.

Connell, J.H. & Slayter, R.O. (1977). Mechanisms of succession in natural communities and their role in community stability and organization. *American Naturalist,* 111: 1119–1144.

DeAngelis, D.L., Post, M.W., & Travis, C.C. (1986). *Positive Feedback in Natural Systems.* New York: Springer-Verlag.

Depew, D.J. & Weber, B.H. (1995). *Darwinism Evolving: Systems Dynamics and the Genealogy of Natural Selection.* Cambridge, MA: MIT Press.

Diamond, J.M. (1975). Assembly of species communities. In *Ecology and Evolution of Communities,* ed. M.L. Cody & J.M. Diamond, Cambridge, MA: Harvard University Press.

Drake, J.A. (1991). Community assembly mechanics and the structure of an experimental species ensemble. *American Naturalist* 137: 1–26.

Drake, J.A., Witteman, G.J., & Huxel, G.R. (1992). Development of biological structure: critical states, and approaches to alternative levels of organization. In *Biomedical Modeling and Simulation*, ed. J. Eisenfeld, D.S. Levine, & M. Witten, pp. 457–463. North Holland: Elsevier.

Drake, J.A., Flum, T.E., Witteman, G.J., Voskull, T., Hoylman, A.M., Creson, C., Kenney, D.A., Huxel, G.R., LaRue, C.S., & Duncan, J.R. (1993). The construction and assembly of an ecological landscape. *Journal of Animal Ecology* 62: 117–130.

Drake, J.A., Flum, T.E., & Huxel, G.R. (1994). On defining assembly space: a reply to Grover and Lawton. *Journal of Animal Ecology* 63: 488–489.

Drake, J.A., Hewitt, C.L., Huxel, G.R., & Kolasa, J. (1996). Diversity and higher levels of organization. In *Biodiversity: A Biology of Numbers and Differences*, ed. K. Gaston, Oxford: Blackwell Scientific. In Press.

Gilpin, M.E. & Soulé, M.E. (1986). Minimum viable populations: the processes of species extinctions. In *Conservation Biology: The Science of Scarcity and Diversity*, ed. M.E. Soulé, pp. 13–34. Sunderland, MA: Sinauer Associates.

Goodman, D. (1987). The demography of chance extinction. In *Viable Populations for Conservation*, ed. M.E. Soulé, pp. 11–34. Cambridge: Cambridge University Press.

Goodwin, B.C. (1987). Developing organisms as self-organizing fields. In *Self-organizing Systems: The Emergence of Order*, ed. F.E. Yates, NY: Plenum Press.

Grover, J. & Lawton, J.H. (1994). Experimental studies on community convergence and alternative states: comments on a paper by Drake *et al*. *Journal of Animal Ecology* 63: 484–487.

Haken, H. (1988). *Information and Self-organization*. Berlin: Springer-Verlag.

Hastings, A., Hom, C.L., Ellner, S., Turchin P., & Godfray, H.C.J. (1993). Chaos in ecology: is mother nature a strange attractor? *Annual Review of Ecology Systematics* 24: 1–33.

Higashi & Burns (1991). *Theoretical Studies of Ecosystems: The Network Perspective*. Cambridge: Cambridge University Press.

Huxel, G.R. (1995). Influences of Community Assembly, Unpublished PhD Dissertation, University of Tennessee, Knoxville, 128 pp.

Johnson, L. (1995). The far-from-equilibrium ecological hinterlands. In *Complex Ecology*, ed. B.C. Patten & S.E. Jorgensen, NJ: Prentice Hall, Englewood Cliffs.

Jorgensen, S.E. (1995). The growth rate of zooplankton at the edge of chaos: Ecological models. *Journal of Theoretical Biology*, 175: 13–21.

Kauffman, S.A. (1993). *The Origins of Order: Self-Organization and Selection in Evolution*. NJ: Oxford University Press.

Keitt, T.H. & Marquet, P.A. (1996). The introduced Hawaiian avifauna reconsidered: evidence for self-organized criticality. *Journal of Theoretical Biology* 182: 161–167.

Lewin, R. (1992). *Complexity: Life at the Edge of Chaos*. Maxwell MacMillian.

Martinez, N.D. (1992). Constant connectance in community food webs. *American Naturalist* 139: 1208–1218.

McIntosh, R.P. (1995). H.A. Gleason's 'Individualistic Concept' and theory of animal communities: a continuing controversy. *Biology Review* 70: 317–357.

Milne, B.T., Johnson, A.R., Keitt, T.H., Hatfield, C.A., David, J., & Hraber, P.T. (1996). Detection of critical densities associated with piñon-juniper woodland ecotones. *Ecology* 77: 805–821.

Nicolis, G. & Prigogine, I. (1989). *Exploring Complexity: An Introduction*. NJ: W.H. Freeman and Company.

Paine, R.T. (1966). Food web complexity and species diversity. *American Naturalist* 100: 65–75.

Paine, R.T. & Levin, S.A. (1981). Intertidal landscapes: disturbance and the dynamics of pattern. *Ecological Monographs*, 51: 145–178.

Pattee, H.H. (1987). Instabilities and information in biological self-organization. In *Self-organizing Systems: The Emergence of Order*, ed. F.E. Yates, NY: Plenum Press.

Plotnick, R.E. & McKinney, M.L. (1993). Ecosystem organization and extinction dynamics. *Palaios* 8: 202–212.

Prigogine, I. (1980). *From Being to Becoming: Time and Complexity in the Physical Sciences*. San Francisco: W.H. Freeman and Co.

Sale, P.F. (1979). Recruitment, loss and coexistence in a guild of territorial coral reef fishes. *Oecologia* 42: 159–177.

Sale, P.S. (1977). Maintenance of high diversity in coral reef fish communities. *American Naturalist* 111: 337–359.

Solé, R.V. & Manrubia, S.C. (1995). Are rainforests self-organized in a critical state? *Journal of Theoretical Biology* 173: 31–40.

Sugihara, G. (1985). Graph theory, homology, and food webs. *Proceedings of the Symposium Applied Mathematics* 30: 83–101.

Tilman, D. (1988). *Plant Strategies and the Dynamics and Structure of Plant Communities*. NJ: Princeton University Press.

Yates, F.E. (1987). General Introduction. In *Self-organizing Systems: The Emergence of Ordered*, ed. F.E. Yates, NT: Plenum Press.

9

Assembly rules as general constraints on community composition

Evan Weiher and Paul Keddy

Introduction

It has been more than 20 years since Jared Diamond focused attention on assembly rules for communities (Diamond, 1975), and his three principal questions about community assembly are still topical and largely unsolved. The questions are: (a) To what extent are the component species of a community mutually selected from a larger species pool so as to 'fit' with each other? (b) Does the resulting community resist invasion, and if so, how? (c) To what extent is the final species composition of a community uniquely specified by the properties of the physical environment, and to what extent does it depend on chance events? These questions are really about how communities, or assemblages, are selected as subsets of a species pool (Fig. 9.1).

For any region and taxa, one could define a species pool of potential members of an assemblage of species. At the largest scale, this would simply be an exhaustive list of all species (i.e., the flora and fauna), while at smaller scales the pool might be a list of birds, beetles, fish, or plant species that could inhabit a given site or habitat. Communities would be composed of some subset of these smaller lists. Assembly rules are explicitly defined constraints on community structure – in other words, assembly rules set limits on which species can be a part of locally coexisting subsets of the species pool (Fig. 9.1). Of all the possible assemblages, some will conform to the rules and will therefore have a greater likelihood of existence. Assemblages that have a large deviation from the assembly rules may exist for short periods of time, but will likely be replaced by assemblages that more closely conform to the contraints.

There has been a fair amount of discussion about the existence of assembly rules. In fact, some of it continues in this volume. These discussions may arise out of basic confusion about the term: are assembly rules a wooly

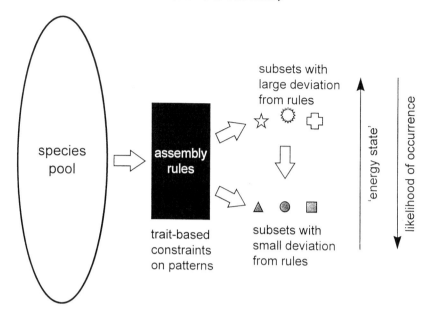

Fig. 9.1. A conceptual model of community assembly from a large species pool. Assemblages that do not conform to the assembly rules are unlikely to persist and can be thought of as having an unstable high energy state. Assemblages that conform to the rules have found low energy, local solutions to the constraints set by the assembly rules.

concept (*sensu* Peters, 1980) or are they a rigorous predictive tool? It has been suggested that it is time to (a) accept that in some places and for some taxa assembly rules do exist, (b) start to define assembly rules and (c) begin testing them for their utility for making predictions (Weiher & Keddy, 1995). Sometimes the very term 'assembly rules' causes confusion. Some argue that assembly rules are only about patterns caused by biotic interactions (e.g., Wilson, this volume), but many different kinds of factors (both biotic and abiotic) can lead to similar patterns (Connor & Simberloff, 1979; Schluter, 1984; Huston, 1997). Some might say that the term was coined by Diamond to mean patterns caused by interactions; however, Diamond currently does not insist upon such a strict definition of assembly rules (Diamond, pers. com.). Assembly rules are simply about constraints on composition, regardless of mechanism. The other main confusion comes from the notion that assembly rules are akin to a mechanistic recipe that describes the necessary steps for building a community. Assembly rules are not recipes for building communities. Rather, they are a set of limits that constrain how species can come together to form assemblages. As limits, or constraints, assembly rules are analogous to the rules of chemical stoichiometry, which limit the ways atoms can come together to form molecules. And, like the rules of basic chemistry, assembly rules are based on functional, characteristic traits, and not on the names given to the entities (be they elements or species).

In this chapter (a) several strategies for finding empirical assembly rules are outlined, (b) some initial attempts at using assembly rules for prediction are described, and (c) an explanation of how the results suggest that ecological communities can be thought of as complex adaptive systems, with properties common to both physical energy dissipating systems and species evolving in an adaptive landscape is given.

Strategies for finding assembly rules

Null models and consistent patterns of assembly

Diamond's interest in patterns of community assembly prompted Dan Simberloff and Ed Connor to investigate the idea using testable null hypotheses about composition. Simberloff and Connor were interested in finding out whether the apparent patterns that Diamond found were more than chance events. Their null model approach (Simberloff & Connor, 1981) allowed them to compare actual communities to simulated communities that were constructed in the absence of species associations, if species were independent of one another. In some respects, they were constructing random communities.

Simberloff and Connor's idea of the community null model both revolutionized community ecology and caused a rather bitter debate over methods and interpretations (e.g., Strong *et al.*, 1984). While the debate lingers (even within this volume), a consensus is emerging. Ecologists are finding ever more evidence supporting assembly rules (Weiher & Keddy, 1995; Gotelli & Graves, 1996; Wilson, this volume). It has been argued (Weiher & Keddy, 1995) that it is time to move on from the simple question of whether assembly rules exist and begin asking what are the assembly rules, how do they vary along gradients and among taxa?

The use of null models has led to several important ideas about the structure of communities. Some of the most important advances have come when the focus has been on traits, rather than on species names. The approach was developed by animal ecologists who were looking for patterns of morphological overdispersion that are consistent with limiting similarity (Ricklefs & Travis, 1980; Moulton & Pimm, 1987; see also Lockwood *et al.*, this volume). Some animal assemblages are overdispersed (e.g., Travis & Ricklefs, 1983; Brown & Bowers, 1984; Dayan & Simberloff, 1994), which means they are assembled according a set of trait-based assembly rules that call for some degree of limiting similarity. In a review of animal assemblages, Miles and Ricklefs (1994) found a positive relationship between species richness and the morphological volume occupied by the assemblage, while the density of species packing (nearest morphological neighbor) remains constant. This means that as animal species are added to assemblages, they tend to enter at the morphological periphery. Again, this corresponds to a view that animal assemblages are structured according to assembly rules that call for limiting similarity.

The prevailing view among plant ecologists has been that environmental factors act like sieves or filters, which remove species from assemblages if they lack certain requisite traits (e.g., Grubb, 1977; Grime, 1979; Box, 1981; Southwood, 1988; Keddy, 1992). This view leads to the idea that coexisting species might be morphologically underdispersed, which is a possibility that has received little attention in the literature on assembly rules (Weiher & Keddy, 1995). Trait dispersion has not been widely studied in plant assemblages. There are, however, a few cases of apparent trait overdispersion. Floral traits have been shown to be overdispersed among congeners (e.g., Armbruster, 1986; Armbruster *et al.*, 1994). There is some tendency for dominant oaks to be from different subgenera (Mohler, 1990), which is also suggestive of limiting similarity. Bastow Wilson has found evidence for guild proportionality, but the patterns are not strong (e.g., Wilson & Roxburgh, 1984; Wilson, this volume).

Investigations of morphological dispersion in temperate riverine (palustrine) wetlands have shown that assemblages can be simultaneously overdispersed in some traits (mostly those associated with size) and underdispersed in others (plasticity and form) (Fig. 9.2; Weiher *et al.*, 1998). This means that wetland plant assemblages conform to both the template-filter model and to the limiting similarity model. In the first case, traits are clumped because species had to pass some sort of environmental filter which likely acts as an external pressure, limiting the morphological volume occupied by an assemblage. In the latter case, the traits are overdispersed due to some sort of internal pressure (like competition), which keeps morphological nearest-neighbor distances large. The results strongly agree with the notion of capacity rules, which limit the presence of species according to their ability to tolerate abiotic conditions, and allocation rules, which limit the presence of species according to their ability to get along with their neighbors (Brown, 1987).

The degree to which wetland plant assemblages are under or overdispersed depends on resources. In order to assess overall morphological dispersion, the first four principal components of 11 measured traits were used. The prin-

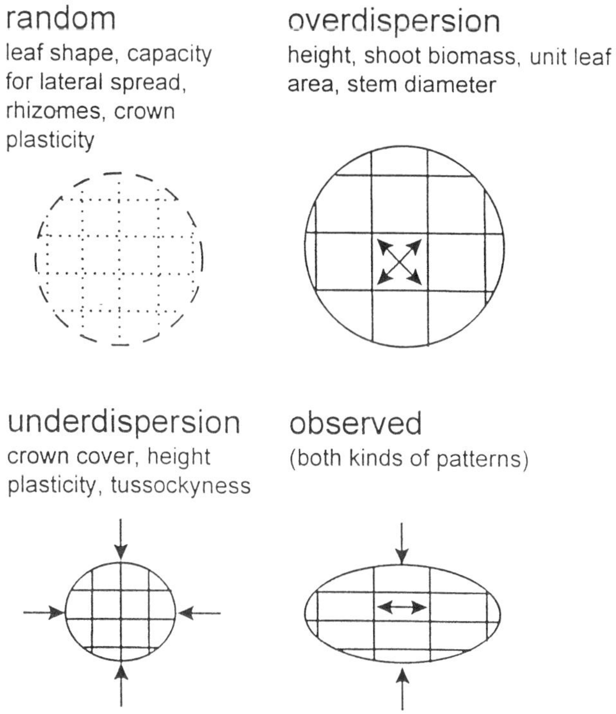

random
leaf shape, capacity
for lateral spread,
rhizomes, crown
plasticity

overdispersion
height, shoot biomass, unit leaf
area, stem diameter

underdispersion
crown cover, height
plasticity, tussockyness

observed
(both kinds of patterns)

Fig. 9.2. Patterns of morphological dispersion for wetland plants. A mixture of over- and underdispersed traits mean that assembly rules constrain community composition in ways that conform to both limiting similarity and environmental filtering (Weiher *et al.*, 1998).

cipal components corresponded to size, leafshape, vegetative reproduction and spread, and plasticity). Mean nearest neighbor Euclidean distance in PC-defined trait space was positively correlated with soil phosphorus concentration (Fig. 9.3; Weiher *et al.*, 1998). We also know that above-ground competition intensity increases with soil resources in these wetlands (Twolan-Strutt & Keddy, 1996) and this supports the idea that competition is driving trait dispersion.

While null models have increased our understanding of community composition, there are still astoundingly few explicit assembly rules. Many of the tests using null models provide very little information about what the assembly rules are. For instance, the checkerboard test (e.g., Stone & Roberts, 1990; Graves & Gotelli, 1993; Weiher *et al.*, 1998) tells if some pairs of species co-occur less often than expected by chance. A significant checkerboard test means that assembly rules are at work, but it tells nothing about the assembly

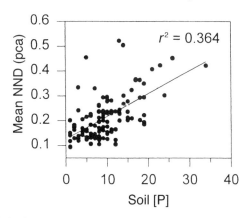

Fig. 9.3. Morphological dispersion, measured as mean nearest-neighbor distance, increases with soil fertility in herbaceous wetland vegetation. Eleven traits were collapsed into four principal components (Weiher *et al.*, 1998).

rules. Tests of trait dispersion may increase our understanding of community composition, but they also do not tell us what the assembly rules are. Wilson's guild proportionality is a test that can potentially supply a simple and powerful assembly rule (Wilson, this volume). When successful, the method provides an explicit rule: one can expect a constant proportion of species in each assemblage to belong to a certain guild. However, there are still few cases where such rules have been found. We failed to reject the null hypothesis for eight guilds of wetland plants, but this was partly due to the problem of multiple hypothesis testing (Weiher *et al.*, 1998). Plant assemblages may simply not be structured according to guild proportionality, but it is too soon to give up on such an elegant approach.

In our opinion, null models can best be used to increase our understanding of community structure, but they rarely provide us with defined assembly rules. Attempts can be made to find significantly low variances in functional groups or traits, and thus explicitly define rules. Unfortunately, more often than not, these rules are hard to find and do not provide much predictive power.

Trait–environment linkages

In plant ecology there has been a rich history of finding associations between plant species or their traits and environmental conditions. The earliest examples come from von Humbolt and the first biogeographers (e.g., see McIntosh, 1985; Bowler, 1993), and from ecophysiologists (e.g., Mooney, 1972; Fitter & Hay, 1987). More recently, there has been interest in how specific plant traits vary along gradients (e.g., Givnish, 1987; Tilman, 1988; Keddy, 1992; Gaudet & Keddy, 1995) and in the relationships between guilds or functional groups and environmental factors (e.g., Cody, 1991; Steneck & Dethier, 1994; Schulze & Mooney, 1994).

It should be made clear that, while some patterns are obvious (such as the lack of spiny succulents in temperate wetlands), others are not. Moreover, many ecological patterns are understood in only a qualitative manner, such as the apparent increase in plant capacity for lateral spread as disturbance rates increase or in the height of dominant vegetation as soil resources increase. These patterns of species replacement are an issue that is central to both ecology in general and for the realm of assembly rules.

In order to investigate trait–environment linkages, plant traits were plotted against various edaphic factors. The riverine wetlands previously described by Day *et al.* (1988), Moore *et al.* (1989), and Weiher *et al.* (1998) were again used. Fertility and elevation are the two main gradients that affect these herbaceous wetlands. The wetlands vary from fertile protected cattail marshes, to wave swept sandy shorelines with high species diversity, to tussocky wet meadows. Trees and shrubs form the upper margin of the wetlands, while at the lower end the wetlands grade into submersed aquatics and *Scirpus acutus*. Plant trait data from Weiher *et al.* (1998) were used

The plan for defining trait–environment linkages was to define how the constraints on plant traits vary along the main environmental gradient (soil resources). To do this, the first focal point focus was on the edges of the morphological space occupied by each assemblage. For each trait, the maximum and the minimum value for each assemblage was determined. In this manner, the morphological edges of the assemblages were defined. To assess the effect of soil resources, each set of assemblage trait maximas and minimas was plotted as a function of soil resources (soil P, N, loss on ignition, pH, and N:P ratio). Nearest-neighbor distances in morphological space were determined in order to measure the dispersion of traits in each assemblage (using Euclidean distances). Lastly, guild proportion were determined (the proportion of species richness from each functional guild). The best relationships were nearly always with soil $[P]\mu g\ g^{-1}$ and this is what has been presented here.

It was hoped that the relationships could be modeled with General Linear Models, but this was not the case and a series of apparently weak, noisy, often non-linear stepped or triangular-shaped 'patterns' were found. Figure 9.4 shows only a selection of the results because some the patterns were repeated for several traits (presumably because some traits are correlated). For brevity, the main types of patterns are shown. Similar patterns were also found for crown and leaf shape mean nearest neighbor distances, for maxima

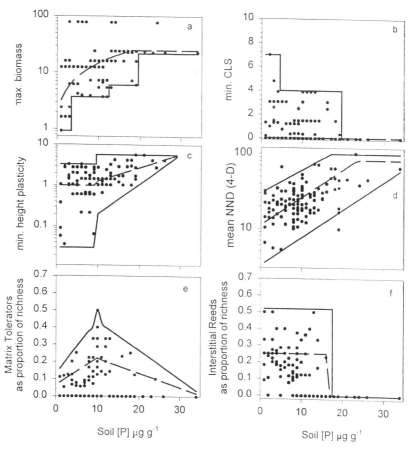

Fig. 9.4. Examples of trait–environment relationships for wetland plant assemblages. The upper or lower limits are putative assembly rules that constrain community composition. The dashed trend lines were used as target values in the simulation model. The dependent variables are: (a) maximum above ground biomass, g, (b) minimum capacity for lateral spread (distance to the farthest attached ramet × number of rhizome apicies, cm), (c) minimum height plasticity (coefficient of variation in height), (d) mean nearest-neighbor distance in traits with strongest associations with the principal component of 11 measured traits (above ground biomass, form index of leaf shape, capacity for lateral spread, and height plasticity), (e) proportion of species richness in the matrix tolerator guild (short plants with high capacity for lateral spread), and (f) proportion of species richness in the interstitial reed guild (compact plants with very small canopy, small or absent leaves).

of height, crown, and leafshape, and for minima in height, crown, and a four dimensional combination of traits that had heaviest loadings on the four principal components of the trait space (biomass, leafshape, capacity for lateral spread, and height plasticity).

There has been growing interest in non-linear relationships among ecological variables (Brown & Maurer, 1987; Keddy, 1994; Brown, 1995). Such patterns often resemble phase diagrams and when plotted, they often appear as filled triangular areas (such as the relationship between average population density and individual body mass for North American birds, Brown & Maurer, 1987) or curves that set upper limits (such as the relationship between standing crop and species density, e.g., Tilman & Pacala, 1993). Conceptually, such patterns likely come from some set of constraints that limit allowable community states. Within such constraints, anything may be possible.

Fourteen patterns were found like those shown in Fig. 9.4 for trait minima, maxima, and nearest-neighbor distances ($n = 33$). If patterns for traits that are significantly correlated are excluded, then eight patterns remain. It seems unlikely that these patterns are the result of chance, but an important next step is to develop methods for testing the significance of these patterns using random assemblages.

Communities as energy dissipating systems

One quite remarkable generality emerges: in nearly every case, the patterns become more tightly constrained as soil fertility increases. It is possible that these patterns could be partly the result of fewer sampling points and lower diversity at high soil [P]. Alternatively, this relationship corresponds to what would be expected if ecological communities are functioning as non-linear energy dissipating systems (Schneider, 1987; Ulanowicz & Hannon, 1987). A general feature of such self-organizing systems is that as the energy available to do work increases, the order or structure of the system tends to also increase (Schneider & Kay, 1995). A simple way of looking at this idea is that it takes work to move the system from an unstructured (high entropy) state to a more ordered state. The idea is also analogous to information theory, where increased 'information flow' results in greater structure (Margalef, 1968). These ideas have, for the most part, been applied to ecosystem-level phenomena such as with energy flows and qualitative changes in energy (Schneider & Kay, 1995; Ulanowicz, 1997). If these ideas are applied to community-level phenomena and patterns, then increased mineral resources likely means increased capacity to do work (production), and as resources increase, the constraints on pattern become tighter. There may be more community structure where there are higher mineral resources. Results from our experiment using wetland microcosms also support this idea (Weiher & Keddy, 1995). In the experiment, a mixture of 20 species were grown in 12 treatments of soil texture, litter, and hydrology at both low and high fertilization. After five years of growth, the high fertilization treatments tended to tightly converge on one community type while the less productive treatments had more variability. The more energy there is to dissipate, the more structure is imposed on the system.

Another observation is that some of the apparent constraints on trait-environment patterns look like a series of steps rather than a linear divide (Figure 9.4 (a), (b), (c), (f)). Initially, these patterns seemed problematic, given the great interest in linear models (e.g., Peters, 1991). However, such stepped patterns are also what one might expect if looking at the problem as a study in non-linear thermodynamics. The effects of added energy to a system often do not cause linear changes in its structure. Rather, energy can usually be added with little or no effect, until some critical point is reached. At that critical point there is a sudden (non-linear) change in the system. A classic example that is often used is the spontaneous organization of Bénard cells in fluids that are heated from the bottom and cooled at the top (Atkins, 1994). When the energy gradient increases to some critical point, hexagonal circulation cells spontaneously appear. When the energy gradient is increased, the system maintains itself until another threshold is reached. At this point the circulation cells spontaneously divide into smaller cells that increase the rate of energy dissipation (Schneider & Kay, 1994). The discontinuities in the trait–environment patterns may also be interpreted similarly. At low soil fertility there may be a single weak attractor in the n-dimensional species space, such that many different outcomes are allowed. Resources can increase to some critical point without changing the rules because the attractor does not change. When that critical point is reached, the system jumps to a new attractor, and the allowable patterns change.

Another expectation from non-linear thermodynamics is that it takes ever-increasing levels of energy to push the system into each successive state of increasing order or structure. Our results hint at this idea as well. At low resource levels it may take only small increases in resources to reach the critical point, which is associated with the first tightening of constraints in Fig. 9.4 (a) and (b). Small additional increases in soil P cause little or no further change until the next point of tightening is reached. Note that each succesive 'step' in the pattern has some tendency to be wider than the previous one.

The ideas presented here owe a great debt to Schneider, Kay, and Ulanowicz who pioneered the notion of applying thermodynamic concepts to complex living systems. Clearly, large concepts are being mapped onto a small set of observations; no explicit hypotheses have as yet been tested. The degree of agreement between the patterns observed and the expectations from non-linear thermodynamics was too great and too interesting to ignore. The value in this is that it suggests non-linear thermodynamics should be considered as a framework for posing hypotheses.

Given the range of trait–environment patterns in Fig. 9.4, there are several methodological choices for seeking assembly rules based on trait–environment relationships. General linear modeling and Loess curves are well established, but their utility in addressing noisy patterns is doubtful. Null models can be used to find statistically significant patterns of consistency, but are probably most applicable to homogeneous patches (Wilson, this volume). There are at least two important steps to take regarding these patterns. One is to develop statistical tests to investigate whether they differ from chance expectation. The other is to test whether such patterns have any use in predicting species composition.

Using trait–environment patterns as assembly rules

So far, different ways of finding patterns in assemblages have been discussed. Implicit in the discussion was the idea that such patterns can be used as assembly rules. Patterns can be used as assembly rules only if accurate predictions can be made with them. It is questionable whether a list of putative constraints on allowable assemblages can actually be used to make useful predictions. While the patterns themselves may be ecologically interesting, they are not very useful unless we can make predictions. It would be good to be able to take a short list of abiotic conditions and predict assemblage composition. It might be useful to predict the most probable assemblages, or quantify the number of equiprobable assemblages, each of which may represent a different endpoint of assembly (cf. Law & Morton, 1993). The predictions might also be in terms of the species that are most likely to occur. Predictions regarding how a known assemblage might change, given certain alterations in either the abiotic conditions or by adding species to the species pool, are also worthwhile goals.

In order to make predictions using trait–environment relationships, we developed an optimization model that is based on the idea of the 'fitness landscape'. Fitness landscapes were developed to understand and model evolution (Wright, 1932). In such models, fitness is defined by an algorithm based on traits or genotypes, and these are either measured or are initially assigned at random. When individuals reproduce, they form novel combinations of alleles and traits, and, over time, the individuals with the highest fitness will tend to predominate by finding fitness peaks in the landscape (Templeton, 1981;

Kauffman, 1995). This kind of approach is very good at finding fitness peaks in any large, complex fitness landscape. There is no guarantee that the global highest fitness will be found because species and natural systems are selected for survival rather than optimized and because they have the baggage of past solutions which limits their ability to find wildly novel phenotypes.

In order to investigate the possibility of making predictions using a large group of weak patterns, it was necessary to use an optimization approach to find assemblages that have a high degree of agreement with the patterns. Over time, it became more and more clear that the analogy of an assemblage as a 'species', evolving over a poorly understood adaptive landscape was very compelling. From a numerical point of view, it probably does not matter if the evolution of species or the assembly of communities is being discussed. In order to adopt the metaphor however, we must dance very close to a Clementsian view of communities. This may raise eyebrows. Therefore, it should be pointed out that is not our contention that community assembly is like the development of an organism, which reaches some stable, adult, climax stage. Rather, community assembly is like the evolution of a species; it is ever adapting to a changing world within the constraints of assembly rules and the limits imposed by its own past.

Species (genotypes) evolve to local fitness peaks on an adaptive landscape. Exactly which local peak is reached is determined in part by historical constraint. It is extremely unlikely that elephants will evolve wings, no matter how much their fitness might increase. Similarly, communities assemble to local fitness peaks on an adaptive landscape. Exactly which local peak is reached is also determined in part by historical constraint (Drake, 1990). Because it is somewhat too Clementsian to talk of assemblage 'fitness', instead assemblage 'deviation' from the trait-based patterns described above will be referred to. Assemblages with low deviations are favored over those with higher deviations. Therefore, assemblage may be described as moving across an assembly landscape and finding local minima (with low deviation).

Low points in assembly space nicely correspond to the idea of low energy states. Therefore, it might be useful to refer to assemblages with high deviations from the observed patterns as also having a high energy state (as in Fig. 9.1). Assemblages at high energy states will move to a lower energy state by altering their composition. Some species may enter, others may fail to persist, and these changes are how assemblages move on the adaptive landscape to find basins of low deviation from the assembly rules.

Assembly rules define the topography of the adaptive landscape for assemblages. The landscape itself is an n-dimensional species space, which for now can be defined according to species presences and absences. In the future, species abundances should also be incorporated. If there are no assembly rules, then there is no topography on the adaptive landscape and all combinations of species are equally favored. Complexity theory suggests that, if the assembly rules call for many trade-offs (that conforming to one rule means deviation from another), the topography will tend to flatten into an apparently ruleless plain (Kauffman, 1993). The greater the topography, the stronger and more consistent are the rules.

Does the topography of the assembly landscape vary with resources? Is it pitted with many local minima or are there smooth transitions? How many minima are there? Are they clumped in one part of the landscape or are there many strong attractor basins? Such questions are basic to understanding community assembly.

As a first step toward addressing these questions, the patterns described above have been used to build a model of community assembly. Please note that a mechanistic simulation model has not been developed. Rather, it is an exploratory model based on a collection of admittedly weak empirical patterns. Our initial questions regarding the model include:

(a) does the model find assemblages that 'fit' the empirical patterns?
(b) do the assemblages make any sense at all, given the natural history of the species?
(c) do the assemblages have a reasonable number of species (i.e., is species diversity an emergent property of the rules)?

An optimization model for community assembly

The model starts with the input of abiotic resource levels and conditions (soil P in μg g^{-1}, and soil percentage organic content as loss on ignition). From this input, 'target' values are determined from the empirical trait-environment patterns (these are shown as dashed lines in Fig. 9.4). Fourteen such patterns were used, as noted above. Tolerance values were set at the edges of the observed patterns. The tolerance values are the putative assembly rules that are used to constrain composition. The model chooses a random set of species with richness between 3 and 25. Three were chosen because of an historical accident (three species are required to calculate an among-species variance) and 25 because it increased the speed of the model (initial runs of the model revealed that assemblages with species richness greater than about 25 either declined to lower values or the assemblages had a very high deviation).

The model then checks whether each trait falls outside the tolerance value (the assembly rule). Whenever it does, a standardized partial deviation was calculated as:

$$(|target - observed|)/target.$$

Partial deviations were standardized in order to minimize the effects of numerical size bias on the amount of deviation from each rule. The partial deviations were summed as the total assemblage deviation. Then the model searches the local species space for the lowest total deviation. This was done by trying all possible single species additions to the assemblage, trying all possible single species deletions from the assemblage, and trying all possible combinations of single species additions and deletions (or more simply, all possible replacements). The assemblage with the lowest total deviation was then adopted as the best fit, and the new local assembly space was searched for a new minimum total deviation. While this search engine may not be the best for finding the global optimum, it has the positive quality of randomly sampling the species space. After 10 to 20 iterations, the search would generally fail to turn up an assemblage with lower total deviation, which means the model had found a local minimum. This assemblage was saved, and the model started with a new random assemblage. The process was repeated until 1000 local minima were found. In general, the model quickly found many minima with low and identical deviation scores. Increasing the number of searches to 10 000 or more simply led to a greater abundance of these putative global minima, but it does not find lower deviation scores. This approach was adopted to sample local minima and to get a general idea of the topography of the adaptive landscape.

The degree to which real assemblages may get stuck in local minima is highly dependent on the number of possible concurrent establishment and extirpation events. In our model, we allowed for the simultaneous deletion of one species with the addition of another. For real plant communities this is not reasonable because the number of simultaneous events is not limited. In wetlands, the majority of seeds arrive in the fall, overwinter, and therefore functionally have a simultaneous arrival time. Similarly, disturbances such as floods, burial, ice scour, or trampling can extirpate many species simultaneously. Because of this, not all local minima found by our model are considered to be ecologically significant minima. If the number of possible simultanous events were increased, then assemblages could escape from many local minima and move on to better ones.

Results of the simulation model

Random assemblages have much higher deviation scores than do the local minima, and the local minima have deviations that approach those of natural assemblages. This means that random assemblages do not conform to the assembly rules and that the search engine is finding assemblages that generally conform to the rules.

The simulated assemblages roughly conform to our expectations based on species natural histories. At low soil phosphorus (1 μg g⁻¹), the best 100 assemblages (with lowest total deviation) were surprisingly reasonable (Table 9.1). These assemblages had the same total deviation score, and all had ten 'core' species which are indeed commonly found on infertile, sandy shorelines. Each assemblage also included three additional species that tend to be interchangeable ecological equivalents of sandy shorelines. A good natural historian would note however that *Phalaris arundinacea* is the wrong grass; *Spartina pectinata* is what would have been expected. The simulated assemblages are compelling, but none exactly matches any observed assemblage. This is partly due to a lack of elevation data, which allows the model to mix upper and lower wetland species. At high soil fertility (30 μg P g⁻¹) the model also predicted startlingly reasonable assemblages (Table 9.1). Cattails and sagittate-leafed species are predicted, as is *Boehmeria cylindrica*, an annual common in cattail stands. Once again, the model chose the wrong grass species. At intermediate soil fertility, a vast number of low deviation assemblages are found, but there are no simple trends to report. One interpretation is that at intermediate fertility species composition is poorly constrained by assembly rules. Another, equally valid interpretation is that our assembly rules are woefully inadequate for predicting these intermediate community types. Robert Ulanowicz (1997) has argued that there is no need to find a precise casual mechanism for everything in nature, and that probablism and gross relationships are just as valuable.

The agreement with real assemblages at both ends of the fertility gradient means that the assembly rules do indeed constrain community compostition and they have some utility in making predictions. Assemblages with equal deviations suggest that alternative equivalent endpoints exist and that at least some species are ecological equivalents.

The simulations were in qualitative agreement with actual patterns of species diversity, but the predictions were generally too high (Fig. 9.5). The ubiquitous 'humped' species richness – productivity relationship with low richness at both ends of the productivity gradient emerged from the model. The pattern also took the form of an upper-limit function, with a filled in

Table 9.1. *Simulated community composition using trait-based assembly rules. These results summarize the best 100 assemblages for each simulation*

$P = 1\ \mu g\ g^{-1}$	Incidence	$P = 30\ \mu g\ g^{-1}$ The four most common assemblages
Core species:		1 *Boehmeria cylindrica*
Carex vesicaria	1	*Carex lenticularis*
Dulichium arundinaceum	1	*Sagittaria latifolia*
Eleocharis acicularis	1	*Sagittaria rigida*
Eupatorium perfoliatum	1	*Typha glauca*
Hypericum boreale	1	
Lindernia dubia	1	2 *Boehmeria cylindrica*
Phalaris arundinacea	1	*Eleocharis erythropoda*
Potentilla palustris	1	*Pontederia cordata*
Scirpus acutus	1	*Sagittaria latifolia*
Scirpus cyperinus	1	*Typha angustifolia*
+ three of the following:		3 *Equisetum fluviatile*
Eleocharis smallii	0.905	*Mentha arvensis*
Epilobium ciliatum	0.571	*Sagittaria latifolia*
Juncus nodosus	0.333	*Rumex verticillatus*
Drosera intermedia	0.238	*Spartina pectinata*
Eriocaulon septangulare	0.190	
Juncus alpinus	0.143	4 *Boehmeria cylindrica*
Juncus subtilis	0.143	*Juncus effusus*
Leersia oryzoides	0.143	*Sagittaria latifolia*
Ranunculus flammula	0.143	*Pontederia cordata*
Taraxicum officinale	0.095	*Typha glauca*
Juncus bufonius	0.048	
Juncus pelocarpus	0.048	

distribution of points. With increasing fertility, the model constrains diversity by setting limits on morphological nearest-neighbor distances. With decreasing fertility, the model constrains diversity by limiting species composition to relatively small species with short nearest-neighbor distances. The results are qualitatively satisfying because they suggest the trait-based assembly rules may be useful for prediction and that species diversity may be an emergent property of assembly rules. However, the model's imprecision shows that we still have much to do.

One way to visualize the topography of the adaptive landscape is to show how deviation varies with soil fertility and species richness (Fig. 9.5). The contours show minimum total deviation scores. The assemblages seek minimal deviation by adding, deleting, or replacing species. Conceptually, assemblages with high deviation scores should not persist and should 'evolve' toward potentially more stable, low deviation endpoints (cf. Fig. 9.1). At both low

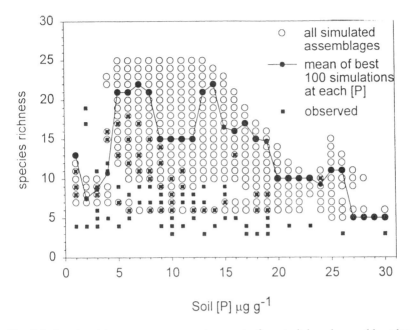

Fig. 9.5. Species richness as an emergent property from trait-based assembly rules. The simulation model overpredicts richness, but there is qualitative agreement with the expected humped and filled-in relationship between richness and productivity.

and high fertility, species richness is constrained to low values (Fig. 9.6). The flat, mildly undulating central region shows that at intermediate fertility species richness is poorly constrained. It also shows that there are many weak attractors, or small depressions, which suggests that there are many possible endpoints. As soil fertility increases, the terrain becomes steeper and more rugged, which is suggestive of a thermodynamic tightening of constraints.

Conclusions

Assembly rules are explicit constraints that limit how assemblages are selected from a larger species pool. Evidence can be sought for assembly rules using a variety of tools, from null models to gradient analysis, but insight is maximized when the focus is on guilds and species traits. When relationships between plant traits and soil fertility were investigated, a series of weak phase-space diagrams were found, which were highly suggestive of a non-linear thermodynamic model of self-organized energy dissipating systems. Community ecologists should make use of these recent conceptual advances in complexity theory and ecosystem ecology.

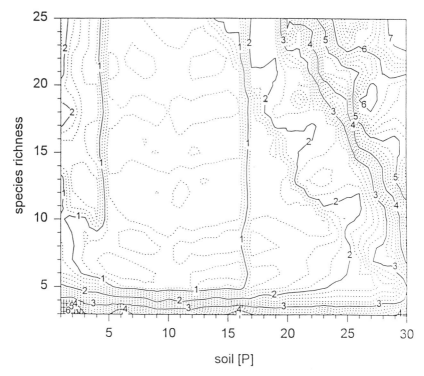

Fig. 9.6. The adaptive landscape for simulated wetland plant communities. Diversity is shown as a function of soil fertility; contours show minimum total deviation scores. In the model, assemblages move vertically to minimal deviations by species addition, deletion, or replacement. Note the flat central region where diversity is poorly constrained.

If assembly rules are going to be useful, then attempts must be made to use them to predict composition. Trait–environment patterns were used as putative assembly rules in an optimization model that used the analogy of a species evolving on an adaptive landscape. A pragmatic search for a modeling tool led communities to be viewed as self-organized entities that change their constituent membership so as to minimize their deviation from a black box of assembly rules. The qualitative success of the model suggests that trait–environment relationships have a place in the search for assembly rules and therefore have some utility in making predictions. As is usual with work on assembly rules, it seemed much easier to find new metaphors and conceptual insights into the workings of ecological communities than to find accurate, reliable predictions.

Acknowledgments

Thanks to James Kay for giving an enlightening seminar. Many thanks to Michael Ell for programming the simulation model.

References

Armbruster, W.S. (1986). Reproductive interactions between sympatric *Dalechampia* species: are natural assemblages 'random' or organized? *Ecology* 67: 522–533.

Armbruster, W.S., Edwards, M.E., & Debevec, E.M. (1994). Floral character displacement generates assemblage structure of Western Australian triggerplants (stylidium). *Ecology* 75: 315–329.

Atkins, P.W. (1994). *The 2nd Law: Energy, Chaos, and Form*. NY: Scientific American Books.

Bowler, P.J. (1993). *The Norton History of the Environmental Sciences*. NY: WW Norton.

Box, E.O. (1981). *Macroclimate and Plant Forms: An Introduction to Predictive Modeling in Phytogeography*. London: Dr W. Junk Publishers.

Brown, J.H. (1987). Variation in desert rodent guilds: patterns, processes, and scales. In *Organization of Communities: Past and Present*. ed. J.H.R. Gee & P.S. Giller, pp. 185–203. Blackwell, London.

Brown, J.H. (1995). *Macroecology*. Chicago, IL: The University of Chicago Press.

Brown, J.H. & Bowers, M. (1984). Patterns and processes in three guilds of terrestrial vertebrates. In *Ecological Communities: Conceptual Issues and the Evidence*. ed. D. Strong, D. Simberloff, L. Abele, & A. Thistle, pp. 282–296. Princeton, NJ: Princeton University Press.

Brown, J. H. & Maurer, B. A. (1987). Evolution of species assemblages: effects of energetic constraints and species dynamics on the diversification of the North American avifauna. *American Naturalist* 130: 1–17.

Cody, M.L. (1991). Niche theory and plant growth form. *Vegetatio* 97: 39–55.

Connor, E.F. & Simberloff, D. (1979). The assembly of species communities: chance or competition. *Ecology* 60: 1132–1140.

Day, R.T., Keddy, P.A., McNeill, J., & Carleton, T. (1988). Fertility and disturbance gradients: a summary model for riverine marsh vegetation. *Ecology* 69: 1044–1054.

Dayan, T. & Simberloff, D. (1994). Morphological relationships among coexisting hereromyids: an incisive dental character. *American Naturalist* 143: 462–477.

Diamond, J.M. (1975). Assembly of species communities. In *Ecology and Evolution of Communities*, ed. M.L. Cody & J.M. Diamond, pp. 342–373. Cambridge, MA: Belknap Press.

Drake, J.A. (1990). Communities as assembled structures: do rules govern pattern? *Trends in Ecology and Evolution* 5: 159–164.

Fitter, A.H. & Hay, R.K.M. (1987). *Environmental Physiology of Plants*. London: Academic Press.

Gaudet, C.L. & Keddy, P.A. (1995). Competitive performance and species distribution in the shoreline plant communities: a comparative approach. *Ecology* 76: 280–291.

Givnish, T.J. (1987). Comparative studies of leaf form: assessing the relative roles of selective pressures and phylogenetic constraints. *New Phytologist* 106: 131–160.

Gotelli, N.J. & Graves, G.R. (1996). *Null Models in Ecology*. Washington, DC: Smithsonian Institution Press.

Graves, G.R. & Gotelli, N.J. (1993). Assembly of avian mixed-species flocks in Amazonia. *Proceedings of the National Academy of Sciences* 90: 1388–1391.

Grime, J.P. (1979). *Plant Strategies and Vegetation Processes*. Chichester, UK: John Wiley.

Grubb, P.J. (1977). The maintenance of species richness in plant communities: the importance of the regeneration niche. *Biological Reviews* 52: 107–145.

Huston, M.A. (1997). Hidden treatments in ecological experiments: re-evaluating the ecosystem function of biodiversity. *Oecologia* 110: 449–460.

Kauffman, S. (1995). *At Home in the Universe*. NY: Oxford University Press.

Keddy, P.A. (1992). Assembly and response rules: two goals for predictive community ecology. *Journal of Vegetation Science* 3: 157–164.

Keddy, P.A. (1994). Applications of the Hertzprung–Russell star chart to ecology: reflections on the 21st birthday of geographical ecology. *Trends in Ecology and Evolution* 9: 231–234.

Law, R. & Morton, R.D. (1993). Alternative permanent states of ecological communities. *Ecology* 74: 1347–1361.

McIntosh, R.P. (1985). *The Background of Ecology*. Cambridge, UK: Cambridge University Press.

Margalef, R. (1968). *Perspectives in Ecological Theory*. Chicago, IL: University of Chicago Press.

Miles, D.B. & Ricklefs, R.E. (1994). Ecological and evolutionary inferences from morphology: an ecological perspective. In *Ecological Morphology: Integrative Organismal Biology*, ed. P.C. Wainwright & S.M. Reilly, pp. 13–41. Chicago, IL: University of Chicago Press.

Mohler, C.L. (1990). Co-occurrence of oak subgenera: implications of niche differentiation. *Bulletin of the Torrey Botanical Club* 117: 247–255.

Mooney, H.A. (1972). The carbon balance of plants. *Annual Review of Ecology* 3: 315–346.

Moore, D.R.J., Keddy, P.A., Gaudet, C.L., & Wisheu, I.C. (1989). Conservation of wetlands: do infertile wetlands deserve higher priority? *Biological Conservation* 47: 203–217.

Moulton, M.L. & Pimm, S.L. (1987). Morphological assortment in introduced Hawaiian passerines. *Evolutionary Ecology* 1: 113–124.

Peters, R.H. (1980). From natural history to ecology. *Perspectives in Biology and Medicine* 23: 191–202.

Peters, R.H. (1991). *A Critique for Ecology*. Cambridge, UK: Cambridge University Press.

Ricklefs, R.E. & Travis, J. (1980). A morphological approach to the study of avian community organization. *Auk* 97: 321–338.

Schluter, D. (1984). A variance test for detecting species associations with some example applications. *Ecology* 65: 998–1005.

Schneider, E.D. (1987). Schrödinger shortchanged. *Nature* 328: 300.

Schneider, E.D. & Kay, J.J. (1994). Complexity and thermodynamics: towards a new ecology. *Futures* 24: 626–647.

Scheider, E.D. & Kay, J.J. (1995). Order from disorder: the thermodynamics of complexity in biology. In *What is Life? The Next Fifty Years*. ed. M.P. Murphy & L.A.J. O'Neill, pp. 161–172. Cambridge, UK: Cambridge University Press.

Schulze, E.D. & Mooney, H.A. (eds.) (1994). *Biodiversity and Ecosystem Function*. NY: Springer-Verlag.

Simberloff, D. & Connor, E.F. (1981). Missing species combinations. *American Naturalist* 118: 215–239.

Southwood, T.R.E. (1988). Tactics, strategies and templets. *Oikos* 52: 3–18.

Steneck, R.S. & Dethier, M.N. (1994). A functional group approach to the structure of algal-dominated communities *Oikos* 69: 476–498.

Stone, L. & Roberts, A. (1990). The checkerboard score and species distributions. *Oecologia* 85: 74–79.

Strong, D., Simberloff, D., Abele, L., & Thistle, A. (eds.) (1984). *Ecological Communities: Conceptual Issues and the Evidence*. Princeton, NJ: Princeton University Press.

Templeton, A.R. (1981). Mechanisms of speciation – a population genetic approach. *Annual Review of Ecology and Systematics* 12: 23–48.

Tilman, D. (1988). Plant strategies and the dynamics and structure of plant communities. Princeton, NJ: Princeton University Press.

Tilman, D. & Pacala, S. (1993). The maintenance of species richness in plant communities. In *Species Diversity in Ecological Communities*. ed. R.E. Ricklefs & D. Schluter, pp. 13–25. Chicago, IL: University of Chicago Press.

Travis, J. & Ricklefs, R.E. (1983). A morphological comparison of island and mainland assemblages of neotropical birds. *Oikos* 41: 434–441.

Twolan-Strutt, L. & Keddy, P.A. (1996). Above- and below-ground competition intensity in two contrasting wetland plant communities. *Ecology* 77: 259–270.

Ulanowicz, R. (1997). *Ecology, The Ascendant Perspective*. NY: Columbia University Press.

Ulanowicz, R.E. & Hannon, B.M. (1987). Life and the production of entropy. *Proceedings of the Royal Society B* 232: 181–192.

Weiher, E. & Keddy, P.A. (1995). Assembly rules, null models, and trait dispersion: new questions form old patterns. *Oikos* 74: 159–164.

Weiher, E., Clarke, G.D.P., & Keddy, P.A. (1998). Community assembly rules, morphological dispersion, and the coexistence of plant species. *Oikos* 81: 309–322.

Wilson, J.B. & Roxburgh, S.H. (1994). A demonstration of guild-based assembly rules for a plant communtiy, and determination of intrinsic guilds. *Oikos* 69: 267–276.

Wright, S. (1932). The roles of mutation, inbreeding, crossbreeding, and selection in evolution. *Proceedings of the XI International Congress of Genetics* 1: 356–366.

10

A species-based, hierarchical model of island biogeography

Mark V. Lomolino

Introduction

MacArthur and Wilson's (1963, 1967) Equilibrium Theory has served as a paradigm for the field of ecological biogeography for nearly three decades (see also Munroe, 1948, 1953; Brown & Lomolino, 1989). The theory provided an insightful alternative to the prevailing, 'static theory' of island biogeography (see Dexter, 1978), which held that insular community structure was fixed in ecological time, resulting from unique immigration and extinction events. Either a species made it to the island in question, or it did not. Once it arrived, the species found adequate resources for survival, and did so in perpetuity, or it failed to establish a population. In contrast, MacArthur and Wilson held that insular community structure was dynamic in ecological time, resulting from recurrent immigrations and extinctions. As an island filled with species, immigration rate declined and extinction rate increased until an equilibrial number of species was attained. At this level, species richness would remain relatively constant, while species composition would continue to change as new species replaced those becoming extinct.

MacArthur and Wilson (1967: 19–21) acknowledged that 'a perfect balance between immigration and extinction might never be reached, ... but to the extent that the assumption of a balance has enabled us to make certain valid new predictions, the equilibrium concept is useful.' The equilibrium theory has indeed served as a worthy paradigm, stimulating much research on structure of isolated communities. It has, however, also been the subject of much criticism (e.g., see Sauer, 1969; Pielou, 1979; Gilbert, 1980), which will not be dwelt on here. Extending MacArthur and Wilson's insights to develop a more general, multi-scale theory of insular community structure is the challenge at hand. Here a model will be presented that is based on MacArthur and Wilson's fundamental assumption that insular community structure results from the combined effects of recurrent immigration and extinction. Like their model, the one developed here is limited in temporal scale to ecological time. Also like their model, it is assumed that immigration and extinction are deterministic with respect to island characteristics. MacArthur and Wilson's model, however, assumed that immigrations and extinctions were stochastic with respect to species characteristics. That is, their model assumed that all species were equivalent. The model presented here assumes that species differ with respect to factors affecting immigration and extinction. Many patterns in insular community structure result from, not despite, interspecific differences in abilities to immigrate to and survive on islands. While the model is species-based, it is hierarchical in that it addresses the potential influences of immigration and extinction across ecological and spatial scales ranging from distributions of individual species within an archipelago, to differences in community assembly, nestedness and species richness patterns among archipelagoes. For the sake of simplicity, this chapter will focus on biogeographic patterns of insular animals and the potential importance of processes occurring within ecological time.

First, a focal species model will be explained; it was originally presented during a symposium on mammalian biogeography in 1984 (Lomolino, 1986). This model describes how the insular distribution of a focal species should be influenced by factors affecting immigration and extinction; specifically, isolation and area, respectively. After describing the expected form of this insular distribution function (IDF), a summary of empirical patterns of insular distributions of selected species will be given. The focal species model serves as the fundamental level of a hierarchical model extending to community and inter-archipelago scales. Linkage between these levels is achieved by exploring how species, islands and archipelagoes covary with respect to factors affecting the two fundamental biogeographic processes, i.e., immigration and extinction.

In some, perhaps many cases, this hierarchical model may provide a relatively simple explanation for apparent anomalies or fundamental ambiguities of biogeography. As Brown (1995) observed in a recent discussion of MacArthur and Wilson's model, 'many advances in insular biogeography come from developing alternative models to explain cases in which the predictions do not hold.' The model presented here may explain many of these special cases. More importantly, rather than providing all the answers, this model should generate some new predictions and identify critical gaps in our understanding of the forces structuring isolated communities. This model should, therefore, provide some useful insights for conservation biology, which has largely become a challenge to conserve native biotas on ever-shrinking 'islands' of their natural habitats. Potential applications of the species-based model for conserving biodiversity will be discussed in a forthcoming paper (see also Lomolino, 1986).

A focal species model of island biogeography

This focal species model is based on three assumptions:

(a) Species are common on those islands where their persistence time exceeds the time between immigrations.
(b) Persistence time of the focal species should increase with island area.
(c) Immigration time should increase with island isolation.

Given this, the minimum area requirements of each focal species should increase with isolation. This pattern can be illustrated on a coordinate system where one axis represents island isolation and the other axis represents island area (Fig. 10.1; see also Alatalo, 1982). The focal species may be present on small islands if they are close enough to a source population such that high immigration rates can compensate for high extinction rates (cf. 'mass effect' – Schmida & Wilson, 1985; Auerbach & Schmida, 1987; 'rescue effect' – Brown & Kodric-Brown, 1977). Conversely, the focal species may be present on distant islands if they are large enough such that low extinction rates compensate for low immigration rates. As a result of these compensatory

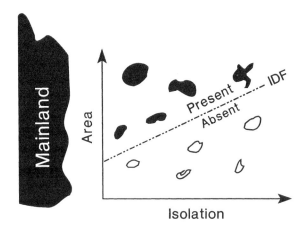

Isolation

Fig. 10.1. A simple, linear Insular Distribution Function (IDF) for a hypothetical species. The IDF delineates those combinations of area and isolation where extinction and immigration rates are equal. The focal species is expected to occur on islands that exceed this ratio (i.e., islands that fall above the IDF – darkened symbols) and absent from those that fall below it (open symbols). The actual form of the IDF is expected to be non-linear (see Fig. 10.3).

effects, insular distribution functions (IDFs; Fig. 10.1), should have a positive slope.

The form of the IDF depends, in turn, on the form of two fundamental relationships: the extinction–area and immigration–isolation functions. Most theoretical work indicates that population persistence increases as an exponential function of island area (MacArthur & Wilson, 1967; Goel & Richter-Dynn, 1974). In contrast, the immigration–isolation relationship has not received nearly as much attention. Nonetheless, it is known that the factors affecting vagilities and, for passive immigrators, persistence and survival during rafting, tend to be correlated with physical traits of individuals such as fat stores, buoyancy and lift, or metabolic rates. These traits tend to be normally or log-normally distributed within species (Calder, 1984; Peters, 1983). The precise form of frequency distributions of these traits is not an issue here. It is only important that immigration abilities (active or passive) of most individuals tend to cluster about some modal value. Simply put, nearly all individuals can immigrate beyond some critical minimum distance, but few can immigrate beyond some maximal distance (Fig. 10.2(*a*). Thus, immigration rate of a focal species should be a negative sigmoidal function of isolation (Fig. 10.2(*b*); see also MacArthur & Wilson, 1967: 56, 127–128). More to the point, the time between immigrations by this species should be a positive sigmoidal function of isolation. Thus, to maintain insular populations, area requirements (and persistence times) of the focal species will increase, again as a positive sigmoidal function of isolation (Fig. 10.3). Because persistence time should be an exponential function of area, increasing more rapidly for the large islands, the slope of the IDF should be more shallow than that of the immigration curve (i.e., increasing less rapidly for the more-isolated islands; Fig. 10.3, solid vs. dashed lines). Yet, the general form of the IDF should still be a positive sigmoidal – increasing relatively slowly for the near islands, more rapidly for those of intermediate isolation, and less rapidly again for the more isolated islands. Because the IDF delineates the combinations of area and isolation where the focal species immigrates as frequently as it suffers extinction, its insular populations should be present on islands that fall above the curve and absent on those that fall below it (Fig. 10.3).

The IDF has two readily interpretable components: an intercept and a generalized, albeit variable, slope. The intercept (A_{min} in Fig. 10.3) is a measure of area required to maintain a population of the focal species on the 'mainland', while the slope is a measure, actually an inverse measure, of its immigration abilities. Given the expected variation in resource requirements and immigration abilities among species, insular distribution patterns may

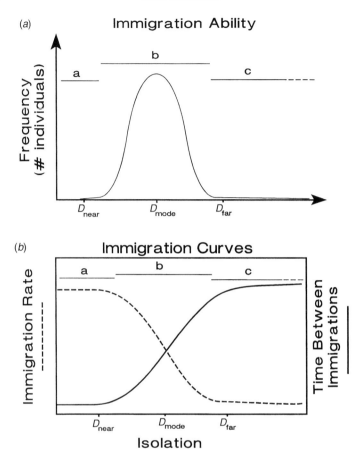

Fig. 10.2. (*a*) Frequency distribution of immigration abilities (distances) among individuals of a hypothetical focal species. Because immigration abilities are influenced by a combination of many physiological, morphological and behavioral traits, they are expected to be distributed as normal or log-normal functions, clustering about some intermediate value (D_{mode}). Here, nearly all individuals can immigrate beyond some minimal distance (D_{near}), while very few can immigrate to the very distant islands (i.e., those beyond D_{far}). (*b*) Given immigration abilities of individuals are distributed as depicted in Fig. 10.2(*a*), then immigration rate (the number of propagules reaching an island per time) of the focal species should decrease as a negative sigmoidal function of isolation (dashed line), while the time between immigrations should be a positive sigmoidal function of isolation (solid line): increasing slowly for the near islands (region a), more rapidly for islands of intermediate isolation (region b), and again more slowly for the more isolated islands (region c).

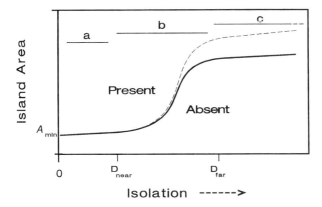

Fig. 10.3. The general form of the insular distribution function (IDF). The IDF delineates those combinations of area and isolation for which immigration equals extinction. The focal species is expected to occur on islands that fall above the IDF. A_{min} represents the minimum area required to maintain populations of this species on the mainland. The dashed line depicts the IDF expected if persistence time was a linear function of area, while the solid line depicts the expected IDF given that persistence time should increase as an exponential function of island area. The influence of isolation on insular distribution of this species should be difficult to detect in regions a and c (i.e., if sampling is limited to the very near or very isolated islands; see Fig. 10.2).

vary widely, even within a group of taxonomically or ecologically similar species (Fig. 4.10 and 10.5). In fact, depending on the range in area and isolation surveyed, the same species may exhibit what appear to be qualitatively different patterns within the same archipelago. Fig. 10.6 illustrates how, given surveys across different ranges in biogeographic variables, the same species may exhibit distributions that appear to be limited by just area, but markedly different areas (boxes A and B in Fig. 10.6), or by just isolation (box C). Alternatively, this species may appear ubiquitous on small islands, but rare on larger islands (boxes D and E). Given a broader range in biogeographical variables sampled, these fundamental ambiguities should clear to reveal the insular distribution function, i.e., the combined influence of immigration and extinction.

The sigmoidal form of the IDF also can account for some other biogeographic ambiguities or anomalies. For example, distributions of some species or groups of species appear to be uninfluenced by isolation. The relaxation faunas of the Great Basin and Bass Straits may be exemplary cases (Fig. 10.7). While the apparent absence of isolation effects for most of these species seems counter to the fundamental assumption of the equilibrium theory, it is

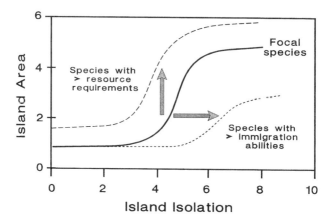

Fig. 10.4. The predicted effects of interspecific differences in resource requirements and immigration abilities on insular distribution functions. In comparison to the IDF of a hypothetical focal species (solid line), IDFs should shift upward for species with greater resource requirements (dashed line) and rightward for species with greater immigration abilities (dotted line).

consistent with predictions of the species-based model. That is, for most non-volant mammals studied, the islands of the Great Basin and Bass Straits may encompass a region where isolation effects are difficult to detect (grey box of Fig. 10.7 (c); see also Fig. 10.3, region c). Here, species distributions should be largely a function of area, but not isolation. In addition, most island biogeography studies are not well designed to detect the effects of isolation. While these natural experiments include islands that vary by three or four orders of magnitude in area, isolation typically varies by just one or two orders of magnitude. Thus, isolation effects may be detectable only when the structure of these isolated communities is compared to that of the less-isolated ones (see Lomolino & Davis, 1997).

In summary, much of what appear to be biogeographic ambiguities or anomalies may stem from the sigmoidal nature of the IDF and the tendency for biogeographic studies to sample just part of its bivariate space. If the IDF was a linear function, the effects of area and isolation would be equally detectable throughout the archipelago. Again, the form of the IDF derives from two fundamental biogeographic functions, the immigration–isolation relationship and the extinction–area relationship. Because these functions are not linear, much of what is studied in biogeography is strongly scale dependent. Thus, to better understand the geographical basis of biodiversity, island biogeography can gain many important insights from comparisons of patterns

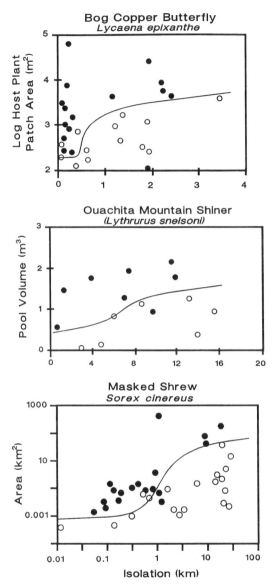

Fig. 10.5. Empirical distribution patterns of three species inhabiting islands, or island-like habitats: the bog copper butterfly (Michaud, 1995), the Ouachita Mountain shiner (Taylor, 1997), and the masked shrew (Lomolino, 1993; darkened symbols depict presence, open symbols depict absence). Distribuitons of each species were significantly associated with both area and isolation ($P < 0.05$; linear regression; sigmoidal IDFs were drawn by inspection).

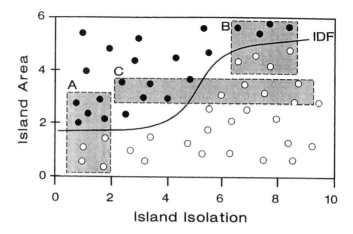

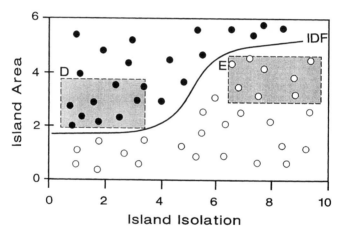

Fig 10.6. Effects of different sampling regimes on apparent patterns of distributions of a hypothetical species (darkened symbols depict presence, open symbols depict absence). Depending on the region sampled, the same species may appear to exhibit a variety of distribution patterns with respect to island area and isolation (see text).

among species and across spatial and ecological scales. These between-scale comparisons are the subject of the following sections.

The influence of immigration and isolation on insular distributions may be studied using multivariate analyses – linear regression, logistic regression or discriminant analysis (see also similar methods of analyzing insular distributions by Schoener & Schoener, 1983; Simberloff & Martin, 1991). In such cases, the dependent variable expresses presence or absence of the focal

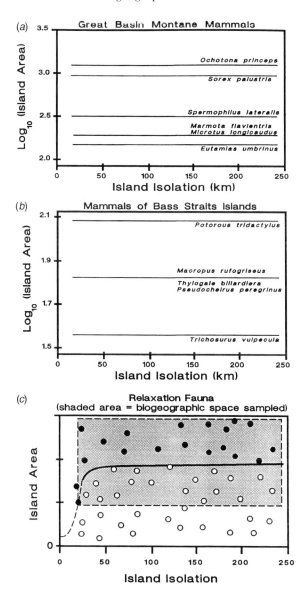

Fig. 10.7. Insular distribution patterns of non-volant mammals on two isolated arch-ipelagoes (*a*) relaxation fauna of montane forest islands of the Great Basin, North America, after Brown, 1978; (*b*) relaxation fauna of the islands of Bass Straits, Aus-tralia, after Hope, 1973. Species occur on islands that fall above the lines (see Lomolino, 1986). Although distributions of these species appear to be uninfluenced by isolation, this may be an artifact of the sampling regime. That is, isolation effects may have been detected if less-isolated islands were included in these studies ((*c*) darkened symbols depict presence, open symbols depict absence).

species (e.g., arbitrarily set at 1 or 0, respectively), and independent variables are isolation and area. Because the IDF is not a linear function, it probably will be necessary to transform one or both independent variables before conducting the analyzes. Given the diversity of possible patterns (Fig. 10.6), the appropriate transformations may best be assessed by inspection. The output of these analyzes can be used to estimate the statistical significance of isolation or area effects and to calculate the slope and intercept of the IDF (for a more detailed description of these methods, see Lomolino, 1986; Lomolino *et al.*, 1989). More general and possibly more robust tests of the influence of isolation and area may be achieved using a randomization approach. Basically, it could be tested whether islands occupied by the target species are significantly clustered with a portion of biogeographic space (i.e., area–isolation plots). To test whether clustering is significant, nearest neighbor distances of observed distributions can be compared to those when the occupancy of the focal species is randomly distributed across the archipelago (see Manly, 1991).

Community structure
Linkage to the focal species model

One way to extend the focal species model to a community level is simply to overlay IDFs of different species on one graph to form what I referred to as a community spectrum (e.g., Fig. 10.8(*a*); Lomolino, 1986). While this descriptive approach has some heuristic value, a more rigorous and insightful model requires explicit linkage among hierarchical levels (see O'Neill *et al.*, 1986; Wiens *et al.*, 1986; Allen, 1987; Hengeveld, 1987). Specifically, to predict patterns in species composition, we need to know how the two components of the IDF, its intercept and generalized slope, covary among species. Alternatively, to predict patterns in species richness, we need to assess the form of frequency distributions of slopes and intercepts.

Again, the intercept of the IDF is a measure of resource requirements and the slope is a measure, actually, an inverse measure of immigration ability. Immigration abilities may either increase with resource requirements, decrease with resource requirements, or the two traits may be uncorrelated among species. For the sake of simplicity, it is assumed here that the latter case will be intermediate with respect to resultant patterns in community structure. Therefore, the following discussion focuses on implications of the two alternative patterns of covariation (positive or negative) of slopes and intercepts on insular community structure.

How should immigration abilities and resource requirements covary?

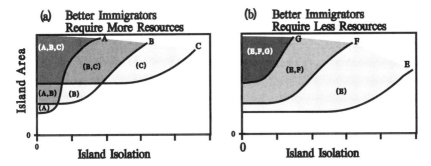

Fig 10.8. Effects of two alternative patterns of covariation of resource requirements and immigration abilities on assembly and nestedness of insular communities. Solid lines depict IDFs for three hypothetical species (A, B, C; or E, F, G) in two archipelagoes. (*a*) Where better immigrators require more resources (larger islands) to maintain their populations, relatively complex assembly of insular communities is expected depending on the size and isolation of islands sampled. (*b*) In contrast, where better immigrators require fewer resources to maintain their insular populations, then assembly sequences should be constant (i.e., first species E, then F, and then G).

Physiological considerations may provide some useful insights here. Within a group of similar species (e.g., non-volant mammals or passerine birds), larger species tend to require more energy, but they also tend to exhibit greater capacities for active migration and greater endurance during stressful conditions (e.g., dehydration or starvation during active or passive immigration; Calder, 1984; Lomolino, 1988, 1989). Therefore, for many faunal groups, resource requirements and immigration abilities may be positively correlated – better immigrators may require more resources to maintain their insular populations. That is, the intercept and slope of the IDF should be inversely correlated. This pattern of covariation, however, is not likely to be a rule for all species groups. Indeed, some faunal groups, such as micro-snails that depend on aerial dispersal for long-distance colonization (see Vagvolgyi, 1975), may exhibit the opposite pattern; i.e., good immigrators may be small and thus require fewer resources to maintain insular populations. This is certainly an area where more empirical studies are sorely needed.

Assembly and nestedness of insular communities

As illustrated in Fig. 10.8, two alternative patterns of covariation have very different implications with respect to assembly of insular communities (Diamond, 1975). Here assembly 'rules' are defined as patterns in how species composition varies along biogeographical gradients (e.g., increasing area or isolation) and gradients in species richness. Positive covariation of slopes and intercepts (i.e., where better immigrators require fewer resources to maintain their populations) implies a remarkably simple and orderly pattern of assembly (species E first, then F, then G; Fig. 10.8(b). In contrast, where better immigrators tend to require more resources, this relatively simple model predicts a diverse collection of communities. Indeed, depending on which ranges in area and isolation are sampled, different groups of these hypothetical islands are expected to be inhabited by almost all possible combinations of species (species A, or B, or C; A and B, B and C, or all three species; Fig. 10.8(a). However, within the limited biogeographical space included in most surveys, even these species may exhibit simple patterns of assembly. For example, if sampling was limited to just the larger islands of Fig. 10.8(a), species accumulation sequences (with decreasing isolation) should always be C first, then B, and then C.

In summary, the species-based, hierarchical model predicts the following assembly rules for insular faunas.

(a) Where better immigrators tend to require fewer resources, regardless of the biogeographic space sampled, communities should accumulate species in order of increasing resource demands and decreasing immigration abilities (i.e., first species E, then F, then G; Fig. 10.8(b).
(b) On the other hand, where better immigrators tend to require more resources (Fig. 10.8(a), accumulation sequences will vary depending on the range in area and isolation sampled (e.g., first A, then B, then C on the near islands; first C, then B, then A on the large islands).

Therefore, the species-based model provides a relatively simple explanation for what may otherwise seem perplexing and contradictory results. Where better immigrators tend to require more resources to maintain their populations, the same species pool may exhibit contradictory patterns of assembly for surveys conducted in different archipelagoes, or different regions of the same archipelago. Only when the results of these studies are combined, does the ambiguity clear to reveal the combined importance of immigration and extinction at the community level. Assembly 'rules' should differ, but in a manner consistent with the resource requirements and immigration abilities of the pool of focal species.

The alternative forms of covariation in resource requirements and immigration abilities also have important implications with respect to nestedness of insular communities. Nestedness, as first described by Darlington (1957) and later developed by Patterson and Atmar (1986), refers to the tendency for more depauperate communities to form proper subsets of richer communities

(Fig. 10.9; see also Patterson & Brown, 1991). Alternatively, nestedness may be viewed as the tendency for communities on smaller or more isolated islands to form proper subsets of those on larger or less isolated islands.

Nestedness of insular communities can result from differential immigration or differential extinction (Fig. 10.9). That is, nestedness implies that immigration and extinction vary in an orderly manner among islands and among species. Of course, both of these fundamental biogeographic processes, immigration and extinction, may contribute to nestedness. If, however, slopes and intercepts of IDFs exhibit negative covariation among species (good immigrators tend to be poor survivors), then immigration and extinction can have confounding effects on nestedness. This should be obvious from inspection of Fig. 10.8(*a*). Sampling islands along gradients of isolation or area can be simulated by moving along horizontal or vertical lines within the bivariate space. Perfect nestedness occurs where, as we move along a gradient of decreasing area or increasing isolation, once a species is omitted, it remains absent from all smaller or more isolated communities. More simply put, archipelagoes comprised of perfectly nested communities are those where their species' IDFs do not intersect (intercepts and slopes exhibit positive covariation). Therefore, perfect or near perfect nestedness is only expected where better immigrators require fewer resources (smaller islands) to maintain their insular populations (Fig. 10.8(*b*), or where biogeographical surveys include only a limited range in area or isolation (e.g., just the near islands of Fig. 10.8(*a*).

Patterns in species richness

Perhaps more than any other ecological pattern, the species–area relationship is of fundamental importance to development of island biogeography theory (Fig. 10.10). Indeed, MacArthur and Wilson (1967:8–9) wrote that 'theories, like islands, are often reached by stepping stones. The species–area curves are such stepping stones'. The primary objective of MacArthur and Wilson's theory, of course, was to explain the general tendencies for species richness to be higher on larger and less-isolated islands. The model presented here, while species based, provides an alternative explanation for patterns in species richness. Given its hierarchical nature, and provided appropriate linkage between species – and community-levels, this model should accurately predict the general form of the species–area and species–isolation relationships.

The species–area relationship

First note that the models presented in Fig. 10.8 predict patterns in richness as well as composition and nestedness. These figures, however, fall short of

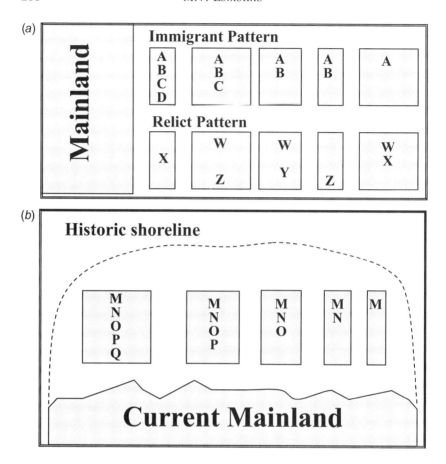

Fig. 10.9. (*a*) Darlington's (1957: 485) 'diagram to compare a simple immigrant pattern, formed by dispersal of several groups of animals from the mainland for different distances along a series of islands, and a relict pattern, formed by partial extinction of an old fauna formerly common to all the islands'. Darlington's immigrant pattern assumes that the species differ in their immigration abilities (A being the best, D being the least vagile immigrator), while his relict pattern assumes that species are equivalent (i.e., extinction is not selective). (*b*) Pattern of nestedness expected for faunas where insular community structure is determined solely by selective extinctions and interspecific differences in resource requirements (resource requirements are highest for species Q and lowest for species M; see Patterson & Atmar, 1986; Patterson & Brown, 1991).

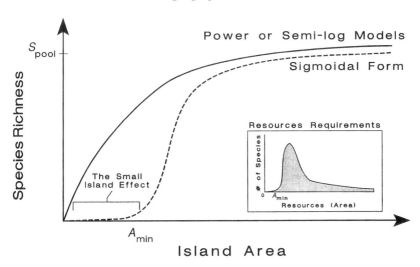

Fig. 10.10. Alternative forms of the species–area relationship. The sigmoidal form is based on observed distributions of body size and on bioenergetic considerations (see text), and is consistent with the small island effect (i.e., a tendency for species richness to be independent of area for the very small islands; i.e., those smaller than A_{min} – see Niering, 1963). Inset depicts the frequency distribution of resource (area requirements) among species (based on the assumption that resource requirements increase in proportion to body size).

describing the form of the species-area and species-isolation relationships. To accomplish this, it is necessary to assess how the intercept and slope of the IDFs are distributed across the species pool. That is, all else being equal, species–area curves are determined by the frequency distribution of the IDF intercepts, which, again, are measures of resource levels required to establish and maintain insular populations. If the intercepts are uniformly distributed among species, then the species–area relationship should be linear. At every increment in area, another species should be added. Of course, it is known that this relationship is less than linear (Fig. 10.10). This suggests that the frequency distribution of IDF intercepts is not uniform, but skewed such that most species have relatively low resource requirements.

Alternatively, inductive means can be used to approach this question, again drawing on the rich literature in physiological ecology. Resource requirements of animals (e.g., energy, water, nutrients and home range size) are positively correlated with body size among species. For most animal taxa, however, the frequency distribution of body size is strongly skewed, with many more smaller than larger species (Brown, 1995; Calder, 1984). Thus, as area increases, the rate at which an island exceeds the resource requirements of additional species should decrease, yielding the general form of the species-area relationship (Fig. 10.10). If pressed, allometric equations could be manipulated to derive the parameters, or at least an exponent for one expression of the species–area relationship – the power model (which as Gould, 1979 pointed out is an allometric formula). This exponent, often referred as z, is a measure of the how the slope of the species-area relationship changes (typically declines) with increasing area. As z-values decrease from 1.0 to 0.0, the slope of the species–area relationship levels off more rapidly (see Lomolino, 1989; Rosenzweig, 1995).

Insights from physiological ecology are again drawn on to predict z-values, here using mass as the common currency. To answer the question of how species should accumulate with increasing area, the question is first asked how species accumulate with increasing mass. That is, how does S_s, the number of smaller species, scale with body mass. Brown and Nicolleto (1991) provide some useful data here. Re-analyzing their data on body mass of North American land mammals, we find S_s scales with mass raised to the 0.234 power ($r^2 = 0.76$). To convert this species–mass to a species–area accumulation function, the question is now asked how resource requirements (RR) scales with mass. Jurgens and Prothero (1991) report that lifetime RR of mammals increases as a function of mass raised to the 0.87 to 0.93 power (for birds the exponent ranges from 0.88 to 0.94, which is not significantly different from 1.0; Jurgens & Prothero, 1991). Home range size of different mammals scales as an approximately linear function of mass (see Lindstedt *et al.*, 1986; see also Harestad & Bunnell, 1979; Schoener, 1968; Calder, 1984; Peters, 1983). Assuming that resource requirements of insular populations increase in a linear fashion with home range size (or roughly proportional to life-time energy budgets), and that the level of insular resources (biomass and productivity) increases as a linear function of island area, area requirements (A) can be substituted for mass (M) to obtain the following.

$$S_s \propto M^{0.23} \text{ (again, after Brown \& Nicolleto, 1991)}$$
$$\text{therefore, } S_s \propto Area^{0.23}$$

These results are eerily close to Preston's (1962) canonical value ($z = 0.26$) for the species–area relationship. This similarity may be more than just a coincidence, given that frequency distributions of body size (the basis of the above derivation) and abundances (the basis of Preston's method) may be causally related: simply put, larger species tend to be more rare. However, it would be foolhardy to make much of the precise values obtained above. The scaling relationships used to generate these results will vary somewhat among taxonomic groups. Similarly, the assumption that insular resources vary as a linear function of area, while intuitively appealing, remains largely untested. Given this, z-values may vary considerably among biotas. The quagmire of predicting how z-values should vary among specific groups or regions will be avoided. However, because the scaling relationships (i.e., exponents in allometric equations) tend to be conservative, the model provides an alternative, species-based explanation for the general tendency for z-values to be conservative, ranging from 0.1 to .5 for most biotas.

The species–isolation relationship

In comparison to the species–area relationship, which is an accumulative function, the species–isolation relationship can be viewed as one of attenuation. How does species richness of insular communities attenuate as isolation increases? Of the two patterns, the species area-relationship has received the lion's share of attention. The species–isolation relationship, in contrast, seems like a neglected sibling. The author is unaware of any debates over its general form, let alone controversies over the meaning of precise, 'canonical' values of its exponents or coefficients.

MacArthur and Wilson (1967: 125–128) predicted that the species–isolation relationship would take one of two forms, depending on the nature of immigration. If immigration is passive with constant directionality, such as that for wind-dispersed, terrestrial propagules, then immigration rate (number of propagules reaching an island per time) should be a negative exponential function of isolation (i.e., immigration should be proportional to e^{-I}). On the other hand, MacArthur and Wilson predicted that, if immigration was active, or passive but on logs or other 'rafts' with normally distributed persistence times, then immigration should be a normal function of isolation (i.e., immigration should be proportional to $e^{-(I \times I)}$).

Observed species–isolation relationships, when significant, tend to be consistent with one of these predictions. MacArthur and Wilson's (1967) reasoning, however, may be problematic because the equilibrium model treated the biota as a collection of homogeneous species. On the contrary, even for taxonomically or ecologically similar faunas, immigration abilities are likely to vary markedly among species. Assuming that this is the more logical alternative, the form of the species–isolation relationship can be derived by asking how immigration abilities should be distributed among species.

The following argument is analogous to that used to derive the form of the IDF. Immigration abilities (passive or active) of most species will probably exceed some minimal distance (D_{near} of Fig. 10.11). This distance depends on the physiological and behavioral characteristics of the species group in question and the nature of the immigration filter (*sensu* Simpson, 1940; see also next section on inter-archipelago comparisons). Beyond this distance, richness should decline at a rate determined by the distribution of immigration abilities among species. Thus, we need to know how the slopes of the IDFs are distributed among species. If they were uniformly distributed, then, as isolation increased, species richness would decline as a linear function of isolation. Again, it is known that this relationship tends to be non-linear. Based on physiological considerations, the frequency distribution of immigration abilities will probably be log-normal or log-skewed, with most species exhibiting relatively limited vagilities.

The above prediction is based on the assumption that immigration abilities of animals, like many species traits, are correlated with body size, which tends to be distributed as a log-normal or log-skewed function (Brown, 1995; Peters, 1983). That is, for a great diversity of taxonomic groups, most individuals tend to be of relatively small size. The species–isolation relationships should therefore, approximate a negative sigmoidal function (Fig. 10.11). Species

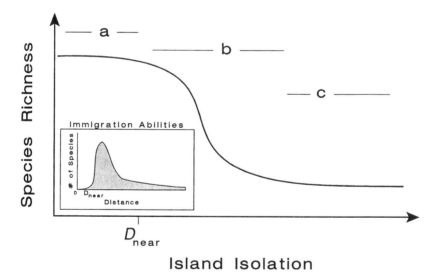

Island Isolation

Fig. 10.11. Species–isolation curve derived from physiological considerations (see text) and frequency distributions of body size for species of a given faunal group. Species richness should remain relatively high until isolation exceeds the immigration ability of the least vagile species (D_{near}, which marks the end of region a). Beyond this distance (region b), species richness should decline rapidly (see Fig. 10.12). Finally, species richness should remain relatively low and appear independent of isolation for the most distant islands (i.e., those in region c). Inset depicts the frequency distribution of immigration abilities (distances) among species (based on assumption that immigration abilities are correlated with body size).

richness should remain high until isolation exceeds the immigration capacity (D_{near}) of the least vagile species. Beyond this minimal critical isolation, species richness should decline rapidly until isolation approximates the immigration range of the modal species. As isolation increases beyond this point, richness should asymptotically approach zero. It would be unwise to make any specific predictions regarding 'canonical' exponents or coefficients of this relationship. Indeed, in comparison to patterns for active immigrators, passive immigrators as a group are likely to have very different physiological relationships, immigration functions and species–isolationship curves. However, because propagule dispersal tends to be log-normally distributed across distance (with most landing near the source), species–isolation relationships for these groups should still approximate a negative sigmoidal in arithmetic space (Fig. 10.11).

In short, there is much to be learned about immigration. At least one important inference, however, can be drawn from the expected pattern. The sigmoidal nature of the immigration function may, at least in part, accounts for the relative paucity of well-documented isolation effects. Biogeographers are unlikely to detect the effects of isolation unless their studies encompass a broad range in isolation. While most studies span three or four orders of magnitude variation in area, they seldom include islands that vary more than an order of magnitude in isolation. If these islands are all relatively close or all distant (i.e., falling within the ranges of isolation where the *s-i* slope is near zero, regions 'a' and 'c' in Fig. 10.11, inset), then isolation effects may be difficult to detect. Aside from this sampling artifact, most species–isolation relationships should belong to the same family of sigmoidal curves whose slopes vary as a function of vagilities of the species and nature of the immigration filter. On pp. 293–295, it will be considered how species–isolation and species–area curves should vary among taxa and among archipelagoes.

Patterns in species richness: summary

The species-based model presented here provides a relatively simple explanation for the species–area and species–isolation relationships. Unlike the equilibrium theory, this model does not assume a balance of immigrations and extinction, but it does assume that species differ and in an orderly pattern. That is, the forms of species–area and species–isolation curves may derive from the general tendency for body size and related physiological and ecological characteristics to be distributed as log-normal or log-skewed functions. Given this and the assumptions, based on bioenergetic considerations, that resource requirements and immigration abilities are correlated with body size, then the species–area relationship should take the form of a positive sigmoidal,

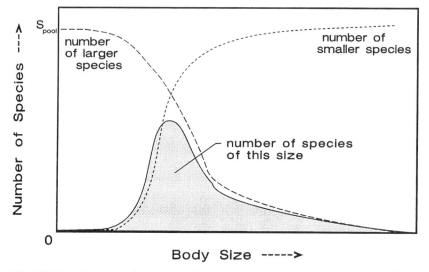

Fig. 10.12. Body sizes of species within a faunal group tend to be log-normally or log-skewed, with most species exhibiting relatively small body size (solid line; see Brown, 1995). Consequently, accumulation functions of the number of species with larger or smaller size tend to be sigmoidal functions (dashed and dotted lines). Because resource requirements and immigration abilities may be correlated with body size, these accumulation functions may have important implications for understanding patterns in species richness of insular faunas (S_{pool} = total number of species in the source, or mainland, pool).

tracking the cumulative number of smaller (less resource intensive) species, while the species–isolation relationship should be a negative sigmoidal, tracking the number of larger (more vagile) species (Fig. 10.12).

The possible sigmoidal form of the species–area relationship is seldom discussed. The conventional models (power and semi-log models) typically provide a good approximation of the species–area relationship of most insular biotas. Indeed, the power and semi-log models are qualitatively similar to the sigmoidal model (Fig. 10.10), with one exception. For all three models, richness increases rapidly for islands of intermediate area, then more slowly and, perhaps, imperceptibly for the larger islands. However, only the sigmoidal model predicts that richness increases slowly for the very small islands as well. Thus, if biogeographic surveys are limited to smaller islands (i.e., those too small to satisfy resource requirements of most species), species richness may appear to be independent of area. This phenomenon, termed the 'small island effect', has only occasionally been observed for some insular biotas (Niering, 1963; see also Dunn & Loehle, 1988; MacArthur & Wilson, 1967:30–33; Wiens, 1962; Whitehead & Jones, 1969; Woodroffe, 1986). It is likely to remain a rarely reported phenomenon as most biogeographic surveys are designed to include islands large enough to satisfy resource requirements of at least a few focal species. Given this, and the ease at which species–area curves can be estimated using power or semi-log models, it is unlikely that biogeographers will switch to a sigmoidal model. Again, these more conventional models typically provide a good approximation of the species–area relationship. However, where surveys encompass a relatively broad range in area and include many small islands, a sigmoidal model may prove superior to conventional models.

The inter-taxa and inter-archipelago scales

An additional tier can be added to the hierarchical model by noting how species-groups vary in ways affecting the fundamental biogeographic processes of immigration and extinction. Biogeographical studies often focus on groups composed of taxonomically and functionally similar species (e.g., amphibians, freshwater fish or passerine birds). Thus, focal groups differ in immigration abilities and resource demands and, more importantly, these differences are predictable. For example, birds and bats tend to be more vagile than non-volant mammals and amphibians. Focal groups comprised of endothermic, large, eusocial or carnivorous species should require more resources (larger islands) than groups composed of ectothermic, small, asocial or herbivorous species.

These differences among groups of species can be modelled as shifts in IDFs. That is, the prediction is that within an archipelago, IDFs should be shifted to the right (reduced slopes) for focal groups composed of more vagile species (Fig. 10.13(a)). Similarly, the prediction is a downward shift in IDFs (reduced intercepts) for faunal groups composed of species with relatively low resource demands (small, ectothermic, and/or herbivorous species; Fig. 10.13(b)).

Alternatively, species or community level patterns may be compared across archipelagoes instead of across taxonomic groups. To model the effects of differences among archipelagoes, we must first assess the relative productivity and filter severity of the different archipelagoes. Here, relative productivity refers to the ability of insular ecosystems to support populations of the focal species group (see applications of the species-energy theory to island bio-geography, Wright, 1983; Wylie & Currie, 1993). By filter severity the reference is to impedance to immigration, which should be associated with the nature of the isolating medium or matrix. For example, Aberg and his colleagues have shown that the distribution of the hazel grouse (*Bonasa*

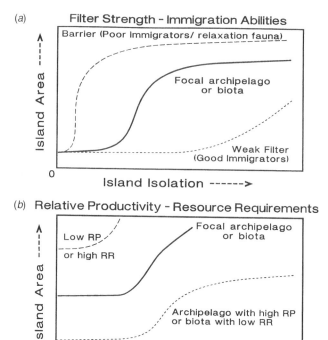

Fig. 10.13. Effects of differences in characteristics of faunal groups (e.g., birds vs. non-volant mammals) and differences among archipelagoes on distribution patterns of insular faunas. The curves depict insular distribution functions (IDFs) of representative species for a given faunal group. (*a*) IDFs should shift leftward (increased slope) in archipelagoes with relatively strong (more isolating) filters or, equivalently, for faunal groups composed of relatively poor immigrators. On the other hand, IDFs should shift to the right in archipelagoes with relatively weak filters or for faunal groups composed of relatively good immigrators. (*b*) IDFs should shift upward (higher intercept) in archipelagoes with higher relative productivity (RP) or, equivalently, for faunal groups composed of species with relatively high resource requirements (RR; e.g., endotherms vs. ectotherms, carnivores versus herbivores). Relative productivity refers to the ability of insular ecosystems to support populations of the focal species, and should be a function of primary productivity and trophic characteristics of the faunal group.

bonasia) among fragmented forest of Sweden is strongly influenced by characteristics of isolating habitats (Aberg *et al.*, 1995). In landscapes dominated by agricultural fields, populations of these birds evidence a strong and highly significant effect of isolation. In contrast, the effect of geographic isolation on grouse populations surrounded by more suitable second growth forests was only marginally significant (Fig. 10.14).

In summary, the prediction is shifts in IDFs among archipelagoes equivalent to those associated with differences among focal groups; downward shifts for archipelagoes with higher relative productivities (equivalent to species groups with lower resource demands), and leftward shifts in IDFs for archipelagoes with more severe immigration filters (or species groups with lower vagilities; Fig. 10.13). Where the isolating matrix is so severe that it forms a barrier to immigration for most species, we then expect a relaxation fauna, with most species limited to the relatively large islands and little evidence of isolation effects (see Fig. 10.7). Again, these predictions point to some potentially fruitful directions for future research, which would focus on the key linkages between community and archipelago (or inter-taxon) scales. For example, do interarchipelago differences in productivity and filter severity covary in nature? That is, do archipelagoes composed of islands with relatively high productivity also tend to be situated in regions with less isolating filters? Do faunal groups composed of relatively good immigrators (e.g., birds versus amphibians; bats vs. non-volant mammals) also tend to be composed of species with high resource demands? Research addressing these questions should provide some important insights into the forces influencing biological variation over a broad range of geographical scales. Such research, however, will be especially challenging as it requires fundamental changes in the nature and scale of ecological research – changes that call for increased emphasis on processes operating at large spatial scales (see Edwards *et al.*, 1994; Brown, 1995).

Species interactions and realized insular distributions

The hierarchical model presented above accounts for a diversity of biogeographic patterns across a range of spatial and ecological levels. There are other interesting ecological patterns, however, that cannot be accounted for by the present version of the model. Checkerboards, or more generally, exclusive distributions of species has interested biogeographers and community ecologists for at least the past three decades (e.g., see Diamond, 1975; chapters in this volume). There remains much controversy over the effects of interspecific interactions at relatively coarse, geographic ranges. At a local or within-island scale, however, many studies indicate that distributions among habitats are strongly influenced by interspecific interactions (e.g., see Connell, 1961; Paine, 1966; Grant, 1972; Menge, 1972; Lubchenco, 1978; Werner, 1979; Munger & Brown, 1981; Schoener, 1983; Brown & Gibson, 1983; Lomolino, 1984). The question remains whether interspecific interactions also influence species distributions at coarser scales (e.g., distributions of insular

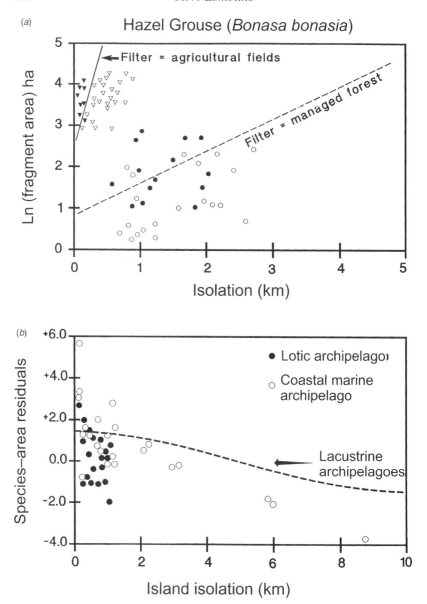

Fig. 10.14. (*a*) Effects of differences in immigration filters on insular distribution functions (IDFs) of Hazel grouse (*Bonasa bonasia*) in fragmented forest of Sweden (Aberg *et al.*, 1995). As predicted by the model of Fig. 10.13, the IDF for this species shifted leftward (increased slope) when forests were surrounded by less hospitable habitats (agricultural fields [triangles] vs. second growth, managed forest (circles); presence and absence is indicated by filled and unfilled symbols, respectively). (*b*) Differences in filter strength should be evidenced by shifts in species richness curves. As pre-

populations among islands; see Simberloff & Connor, 1981; Connor & Simberloff, 1979).

While the species-based model in its present form does not include the potential influence of interspecific interactions, it does describe the biogeographical space (ranges of isolation and area) over which those interactions may occur. Borrowing from niche theory, this biogeographic space can be referred to as the fundamental (vs. realized) distribution. If interspecific interactions are intense enough to reduce persistence of insular populations then, at least for some species, realized insular distributions may be substantially less than their fundamental distributions.

For now, this exercise can be simplified by focusing on just two hypothetical species. To predict the effects of interspecific interactions it first is necessary to define the fundamental insular distributions of these species. Specifically, how do these species differ in vagilities and resource requirements? Second, it is necessary to consider the nature of the interaction; predation, amensalism, commensalism, parasitism, mutualism or competition? In the latter case, is the competitive interaction symmetrical, or does one species tend to dominate in interference or exploitative interactions?

Rather than consider all possible permutations of IDFs and interspecific interactions, this exercise will be further simplified by limiting it to the potential effects of asymmetrical competition. Figure 10.15 illustrates the IDFs of two hypothetical species with asymmetrical competition. Assume that these interactions are strong enough to result in competitive exclusion from islands where the fundamental IDFs overlap. Also assume that species B is a better immigrator than A. Regardless of which species dominates, we expect exclusive distributions on all but the largest, mainland-like islands, which should afford ecological refugia for inferior competitors. If B dominates, then the competitively inferior species (A) should be limited to the very small, near islands and the very large islands. On the other hand, if A is the dominant competitor, then species B, the more vagile species, should be limited to the distant islands (and again, the very large, mainland-like islands as well).

dicted by the species-based model, species richness of non-volant mammals on islands of the Great Lakes Region of North America declines more rapidly for islands in archipelagoes with more severe (less hospitable) immigration filters (for these mammals which either swim or colonize islands by traveling across ice, filter severity is highest for the lotic archipelagos, intermediate for the coastal marine archipelago, and lowest for the lacustrine archipelago, which is characterized by relatively weak currents and more prolonged periods of ice-cover; see Lomolino, 1994).

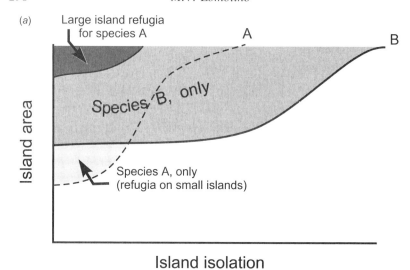

Fig. 10.15. Effects of interspecific interactions on realized insular distribution functions of two hypothetical species. (*a*) Here the better immigrator, species B, tends to dominate interspecific interactions, but requires larger islands to maintain its populations. As a result, species B is widely distributed across the archipelago, while populations of species A (the ecologically subordinate and less vagile species) should be restricted to relatively small islands (i.e., those too small to maintain populations of species B) or the relatively large islands (i.e., those large enough so that they afford ecological refugia for species A). (*b*) Here the better immigrator, species B, is ecologically subordinant (i.e., A is the dominant competitor or predator) and requires larger islands to maintain its populations. As a result, species A (the relatively poor immigrator) is restricted to the less-isolated islands, while populations of species B

Case studies: effects of competition and predation on IDFs

Although examination of the archipelago scale effects of interspecific inter-actions has only just begun, there is some evidence that isolated habitats often serve as ecological refugia. In his recent analysis of distributions of fresh-water fish among pools of the Red River in Oklahoma, Taylor (1997) reports that distributions of small mouth bass (*Micropterus dolomieu*) and creek chub (*Semotilus atromaculatus*) appear to be strongly affected by asymmetrical competition (with bass being competitively dominant). The distributions of these two species are nearly exclusive, with the exception of one co-occurrence in a relatively large, near pool (Fig. 10.16).

The effects of predation on insular distributions may be similar to that of intense asymmetrical competition, and here again there is some empirical evidence that distant islands serve as refugia. Meadow voles and deermice on islands of Lakes Huron and Michigan are largely restricted to the more isolated islands, apparently as a result of predation by shrews, which are smaller and less vagile than their vertebrate prey (Fig. 10.17; see also Lomolino, 1984, 1989). These species also co-occur on near islands if they are relatively large. Experimental introductions of shrews (predators on young mice) onto islands of another archipelago of the Great Lakes Region demon-strated that predation by short-tailed shrews (*Blarina brevicauda*) may increase extinction rates of at least some insular rodents (Lomolino, 1984). Taking this a step further, after these prey are extirpated, predators may go extinct, especially on small islands where alternative resources are scarce. Thus, communities on small islands within the immigration range of shrews may undergo cycles of colonization by prey, colonization by predator, extinc-tion of prey followed by extinction of predator and reinitiation of the cycle. Unless there is some factor causing synchronization of this ecological cycle among islands, a temporally dynamic checkerboard pattern may be expected on islands of intermediate isolation – each in different phases of the cycle (e.g., see Fig. 10.17, islands from 0.5 to 5 km isolation). More generally, where fundamental ranges overlap, competition and predation may alter assembly sequences of insular biotas, increasing the number of 'holes' or departures from perfect nestedness.

As a final case study, consider Jared Diamond's (1975) seminal paper on assembly rules and bird communities of New Guinea satellite islands. The

should be restricted to the more isolated islands (i.e., those lacking A) or the rela-tively large islands (i.e., those large enough so that they afford ecological refugia for species B).

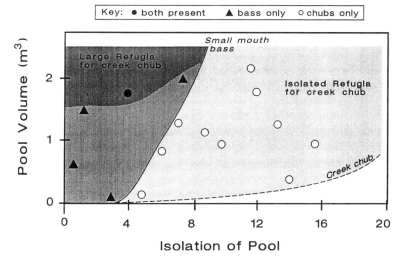

Isolation of Pool

Fig. 10.16. Distribution functions of two species of freshwater fish among pools of the Red River in Oklahoma (Taylor, 1997). Small mouth bass (*Micropterus dolomieu*) appear to dominate in interspecific interactions with the creek chub (*Semotilus atromaculatus*), but appear more limited in its dispersal abilities. As a result, the distributions of these species is nearly exclusive, with populations of small mouth bass being restricted to the less-isolated pools, while all but one population of the creek chub are restricted to the most-isolated pools. The one exception is the occurrence of both species on a relatively large, less-isolated pool (i.e., one large and near enough to maintain populations of creek chubs despite the effects of competition and possible predation from small mouth bass).

purpose of that paper was to develop the concept of alternative stable communities, which as Diamond observed, 'was brilliantly explored by Robert MacArthur (1972) in *Geographical Ecology*'. Building on MacArthur's theoretical discussion, Diamond developed a set of empirical procedures, including incidence functions and analysis of checkerboard patterns, to investigate how community structure is influenced by characteristics of the islands, by species interactions or by chance (see also Connor & Simberloff, 1979, 1983; Simberloff & Connor, 1981; Diamond & Gilpin, 1982; Gilpin & Diamond, 1982, 1984). The purpose here is to complement Diamond's work by providing an explicit spatial reference for checkerboard patterns. This may be accomplished by plotting distributions of guild members across biogeographical space (i.e., area–isolation plots).

There are at least two, fundamentally different checkerboard patterns. On the one hand, if distributions of ecologically similar species were influenced by intense competition, but not by island characteristics (e.g., area and

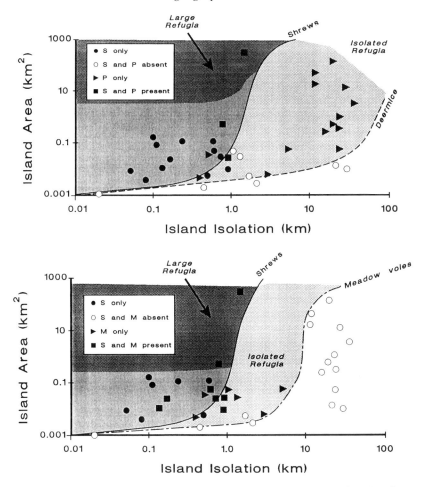

Fig. 10.17. Insular distribution patterns of non-volant mammals on islands of two lacustrine archipelagoes of Lake Huron and Lake Michigan (S = shrews [*Blarina brevicauda* and *Sorex cinereus*], P = deermice (*Peromyscus maniculatus*) and M = meadow voles (*Microtus pennsylvanicus*)). Because shrews tend to be relatively poor immigrators, their populations are generally restricted to the less-isolated islands. By preying on meadow voles and deermice (especially juveniles), shrews may be able to exclude populations of these rodents from relatively small, less-isolated islands. On the other hand, the ecologically subordinant but more vagile deermice and meadow voles may find refugia on the more isolated or larger islands.

isolation), then we would expect species to be exclusively, but uniformly distributed across biogeographical space (Fig. 10.18(*a*)). On the other hand, if distributions are influenced by competition as well as by differences in vagilities and resource requirements, then a complex set of insular distributions is

(a)

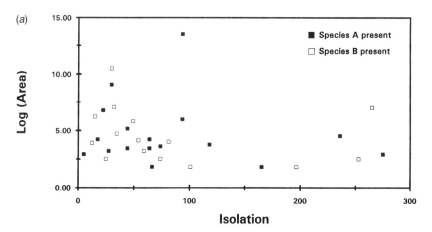

(b) **Fruit-pigeon Guild**

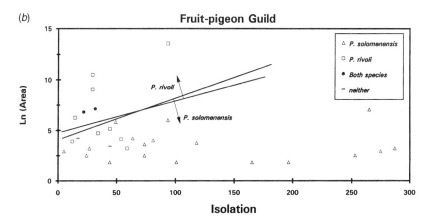

(c) **Cuckoo-dove Guild**

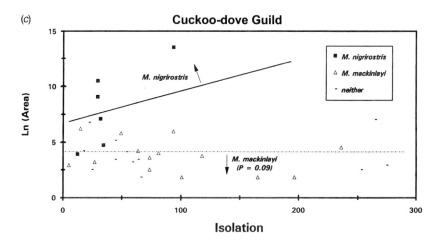

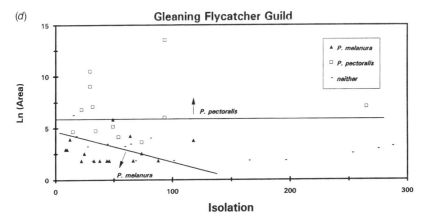

Fig. 10.18. (*a*) A spatial checkboard distribution of two hypothetical species. In this case, both species are uniformly distributed across the archipelago (i.e., their distributions are independent of isolation and area), but they never occupy the same island. This pattern would be expected where competition is intense, but species are essentially equivalent with respect to immigration abilities and resource requirements. (*b*) *thru* (*d*) In contrast to the above, hypothetical pattern, analysis of insular distribution patterns for three of Diamond's (1975) avian guilds reveals that exclusive distribuitons are often achieved by species segregating their realized distributions over different ranges of isolation and area (Perault and Lomolino, unpublished report, 1996; see also Figs. 10.16 and 10.17). Regression analysis was used to estimate linear insular distribution functions of these species (see Lomolino, 1986). Insular distributions were significantly ($P < 0.05$) associated with island area and isolation for *Ptillinopus rivoli, P. solomenensis, Macropygia nigrirostris*, and *Pachcephala pectoralis*. Insular distributions were significantly associated with just area for *P. melanura*, and marginally ($P < 0.09$) associated with area for *M. mackinlayi*.

expected similar to those illustrated in Fig. 10.18(*b*) to (*d*). In such cases, competitors may be restricted to only a portion of their fundamental distribution (i.e., biogeographic refugia).

Examination of insular distributions of three of Diamond's avian guilds has taken place: the fruit-pigeon, cuckoo-dove and gleaning flycatcher guilds (D. Perault and M.V. Lomolino, unpublished data). At first glance, distributions of members of the fruit-pigeon guild (*Ptilinopus solomenensis* and *P. rivoli*; Fig. 10.18(*b*) may seem consistent with non-interactive, species-based model. However, as the arrows in Fig. 10.18(*b*) indicate, the distribution of *P. rivoli* is biased toward large, near islands, while that of *P. solomenensis* is biased toward small, distant islands. For the cuckoo-dove guild (Fig. 10.18(*c*)), *Macropygia nigrirostris* occurs on near and large islands, while *M. mackinlayi* tends to be restricted to smaller islands. Finally, *Pachycephala pectoralis* of the gleaning flycatcher guild (Fig. 10.18(*d*)) occurs on the largest

M.V. Lomolino

islands while *P. melanura* is generally restricted to near, but small islands. Therefore, at least for these guilds, insular distributions appear to be influenced by the combined effects of interspecific interactions and interspecific differences in resource requirements and vagilities.

Including the effects of interspecific interactions will certainly add a substantial degree of complexity to the model, but it seems to be a more accurate account of the complexity of nature. Moreover, this exercise identifies another potentially illuminating area of future research. Do vagilities or resource demands covary with ecological dominance? In the examples discussed above, immigration abilities appeared to be lower for the competitively dominant species (see also MacArthur, 1972; Tilman & Wedin, 1991). Is this a general pattern in nature? Given that there may emerge some pattern of covariation in vagilities and 'interspecific dominance' among species, does this pattern vary with the nature of interspecific interactions (e.g., negative covariation for competitors, positive covariation for predators)? Research directed at these questions may provide illuminating linkages between the fields of biogeography and community ecology.

Summary of the hierarchical model

The species-based, hierarchical model developed in the previous sections can account for patterns spanning three biogeographical scales. These include the following.

(a) *Species level patterns:* distribution of a focal species as a function of area and isolation (i.e., insular distribution functions).

(b) *Archipelago (inter-community) level patterns:* differences in community structure (especially species richness and species composition) among islands.

(c) *Inter-archipelago patterns:* differences in intercommunity patterns (e.g., differences in species–area relationships, species–isolation relationships and nestedness) among archipelagoes. Alternatively, these patterns may also be compared among taxa (e.g., comparing species–isolation relationships of bats vs. birds).

Thus, the model can be used to study a diversity of ecological patterns of isolated systems. The model, however, remains relatively simple. Because the empirical cases discussed here are limited to animals, the model remains untested for plants and other taxa. In addition, the model is limited to events occurring in ecological time and it is entirely deterministic, focusing on characteristics of species or islands that affect immigration and extinction.

MacArthur and Wilson's equilibrium theory also assumed immigration and extinction were deterministic with respect to island characteristics. On the other hand, their equilibrium theory assumed that these processes were stochastic with respect to species characteristics. That is, all species were assumed equivalent. For this reason, the equilibrium theory remains neutral with respect to patterns in species composition.

Linkages: species to archipelago levels

The true challenge of the hierarchical model presented here lies in identifying and understanding the key linkages among biogeographical scales. At a fundamental level, the linkages among components of insular diversity, in ecological time, are immigration and extinction. Linkages among the three biogeographic levels listed above will therefore concern the scale-apppropriate characteristics of species and ecosystems that influence these fundamental biogeographic processes. Four such linkages are requisite for moving from the species to archipelago levels.

(a) *The frequency distribution of resource requirements and immigration abilities among species;*
(b) *Patterns of covariation of resource requirements and immigration abilities among species;*
(c) *Effects of interspecific interactions on establishment and persistence of insular populations;*
(d) *Correlations between interspecific dominance and immigration abilities or resource requirements.*

Linkages: archipelago to inter-archipelago

Moving from within to between archipelago levels involves at least the following linkages.

(a) *Differences and possible covariation in strength of immigration filters and productivity among archipelagoes;*
(b) *Differences and possible covariation in representative vagilities and resource requirements among species pools.*

Conclusions

The requisite information to explore many of these linkages seems to be available. On the other hand, for those linkages where adequate data is lacking, areas of potentially important research have now been identified. This information may enable comparisons among biotas at regional to global scales, scales where biogeographers are seldom blessed with the luxury of having identical ecosystems and species pools. By acknowledging the importance of these linkages, some otherwise perplexing ambiguities may be resolved. Rather than ignoring differences among archipelagoes and faunal groups, our questions may actually focus on these differences. In a characteristically intriguing paper, Janzen (1967) asked whether 'mountain passes are higher in the tropics', arguing that, in comparison to species inhabiting temperate ecosystems, those of the tropics have lower vagilities, or at least lower propensities for immigration across mountains. If this is a genuine tendency, how should patterns in richness and nestedness vary between faunal groups of tropical and temperate regions? To adequately address such questions, differences among archipelagoes as well as differences among regional biotas need to be studied.

As biogeography and ecology continue to develop as scientific disciplines, the scope of our studies needs to be broadened. The interface between biogeography and local-scale ecology is the domain of an exciting new discipline, macroecology, which focuses on variation of population and community level parameters at geographical scales (Brown, 1995). The diverse, multi-scale nature of this new discipline, and biodiversity research in general, calls for hierarchical approaches and others that can capitalize on natural experiments, especially those 'designed' at landscape to regional scales (see Edwards *et al.*, 1994; Brown, 1995).

Although not an equilibrium theory, the model presented here, like much of metapopulation theory (see Gilpin & Hanski, 1991), was derived from a fundamental premise of MacArthur and Wilson's theory – that insular community structure is dynamic and results from the combined effects of immigration and extinction. Unlike the equilibrium theory which assumes species are equivalent, the current model assumes that many patterns in insular community structure result from, not despite, differences among species. The hierarchical model, while encompassing three biogeographic scales, distills down to a relatively simple, deterministic model: one focusing on the species and ecosystem characteristics affecting immigration and extinction. Patterns in insular community structure derive from the non-random variation among species and islands that affect these fundamental biogeographic processes. It is not being denied that immigration and extinction also are influenced by stochastic factors, but such factors cannot account for the *patterns* addressed here.

Finally, note that my focus has been limited to just a subset of the earth's ecosystems – islands and other ecosystems distributed as isolated patches in 'seas' of other habitats. On the other hand, studies of these systems have had a disproportionately important influence on the development of ecology, evolutionary biology and conservation biology. Hopefully, the model presented here, in combination with insights from metapopulation theory and other disciplines will contribute to our ability to understand the forces structuring isolated communities and, ultimately, conserve endangered biotas, many of these persisting on true islands or ever shrinking fragments of native habitats.

Acknowledgments

This work was partially supported by a grant from the National Science Foundation (DEB-9322699). David Perault, James H. Brown, Russell Davis, Rob Channell, Gregory A. Smith, and Evan Weiher provided useful comments on an earlier version of this manuscript.

References

Aberg, J., Jansson, G. Swenson, J.E, & Anglestam, P. (1995). The effect of matrix on the occurrence of hazel grouse (*Bonasa bonasia*) in isolated habitat fragments. *Oecologia* 103: 265–269.

Alatalo, R.V. (1982). Bird species distributions in the Galapagos and other archipelagoes: competition or chance? *Ecology* 63: 881–887.

Allen, T.F.H. (1987). Hierarchical complexity in ecology: a non-euclidean conception of the data space. *Vegetatio* 69: 17–26.

Auerbach, M.J. & Schmida, A. (1987). Spatial scale and the determinants of plant species richness. *Trends in Ecology and Evolution* 2: 238–242.

Brown, J.H. (1978). The theory of insular biogeography and the distribution of boreal birds and mammals. *Great Basin Naturalist Mem.* 2: 209–227.

Brown, J.H. (1995). *Macroecology*. Chicago, IL: University of Chicago Press.

Brown, J.H. & Gibson, A.C. (1983). *Biogeography*. Mosby Company, St Louis.

Brown, J.H. & Kodric-Brown, A. (1977). Turnover rates in island biogeography: effect of immigration on extinction. *Ecology* 58: 445–449.

Brown, J.H. & Lomolino, M.V. (1989). On the nature of scientific revolutions: independent discovery of the equilibrium theory of island biogeography. *Ecology* 70: 1954–1957.

Brown, J.H. & Nicolleto, P.F. (1991). Spatial scaling of species composition: body masses of North American land mammals. *American Naturalist* 138: 1478–1512.

Calder, W.A., III. (1984). *Size, Function and Life History*. Cambridge, MA: Harvard University Press.

Connell, J.H. (1961). The influence of interspecific competition and other factors on the distribution of the barnacle *Chthalamus stellatus*. *Ecology* 42: 710–723.

Connor, E.H. & Simberloff, D. (1979). The assembly of species communities: chance or competition? *Ecology* 60: 1132–1140.

Connor, E.H. & Simberloff, D. (1983). Interspecific competition and species co-occurrence patterns on islands: null models and the evaluation of evidence. *Oikos* 41: 455–465.

Darlington, P.J. (1957). *Zoogeography: The Geographical Distribution of Animals.* New York: John Wiley.

Dexter, R.W. (1978). Some historical notes on Louis Agassiz's lecutres on zoogeography. *Journal of Biogeography* 5: 207–209.

Diamond, J.M. (1975). Assembly of species communities. In *Ecology and Evolution of Communities*, ed. M.L. Cody & J.M. Diamond, pp. 342–444. Cambridge, MA: Harvard University Press.

Diamond, J.M. & Gilpin, M. (1982). Examination of the 'null' model of Connor and Simberloff for species co-occurrences on islands. *Oecologia* 52: 64–74.

Dunn, C.P. & Loehle, C. (1988). Species–area parameter estimation: testing the null model of lack of relationships. *Journal of Biogeography* 15: 721–728.

Edwards, P.J., May, R.M. & Webb N.R. (1994). *Large-scale Ecology and Conservation Biology*. London: Blackwell Scientific Publications.

Gilbert, F.S. (1980). The equilibrium theory of island biogeography: fact or fiction? *Journal of Biogeography* 7: 209–235.

Gilpin, M. & Diamond, J.M. (1982). Factors contributing to non-radomness in species co-occurrence on islands. Oecologia 52: 75–84.

Gilpin, M. & Diamond, J.M. (1984). Are species co-occurrences on islands non-random, and are null hypotheses useful in community ecology? In *Ecological Communities: Conceptual Issues and the Evidence*. ed. D.R. Strong, D. Simberloff, L.G. Abele, & A.B. Thistle, pp. 297–315. NJ: Princeton University Press.

Gilpin, M. & Hanski, I. (1991). *Metapopulation Dynamics: Empirical and Theoretical Investigations*. London: Academic Press.

Goel, N.S. & Richter-Dynn, N. (1974). *Stochastic Models in Biology*. NJ, USA: Academic Press.

Gould, S.J. (1979). An allometric interpretation of species-area curves: the meaning of the coefficient. *American Naturalist* 114: 335–343.

Grant, P.R. (1972). Interspecific competition among rodents. *Annual Review of Ecology and Systematics* 3: 79–106.

Harestad, E.L. & Bunnell, E.L. (1979). Home range and body weight – a re-evaluation. *Ecology* 60: 389–402.

Hengeveld, R. (1987). Scales of variation: their distinction and ecological importance. *Annales Zoologica Fennici* 24: 195–202.

Hope, J.H. (1973). Mammals of Bass Strat Islands. *Proceedings of the Royal Society of Victoria* 85: 163–195.

Janzen, D.H. (1967). Why mountain passes are higher in the tropics. *American Naturalist* 101: 233–249.

Jurgens, K.D. & Prothero, J. (1991). Lifetime energy budgets in mammals and birds. *Comparative Biochemistry and Physiology* 100A: 703–709.

Lindstedt, S.L., Miller, B.J. & Buskirk, S.W. (1986). Home range, time, and body size in mammals. *Ecology* 67: 413–418.

Lomolino, M.V. (1984). Immigrant selection, predatory exclusion and the distributions of *Microtus pennsylvanicus* and *Blarina brevicauda* on islands. *American Naturalist* 123: 468–483.

Lomolino, M.V. (1986). Mammalian community structure on islands: immigration, extinction and interactive effects. *Biological Journal of the Linnoean Society* 28: 1–121, and in *Island Biogeography of Mammals*, Chapter 1. NY: Academic Press.

Lomolino, M.V. (1988). Winter immigration abilities and insular community structure of mammals in temperate archipelagoes. In *Biogeography of the Island Region of Western Lake Erie.* Ohio University Press.

Lomolino, M.V. (1989). Bioenergetics of cross-ice movements of *Microtus pennsylvanicus, Peromyscus leucopus* and *Blarina brevicauda. Holarctic Ecology* 12: 213–218.

Lomolino, M.V. (1993). Winter filtering, immigrant selection and species composition of insular mammals of Lake Huron. *Ecography* 16: 24–30.

Lomolino, M.V. (1994). Species richness patterns of mammals inhabiting nearshore archipelagoes: area, isolation, and immigration filters. *Journal of Mammalology* 75: 39–49.

Lomolino, M.V. & Davis, R. (1997). Biogeographic scale and biodiversity of mountain forest mammals of western North America. *Global Ecology and Biogeography Letters* 6: 57–76.

Lomolino, M.V., Brown, J.H. & Davis, R. (1989). Island biogeography of montane forest mammals in the American Southwest. *Ecology* 70: 180–194.

Lubchenco, J. (1978). Plant species diversity in a marine intertidal community: importance of herbivore food preference and algal competitive abilities. *American Naturalist* 112: 23–39.

MacArthur, R.H. (1972). *Geographical Ecology: Patterns in the Distribution of Species.* NJ, USA: Princeton University Press.

MacArthur, R.H. & Wilson, E.O. (1963). An equilibrium theory of insular zoogeography. *Evolution* 17: 373–387.

MacArthur, R.H. & Wilson, E.O. (1967). *The Theory of Island Biogeography.* NJ: Princeton University Press.

Manly, B.F.J. (1991). *Randomization and Monte-Carlo Methods in Biology.* NY: Chapman and Hall.

Menge, B.A. (1972). Competition for food between two intertidal starfish species and its effect on body size and feeding. *Ecology* 53: 635–644.

Michaud, J.M. (1995). Biogeography of the bog copper butterfly (*Lycaena epixanthe*) in southern Rhode Island Peatlands: a metapopulation perspecitve. MS thesis, Department of Natural Resources, University of Rhode Island, USA.

Munger, J.C. & Brown, J.H. (1981). Competition in desert rodents: an experiment with semipermeable exclosures. *Science* 211: 510–512.

Munroe, E.G. (1948). The geographical distribution of butterflies in the West Indies. PhD dissertation, Cornell University, New York, USA.

Munroe, E.G. (1953). The size of island faunas. In *Proceedings of the 7th Pacific Science Congress of the Pacific Science Association.* volume IV, Zoology. pp. 52–53. Auckland, New Zealand: Whitcome and Tombs.

Niering, W.A. (1963). Terrestrial ecology of Kapingamarangi Atoll, Caroline Islands. *Ecological Monographs* 33: 131–160.

O'Neill, R.V., DeAngelis D.L., Waide, J.B. & Allen, T.F.H. (1986). *A Hierarchical Concept of Ecosystems.* Princeton, NJ: Princeton University Press. 254 pp.

Paine, R.T. (1966). Intertidal community structure: experimental studies between a dominant competitor and its principal predator. *Oecologia* 15: 93–120.

Patterson, B.D. & Atmar, W. (1986). Nested subsets and the structure of insular mammalian faunas and archipelagoes. In *Island Biogeography of Mammals,* ed. L. Heaney & B.D. Patterson, pp. 65–82. *Biological Journal of the Linnaean Society* 28 (1&2) and London: Academic Press.

Patterson, B.D. & Brown, J.H. (1991). Regionally nested patterns of species composition in granivorous rodent assemblages. *Journal of Biogeography* 18, 395–402.

Peters, R.H. (1983). *The Ecological Implications of Body Size*. Cambridge: Cambridge University Press.

Pielou, E.C. (1979). *Biogeography*. NY: John Wiley.

Preston, F.W. (1962). The canonical distribution of commonness and rarity: Part I. *Ecology* 43: 185–215.

Rosenzweig, M.L. (1995). *Species – Diversity in Space and Time*. NY: Cambridge University Press.

Sauer, J. (1969). Oceanic islands and biogeographic theory: a review. *Geographical Reviews* 59: 582–593.

Schmida, A. & Wilson M.V. (1985). Biological determinants of species diversity. *Journal of Biogeography* 12: 1–20.

Schoener, T.W. (1968). Sizes of feeding territories among birds. *Ecology* 49: 123–141.

Schoener, T.W. (1983). Field experiments on interspecific competition. *American Naturalist* 122: 240–285.

Schoener, T.W. & Schoener, A. (1983). Distribution of vertebrates on some very small islands. I. Occurrence sequences of individual species. *Journal of Animal Ecology* 52: 209–235.

Simberloff, D. & Connor, E.F. (1981). Missing species combinations. *American Naturalist* 118: 215–239.

Simberloff, D & Martin, J. (1991). Nestedness of insular avifaunas: simple summary statistics masking complex species patterns. *Ornis Fennica* 68: 178–192.

Simpson, G.G. (1940). Mammals and landbridges. *Journal of the Washington Academy of Science* 30: 137–163.

Taylor, C.M. (1997). Fish species richness and incidence patterns in isolated and connected stream pools: effects of pool volume and spatial position. *Oecologia* (in press).

Tilman, D. & Wedin, D. (1991). Plant traits and resource reduction for five grasses growing on a nitrogen gradient. *Ecology* 72: 685–700.

Vagvolgyi, J. (1975). Body size, aerial dispersal, and origin of Pacific Island snail fauna. *Systematic Zoology* 24: 465–488.

Werner, E.E. (1979). Foraging efficiency and habitat switching in competing sunfishes. *Ecology* 60: 256–264.

Whitehead, D.R. & Jones, C.E. (1969). Small islands and the equilibrium theory of insular biogeography. *Evolution* 23: 171–179.

Wiens, H.J. (1962). *Atoll Environment and Ecology*. New Haven, CT: Yale University Press.

Wiens, J.A., Addicott, J.F., Case, T.J. & Diamond, J.M. (1986). The importance of spatial and temporal scale in ecological investigations. In *Community Ecology*, ed. J.M. Diamond & T.J. Case, pp. 145–153. NY: Harper and Row.

Woodroffe, C.D. (1986). Vascular plant species-area relationship on Nui Atoll, Tuvalu, Central Pacific: a reassessment of the small island effect. *Australian Journal of Ecology* 11: 21–31.

Wright, D.H. (1983). Species energy theory: an extension of species-area theory. *Oikos* 41: 496–506.

Wylie, J.L. & Currie, D.J. (1993). Species-energy theory and patterns of species richness: I. Patterns of bird, angiosperm, and mammals species richness on islands. *Biological Conservation* 63: 137–144.

11

Interaction of physical and biological processes in the assembly of stream fish communities

Elizabeth M. Strange and Theodore C. Foin

Introduction

The concept of assembly rules originally was developed by Diamond (1975) to describe regular patterns of species occurrence on islands, and has since been generalized to include all processes influencing community development, the emphasis on colonization and extinction having lessened somewhat (Drake, 1990a; Nee, 1990). The contemporary meaning of community assembly also de-emphasizes the term 'rule' in favor of less-fixed combinations of interactions and outcomes that are known to occur within different communities (Post & Pimm, 1983; Wilbur & Alford, 1985; Gilpin *et al.* 1986; Robinson & Dickerson, 1987; Robinson & Edgemon, 1988; Barkai & McQuaid, 1988; Drake, 1988, 1990b, 1991; Drake *et al.*, 1993; Case, 1991; Keddy, 1992; Weiher & Keddy, 1995a, b). Community assembly now encompasses the ecology and short-term, observable evolution of the community from initial colonization to species saturation, and logically includes a number of topics that are referred to by other names, such as trophic cascades (Carpenter & Kitchell, 1993), top-down vs. bottom-up regulation of trophic webs (Matson & Hunter, 1992), and the regulation of biodiversity (Rosenzweig, 1995). The scope of the processes involved and the applicability of assembly rules to all communities makes this one of the central problems in ecology.

Although the physical environment is important in a number of ways, there has been more focus upon biological interactions in assembly processes. The traditional view is that environmental gradients act principally to filter some species out of the pool of potential immigrants (Drake, 1990b). In most natural communities, however, physical conditions may do much more than act as a coarse filter of unsuitable species; they may exert subtle, quantitative effects that play a continuing role in shaping community composition and dynamics (Chesson, 1986).

311

There is a need to incorporate physical variation into the picture of community assembly to account for interactions between physical variation and biotic processes within specific systems. Because such interactions will depend on both the nature and rates of interaction, if general rules of community assembly are to emerge, they will likely do so by comparative analysis of a number of different kinds of communities. Stream fish communities are particularly appropriate for the study of how abiotic factors influence community assembly because they have been studied intensively and their biological interactions defined, and because flow conditions are known to be critically important and are readily quantified. For example, native topminnows in the American southwest can survive predation by introduced mosquito fish (*Gambusia* spp.) if floods are frequent enough to reduce populations of *Gambusia* periodically. In the absence of floods, mosquito fish typically eliminate topminnows within 1–3 years. *Gambusia* is known to be poorly adapted to flood conditions, and observations in streams suggest that floods are a key factor in coexistence (Meffe, 1984, 1985; Minckley & Meffe, 1987).

In this chapter, models of interactions are developed for stream fishes in what is known as the Trout Zone Assemblage (Moyle, 1976), typical of the headwaters of cold montane streams throughout North America, and results are used to interpret the development of other communities. In northern California streams, high scouring discharges can occur in winter and spring as a result of rainfall and snowmelt, and cause high mortality of young-of-year (Erman *et al.*, 1988). This contributes to differential recruitment success among species that spawn in different seasons, and as a consequence hydrology exerts a strong control over community structure and organization in these systems (Seegrist & Gard, 1972; Gard & Flittner, 1974; Strange *et al.*, 1992; Strange, 1995).

Analysis of fluctuations in fish abundance over a 15-year period in Martis Creek, a small coldwater stream in the Truckee River drainage of northern California, shows that the impact of invasion by the piscivorous European brown trout (*Salmo trutta*) on native fishes is mediated by the timing of floods and droughts (Strange *et al.*, 1992; Strange, 1995). The winter-spawning brown trout is favored in years when winter floods are absent but spring floods reduce recruitment of native fishes, all of which are spring spawners. On the other hand, when floods occur in winter but not in spring, the spring-spawning native-species are favored and the winter-spawning brown trout shows poor recruitment. During times of drought, relative abundances shift in favor of native species that show greater physiological adaptation to low flow conditions than brown trout, which is a native of Europe and is presumably not adapted to drought conditions characteristic of northern California.

This shifting pattern of relative abundances is well known, but is almost impossible to investigate experimentally. In order to examine the influence of varying flows on relative abundance, to determine what conditions would cause local extinction, and to define conditions which would favor establishment of invading species, an age-structured population model was developed for the fish community of these types of streams. The model was used to set up specific strengths and types of interactions between species and, through variation in inputs, investigated how stream flow can drive variation in stream fish densities and assemblage structure.

Assembly was modeled by introducing different stream fish types, one at a time, under a variety of physical and biological conditions. Although each species is represented by a Leslie matrix formulation, the model has two important additional features. First, the community is represented by linking the individual matrices through the survival vectors rather than through a community matrix. Second, the linkage of survival vectors is explicitly related to flow regimes, competitive interactions, and predation. By expressing species survival rates as a function of density, greater biological realism is achieved, and the model is capable of a range of dynamical behavior not possible with the linear formulations of the community matrix. Using an experimental format that permitted variation of single factors and realistic combinations, we were able to examine how flow-driven variations in stream fish densities combine with ongoing competition and predation in the assembly of stream fish assemblage structure. Validation of the model was accomplished by comparing our predictions with results of field studies of stream fish assemblages in two coldwater streams in the Sierra Nevada region of northern California.

Together, model analysis and field observations allowed us to analyze how stream fish assembly dynamics are influenced by interactions between physical and biological conditions and to determine what general assembly rules might be operating. Our findings also suggest some general conclusions regarding the interactive roles of physical variation and biotic processes in the assembly of communities in other physically variable environments.

The simulation model
Stream fish life history types and species

In order to maximize the generality of the model, a system was established using four common stream fish life history types. These combine theoretical studies of fish life histories (Winemiller, 1992; Winemiller & Rose, 1992) and demographic properties of fish species typical of northern California

Table 11.1. *Summary of demographic characteristics of model life history types*

Life history type	Demographic characteristics
Catostomid	matures age 4 $r = 0.2475$ under 25 cfs maximum annual fecundity = 6000 eggs mean generation time = 6.2 years maximum age = 7
Cyprinid	matures age 1 $r = 0.8082$ under 25 cfs maximum annual fecundity = 400 eggs mean generation time = 2.15 years maximum age = 2
Cottid	matures age 1 $r = 0.6279$ under 25 cfs maximum annual fecundity = 350 eggs mean generation time = 2.3 years maximum age = 4
Salmonid	matures age 2 $r = 0.4109$ under 25 cfs maximum annual fecundity = 1600 eggs mean generation time = 3.7 years maximum age = 6

coldwater streams (Moyle, 1976). The four life history types can be identified with real species: (a) a salmonid, characterized by maturity at 2 years, moderate fecundity, and a reproductive span of 5 years; (b) a cyprinid, characterized by maturity after 1 year, low fecundity, and a reproductive span of 2 years; (c) a catostomid, characterized by late maturity, high fecundity, and a reproductive span of 4 years; and (d) a cottid, characterized by maturation after 1 year, low fecundity, and a reproductive span of 4 years. The salmonid–catostomid–cyprinid–cottid fish assemblage is characteristic of coldwater streams throughout North America (Moyle & Herbold, 1987). Demographic characteristics of model life histories are summarized in Table 11.1.

By changing parameter values, we could make each life history correspond to a specific species for purposes of validation against field data. Parameter values for six species were supplied for field validation. The species are the introduced European piscivore, brown trout (*Salmo trutta*), rainbow trout (*Onchorynchus mykiss*), which was introduced to California's Sierra Nevada region, and several species native to the Truckee River drainage, including two cyprinids, Lahontan redside (*Richardsonius egregius*) and Lahontan speckled dace (*Rhinichthys osculus robustus*), the cottid Paiute sculpin (*Cottus*

beldingi), and the catostomid Tahoe sucker (*Catostomus tahoensis*). The
model assumes that the four basic life history types are spring-spawners that
show weak year-classes following spring floods. A salmonid predator, based
on the European brown trout, is assumed to spawn in the winter and show
poor recruitment following winter floods.

Overall structure of the simulation model

The model consists of linked, age-structured Leslie matrices, one for each
life history, that simulate annual changes in the population sizes of interacting fish species as survival is modified by stream discharge, competitive interactions, and predation. For each yearly integration, the model calculates the
net change in the number of individuals in each age/size class of each species
as a function of age-specific growth and mortality rates. Recruitment varies
as a function of female fecundity and the survival of early life stages. Specified patterns of high stream discharge during early development, interspecific
competition, and predation all contribute to the rate of first year mortality.
Adult survival is reduced principally by low flow conditions.

Relationship of first year survival to stream discharge

Curves fit to field data by regression analysis show a negative exponential
relationship between recruitment of a year-class and mean monthly discharge
during May for spring-spawners and during January for winter-spawners
(Strange *et al.*, 1992). The model varies first year survival as a continuous
negative exponential function of mean monthly May discharges above 25 cfs
for spring-spawners and as a continuous function of mean January discharges
above 25 cfs for winter-spawners. Differential sensitivity to extremely high
or low stream discharge was estimated by differences in egg size and rates
of first year growth (Winemiller, 1992; Winemiller & Rose, 1992). It was
assumed that 25 cfs was the critical threshold for changes in survival rate;
the larger the egg size and the higher the potential growth rate, the less the
reduction in survival above 25 cfs. Because salmonids have relatively large
eggs and rapid first year growth, they were assumed to show 25% higher survival as stream discharge increased than the moderate-sized catostomid life
history. The small-bodied cyprinid and cottid life histories were assumed to
show 25% lower survival than the catostomid.

Optimal temperatures for salmonids and cottids are generally lower than
for catostomids and cyprinids, and salmonids and cottids are less tolerant of
higher water temperatures resulting from low flows (Moyle, 1976). Thus, the

model assumes that first-year survival under prerecruitment discharges less than 25 cfs is highest for the catostomid and cyprinid life histories and 25% less for the salmonid and cottid life histories based on North American species.

Relationship of adult survival in response to droughts

Adult survivorship is also reduced under low flows. In the model, survival rates of fish older than one year are reduced as a continuous function of mean annual flows less than 25 cfs. Mean annual flow is calculated as the average of the mean January and mean May discharges at each yearly time step. As was the case for first-year survival under low flows, the model assumes that percentage decrease in adult survival under discharges less than 25 cfs is least for the catostomid and cyprinid life histories and 25% more for the salmonid and cottid life histories. Decreases in adult survival were assumed to be 50% more for the European piscivorous salmonid based on brown trout.

Competition during the first year of life

Density-dependent mortality is characteristic of young-of-year fish (Wootton, 1990), and models of fish populations generally assume that only this age group contributes to density effects (e.g., Levin & Goodyear, 1980; DeAngelis et al., 1980). This was the basis for using an inverse logistic function to represent competitive interactions that modify survival to age 1 based on stream discharge. The densities of each species and its competitors were used as inputs to reduce survival below that due to streamflow alone. The percent reduction ranges from 0 (under no competition at low densities) to 90% (under maximum competition at highest densities). Species densities at each time step are given by the number of eggs spawned by each species plus the number of eggs spawned by competing species modified by weighting factors representing the strength of the competitive effect.

Predation intensities

The model contains only one piscivorous species. Adults of the piscivorous salmonid prey on young-of-year of other species based on experimental studies of brown trout (Garman & Nielsen, 1982). To model predator functional response, first year survival of prey species based on stream discharge was modified as a function of the ratio of predator to prey species for each prey species. We assumed that the piscivorous salmonid consumes prey species according to a type III functional response, and we used an inverted logistic function to model the relationship between prey survival and predator consumption. The percentage reduction in prey survival ranges from 0 (under no predation) to 90% (under maximum predation). The numerical response of the predator was determined to be an asymptotically increasing function of the total number of prey consumed per predator up to the maximum consumption rate. This function gives a percent increase in survival of piscivorous salmonids > 150 mm standard length (age 2 and over in our model) based on data from Garman and Nielsen (1982). The survival rate for each predator size class is the baseline survival for that age/size class in the model in the absence of predation; it increases up to a maximum of 90% under maximum consumption.

Parameter estimation

Pre-recruitment stream flows were sampled randomly from discharge distributions based on the historical record of discharges in our reference stream (USGS gaging station no. 10339400, Martis Creek, California, 1972–1994). The time series included a range of discharges considered representative of the natural regime and included the most extreme high and low discharges on record. Rates of survival and reproduction for each life history type were estimated from field data (Strange, 1995) and information in the general stream fish literature for representative species (Moyle, 1976). Survival rates associated with the function relating early survival to stream discharge are based on estimates derived from our field data (Strange, 1995) and estimates in the stream fish literature for early survival of stream fishes under a range of stream discharges (Allen, 1951; Cooper, 1953; Shetter, 1961; Latta, 1962; McFadden *et al.*, 1967; McFadden, 1969; Mortensen, 1977; Nehring, 1988). Baseline survival was estimated as 20% for age 1, 30% for age 2, 50% for age 3, and 70% for ages up to the maximum age/size class, for which survival is 0 (Needham *et al.*, 1945; McFadden & Cooper, 1962; McFadden *et al.*, 1967; Hunt, 1969). Age/size classes are based on length-frequency histograms for representative species in our study stream and the general literature (Moyle, 1976). Fecundity of each life history type is the mean for the average-sized mature female in each mature age/size class and based on fecundity estimates for representative species (Moyle, 1976).

Simulation experiments

Our simulation experiments were designed to reveal how interactions between physical and biological conditions vary species' relative success and outcomes of assembly processes. Model life history types were introduced one at a time under low abundance (50 individuals), with invasions occurring every ten years. Experimental treatments varied invasion sequence, species' interactions, and the physical regime. Simulations examined outcomes for invasion first by early-maturing life history types with short generation times compared to entry first by the late-maturing catostomid life history, which has the longest generation time. Results were compared for the different sequences in the case of invasion by competing species that spawn in the spring and for a predator–prey system in which the predator spawns in winter and the prey species spawn in spring. For the competition experiments, the cyprinid and cottid life history types were modeled as strong competitors of each other but weak competitors of the catostomid and salmonid life histories. Similarly, the salmonid and the catostomid were assumed to be strong competitors of each other and weak competitors of the cyprinid and cottid. Symmetric competition was assumed between the salmonid and the catostomid. Based on a study of competitive interactions between a cottid and cyprinid in a northern California stream (riffle sculpin, *Cottus gulosus*, and speckled dace, *Rhinichthys osculus*, Baltz *et al.*, 1982), the cottid is a better competitor. For the predator–prey system, the predator life history was a winter-spawning piscivorous salmonid, and prey species included the spring-spawning cyprinid, cottid, and catostomid. Under predation, prey species were assumed to show intraspecific competition only. Details of the simulation experiments are presented below.

Relative success of invader life histories under a range of physical and biological conditions

To define conditions that will alter relative success, we examined how the success of different invader life histories varied for contrasting invasion sequences and under different physical and biological conditions. Physical conditions were varied by alternating floods and droughts at specified frequencies (# of floods or droughts over 100 years from the time of the first invasion) with a moderate streamflow (50 cfs) based on average discharges observed in our study stream. For each invasion sequence, the percentage change in relative abundances was determined under the different physical conditions for a competition community and a predator-prey system compared to a baseline model without competition or predation. Floods occurred during a species' month of hatching and reduced survival to age 1. For the competition community, all invading species were assumed to spawn in spring and show poor recruitment in response to spring floods. For the predator–prey system, floods occurred only in winter and therefore only affected recruit-

Table 11.2. *Outline of simulation experiments*

Invasion sequence	Biological regime	Physical regime
COT-CYP-SAL-CAT	CAT-SAL strong competitors, COT-CYP strong competitors, and COT,CYP weak competitors of CAT,SAL	no droughts, 10% droughts, 20% droughts, 30% droughts vs. no spring floods, 10% spring floods, 20% spring floods, 30% spring floods vs. 50 cfs
CAT-SAL-CYP-COT	same	same
COT-CYP-SAL-CAT	SAL predator of COT, CYP, CAT; prey show intraspecific competition only	no droughts, 10% droughts, 20% droughts, 30% droughts vs. no winter floods 10% winter floods 20% winter floods 30% winter floods vs. 50 cfs
CAT-SAL-CYP-COT	same	same

COT = COTtidae, CYP = CYPrinidae, SAL = SALmonidae, CAT = CATostomidae.

ment of the winter-spawning predator. A flood was defined as 200 cfs, based on maximum discharges observed in our study stream. A drought was defined as a streamflow during the month of hatching of 10 cfs and an average annual flow of 10 cfs. Experimental treatments are outlined in Table 11.2.

Simulation of invasion into the stream fish community

To illustrate how specific interactions between physical and biological conditions can influence assembly outcomes, assembly of the competition community and the predator–prey system was examined under regimes that differed in the timing of floods and droughts. Floods or droughts were introduced at the time particular life histories invaded and population trajectories followed for up to 50 years after the last invasion. Streamflow regimes were based on a lognormal distribution of flows based on the historical record of discharges in our study stream. For the competition community, a decade of drought (May and January mean = 15 cfs, cv = 0.3) during the period of cottid and salmonid invasion was first simulated. Next, a decade of high flows (May mean = 150 cfs, cv = 0.1) was simulated during the period of cyprinid and catostomid invasion. The same treatment was applied to each invasion sequence. For the predator–prey system, a period of high flows in winter (January mean = 150 cfs, cv = 0.1) was introduced at the time of invasion of the piscivorous salmonid under each invasion sequence.

Model validation

Model predictions were compared with field data for two coldwater streams in the Truckee River drainage of northern California to examine how the physical regime may mediate processes of community assembly in the case of invasion by the piscivorous European brown trout. One stream is our reference stream, Martis Creek (Moyle & Vondracek, 1985; Strange *et al.*, 1992; Strange, 1995), and the other is nearby Sagehen Creek (Seegrist & Gard, 1972; Gard & Seegrist, 1972; Gard & Flittner, 1974). Modeled species included two salmonid species which are not native to the Truckee drainage, brown trout (*Salmo trutta*) and rainbow trout (*Oncorhynchus mykiss*), two native cyprinids, Lahontan redside (*Richardsonius egregius*) and Lahontan speckled dace (*Rhinichthys osculus robustus*), a native cottid, Paiute sculpin (*Cottus beldingi*), and a native catostomid, Tahoe sucker (*Catostomus tahoensis*). Native Truckee drainage species and rainbow trout spawn in spring, whereas brown trout is a winter-spawner. Census data for each of the six species and stream discharges for each stream were used as model inputs. Stream discharge data were obtained for Martis Creek from records for USGS stream gage no. 10339400 and for Sagehen Creek from records for USGS stream gage no. 10343500. Initial population sizes were the actual censused population sizes for the first year of each census.

Population limits for the competition and predation functions were adjusted according to the range of population sizes for the different species observed in each stream. Predation rates and other parameter values were the same as for the base model used in the simulation experiments.

Results

Changes in relative success under different physical and biological conditions

Simulation results show how the relative success of interacting stream fish species can vary during community assembly depending on order of invasion, type of biotic interaction, and differences in life history response to streamflow conditions (Fig. 11.1, 11.2).

Competition

In the case of competing salmonid and catostomid life history types, the salmonid shows increased advantage as floods increase because of its earlier maturity and shorter generation time. However, as droughts become more frequent, dominance shifts in favor of the catostomid because of the salmonid's lower tolerance of drought conditions. Thus, although the salmonid has an advantage if it invades before the catostomid, advantage is lost under frequent droughts (Fig. 11.1). If the salmonid invades after its competitor, advantage will be lost even under moderate drought frequencies (Fig. 11.2).

For the competing cyprinid and cottid life histories, relative advantage shifts to the weaker competitor, the cyprinid, when either floods or droughts become more frequent. The cyprinid is favored under floods because of a higher population growth rate and is favored under droughts because the cottid shows greater sensitivity to drought conditions. Thus, the advantage of invading first is lost by the cottid as floods and droughts increase (Fig. 11.1). When the cyprinid invades first, it increases in relative abundance as floods and droughts increase and the cottid declines (Fig. 11.2).

Such shifts in the relative success of competing species under changing physical conditions strongly influence assembly outcomes. The cyprinid and catostomid dominate assemblage structure under frequent droughts because of their higher tolerance of drought conditions, whereas the salmonid and cyprinid are favored under frequent floods. Advantage is gained by the cyprinid under floods because it matures earlier and shows a shorter generation time so it can maximize reproductive success in good years. The longer-lived salmonid shows greater advantage as flood frequency increases because it spreads reproductive value over many age classes and is therefore better able to withstand many years of poor recruitment. In contrast, the cottid shows advantage only if floods and droughts are infrequent and if it invades before its competitor.

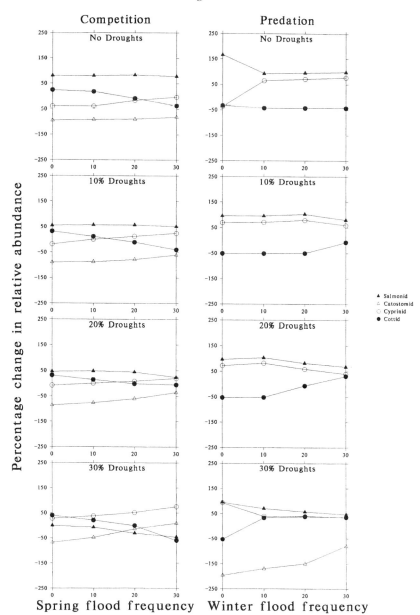

Fig. 11.1. Percentage change in relative abundances of model life history types under competition and predation for different flood and drought frequencies alternated with 50 cfs for invasion sequence cottid (invades year 0), cyprinid (invades year 10), salmonid (invades year 20), and catostomid (invades year 30). For the competition experiments, the catostomid and salmonid show symmetric competition and the cottid is a stronger competitor than the cyprinid. For the predation experiments, the

Predation

Results for the predator-prey system indicate how storm events that impact predator recruitment can function to regulate prey populations. When a winter-spawning salmonid predator invades late in the invasion sequence, early invading prey species that spawn in spring can persist and increase in relative abundance as long as frequent droughts or winter floods limit increase of the predator population (Fig. 11.1). Even a species that invades after the predator can persist if storm events that limit predator recruitment become more frequent. By contrast, if the salmonid predator invades when droughts are infrequent, extinction of prey species that invade later will occur, and a prey species that invades first will show advantage only as both droughts and floods become more frequent (Fig. 11.2). Results for both invasion sequences show that all species can coexist under droughts if floods occur that limit only the predator population.

Assembly scenarios

Results of sample assembly scenarios show that when floods or droughts occur at the time particular species invade there are significant consequences for assembly dynamics that can overwhelm the effects of invasion order (Fig. 11.3–11.5). If the cottid and salmonid life histories invade under droughts, the system is dominated by the catostomid and cyprinid life histories, irrespective of the order in which the salmonid or cottid invade, because the catostomid and cyprinid strategies are advantageous under such conditions (Fig. 11.3). Similarly, if increases of the catostomid and cyprinid are limited by high flows at the time of their invasion, the cottid and salmonid are able to benefit (Fig. 11.4). If the cottid and salmonid invade under such conditions, and before their catostomid and cyprinid competitors, they will be able to dominate the assemblage (Fig. 11.4, top). Even when the catostomid and cyprinid invade first, the cottid and salmonid are able to increase (Fig. 11.4, bottom). When the short-lived cyprinid invades under extreme high flows, invasion fails irrespective of the order of invasion relative to its cottid competitor. By contrast, the late-maturing, long-lived catostomid, which spreads reproductive value over many age classes, can invade successfully even though high flows occur at the time of invasion.

Fig. 11.1. (continued)
salmonid is a winter-spawning predator and the prey species spawn in spring and show intraspecific competition only. Percentage change in relative abundance is the change in relative percent based on mean populations over 100 years after the first invasion calculated as (experimental treatment-baseline)/baseline × 100.

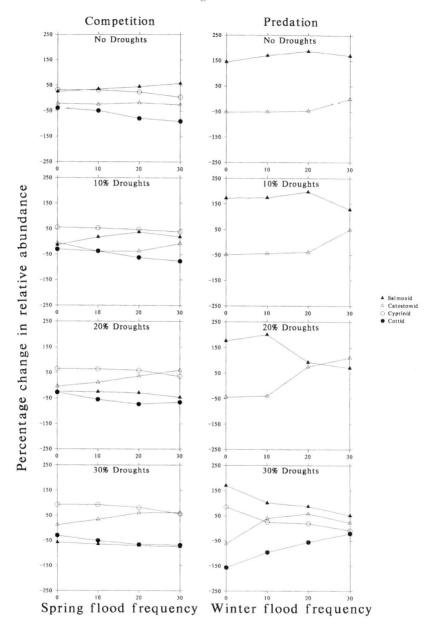

Fig. 11.2. Percentage change in relative abundances of model life history types under competition and predation for different flood and drought frequencies alternated with 50 cfs for invasion sequence catostomid (invades year 0), salmonid (invades year 10), cyprinid (invades year 20), and cottid (invades year 30). Relative competitive abilities, predator dynamics, and abscissa are as in Fig. 11.1.

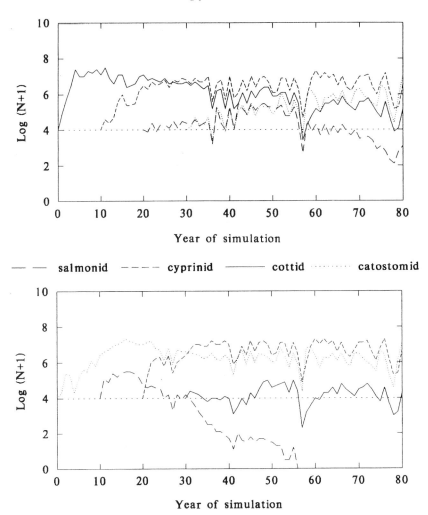

Fig. 11.3. Simulated community assembly of competing species for invasion sequence cottid–cyprinid–salmonid–catostomid (*top*) and invasion sequence catostomid–salmonid–cyprinid–cottid (bottom) under the average regime for Martis Creek, California with periods of drought. A decade of extreme low flows (May and January mean = 15 cfs, cv = 0.3) was introduced at the time of invasion of the cottid and salmonid. Relative competitive abilities as in Fig. 11.1.

In the case of the predator–prey system, the flow regime at the time of predator invasion also strongly influences the resulting assemblage (Fig. 11.5). When the winter-spawning piscivorous salmonid invades under extreme high flows during its spawning season, predator and prey species can coexist if

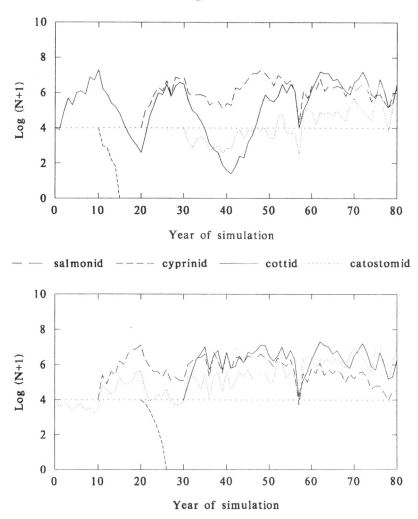

Fig. 11.4. Simulated community assembly of competing species for invasion sequence cottid–cyprinid–salmonid–catostomid (top) and invasion sequence catostomid–salmonid–cottid–cyprinid (bottom) under the average regime for Martis Creek, California with periods of high flows. A decade of spring high flows (May mean = 150 cfs, cv = 0.1) was introduced at the time of invasion of the cyprinid and catostomid. Relative competitive abilities as in Fig. 11.1.

prey species show comparatively high numbers at the time of predator invasion, either because they invade before the predator or before predator numbers increase (Fig. 11.5, top). However, if the predator is able to recover from deleterious high flows before prey species invade, predator invasion can lead to prey extinction (Fig. 11.5, bottom).

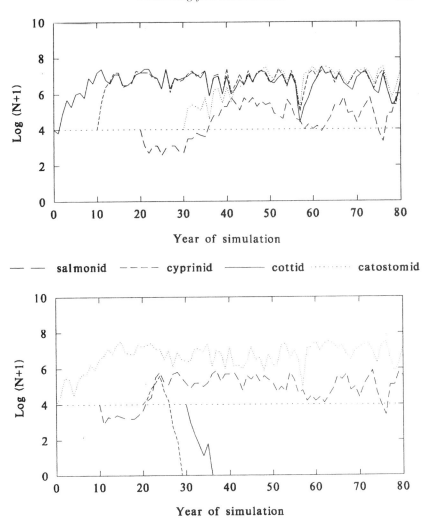

Fig. 11.5. Simulated community assembly of a predator–prey system for invasion sequence cottid–cyprinid–salmonid–catostomid (top) and catostomid–salmonid–cyprinid–cottid (bottom) under the average regime for Martis Creek, California with a decade of high flows in winter (January mean = 150 cfs, cv = 0.1) at the time of invasion of the salmonid predator. The cottid, cyprinid, and catostomid spawn in spring and are prey of the winter-spawning salmonid. Prey species show intraspecific competition only.

Model predictions and field observations: Martis Creek

Model predictions generally show a good fit to census data for both Martis Creek and Sagehen Creek. In the case of Martis Creek (Fig. 11.6), the model satisfactorily predicts the major trends in species' abundances over 15 years. From 1979–1983, species native to Martis Creek (Tahoe sucker, Lahontan redside, Paiute sculpin, and Lahontan speckled dace) dominated the assemblage and showed relatively stable ranked abundances. During this time winter and spring discharges were dissimilar and varied from year to year, and brown trout and spring-spawning species coexisted, with brown trout showing relatively low numbers. In 1983, severe spring floods decimated populations of spring-spawning species while the winter-spawning brown trout increased under favorable winter flows. For several years following the spring floods of 1983, winter and spring flows were favorable but native species failed to recover from flood-related declines while brown trout showed a dramatic increase. Paiute sculpin remained rare and Lahontan redside became locally extinct. From 1987 to 1992, drought conditions prevailed in Martis Creek, and brown trout declined while remnant populations of native species increased. The model also accurately predicts this decline of brown trout and recovery of Tahoe sucker and speckled dace during the last years of our study.

The model does not predict the dramatic reversal in rainbow trout and brown trout abundance in 1987, which followed severe winter floods in 1986. This is probably a census error. We didn't census fish populations in Martis Creek in 1986, and there may have been important changes in that year which are not included in our model. The model also underestimates the decline in Tahoe sucker from 1983–1990. This may be a true underestimation, or we may have failed to adequately census Tahoe sucker during those years. Despite these shortcomings, the accurate prediction of overall trends suggests that the model captures key dynamics.

Model predictions and field observations – Sagehen Creek

The model provides an excellent fit to census data for Sagehen Creek (Fig. 11.7). Both model predictions and field data show relative stability of the Sagehen Creek fish populations over the census period (1952–1961). This period included two major floods, one in May 1952 and another in December 1955. Average annual flows remained favorable and stable. Differences in the seasonal timing of floods combined with overall stability of annual flows appears to have prevented any major shift in the relative abundances of brown trout and prey or competing species in Sagehen Creek.

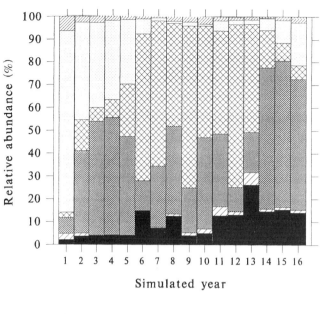

Simulated year

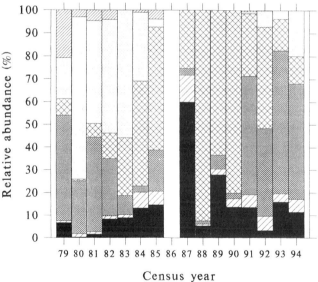

Census year

▨ Lahontan redside (cyprinidae)	▨	Tahoe sucker (catostomidae)
☐ speckled dace (cyprinidae)	▨	Paiute sculpin (cottidae)
▨ brown trout (salmonidae)	■	rainbow trout (salmonidae)

Fig 11.6. Model predictions (*above*) compared to census data (*below*) for the fish species of Martis Creek, California, 1979–1994. No census was taken in 1986. Data are relative percentages based on total numbers.

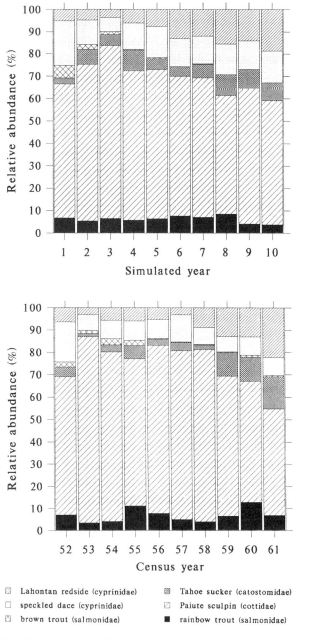

Fig 11.7. Model predictions (*above*) compared to census data (*below*) for the fish species of Sagehen Creek, California, 1952–1961. Census data are from Gard and Flittner (1974) and Gard and Seegrist (1972). Data are relative percentages based on total numbers.

Discussion

Assembly and invasion of stream fish communities

Our results underscore the important role of physical variation in community organization and assembly processes. Results suggest that invader success will vary depending on how physical conditions at the time of invasion affect the ratio of invader to resident species and the outcomes of their interactions. As a result, the outcome of assembly from the same species pool will vary depending on specific interactions between the physical regime and underlying biotic factors. Invasion sequence is important, but the historical context of changing physical conditions affecting rates of biotic processes ultimately determines assemblage structure.

In the case of the Trout Zone Fish Assemblage of the Sierra Nevada, our evidence indicates that the community is readily invaded by brown trout when this species has strong recruitment years coupled with low recruitment of native stream fishes. Such conditions shift relative abundances in favor of brown trout, whose predatory activity can affect recruitment of the other fishes significantly. For the same reason, the native community can resist brown trout invasion if brown trout recruitment is weak and their own recruitment is strong. Thus, the influence of changing physical conditions on population densities is critical in determining any particular assembly outcome.

The ability of our model to predict outcomes, not only in our study stream but in one that has the same species pool but is otherwise dynamically independent, is evidence that the model is a reasonable simulation of the dynamics of the real system. The one exception that is not accounted for very well is rainbow trout in Martis Creek, but even in this case it is not clear whether the differences between actual and predicted abundance are flaws in the model or problems with the census. Otherwise, we are satisfied that the model verifies our understanding of assembly processes in such stream fish systems.

If top-down regulation and predator–prey ratios were the dominant influences on community assembly and invasion success, the order of entry into the community would have been expected to be more important than it was in our simulations. Although there are some effects of invasion order (e.g., the success or failure of the short-lived cottid and cyprinid forms), the model suggests that variation in stream flows (especially in the form of floods and droughts) is the dominant factor determining the relative advantage and invasion potential of each of the fishes in the system.

This has important implications for studies of biological invasions. Previous studies have suggested that if the habitat is unfavorable for invading species, communities will show 'environmental resistance' to invasion (Baltz

& Moyle 1993), whereas highly interactive, species-rich communities will show 'biotic resistance' (Post & Pimm, 1983; Pimm, 1989; Case, 1991). However, our findings make clear that ongoing physical variation will continually alter ratios of invader and resident species and outcomes of their interactions if species respond in different ways to changing physical conditions. As a result, community susceptibility to invasion will vary, and no community will be 'invasion-resistant' under all conditions.

Life history–environment interactions

Our findings also suggest that studies of life history–environment interactions could benefit by shifting focus from defining optimal traits or strategies based on average physical conditions to predicting how interactions between physical and biotic conditions will alter relative success as the environment varies. Previous models of life history evolution have focused on average environments or habitats and attempted to predict a single optimal strategy in stable vs. variable environments. For example, the theory of r/K selection proposes that high equilibrium population size (high K) is favored in stable environments whereas a high intrinsic growth rate (high r) is favored in variable environments (MacArthur, 1962; MacArthur & Wilson, 1967; Pianka, 1970). Habitat-based models suggest that habitat characteristics serve as a 'template' for determining life history success, favoring certain traits and eliminating others (Grime, 1974, 1977; Southwood, 1977; Begon, 1985; Silby & Calow, 1985). However, Stearns (1992) has argued that a major weakness of these models is that they attempt to tie life history directly to habitat, whereas what is needed is an understanding of the link between habitat, mortality regimes, and life history – that is, the mechanisms linking habitat and life history. In terms of community assembly, our results suggest that, if different life histories are favored at different times as physical conditions change, no single strategy will show overall advantage in a variable environment, or uniform advantage across environmental gradients.

This suggests that models of community assembly in variable environments should seek to identify not only those conditions which favor particular life histories, but also conditions that will alter relative success. Following van der Valk (1981), Keddy (1992) suggests that 'the environment acts as a filter removing species which lack traits for persisting under a particular set of conditions'. In a study of experimental wetlands, Weiher and Keddy (1995a) found that environmental filters can constrain assembly and result in a common end state even though pathways may differ. Our results suggest that, in systems in which the environmental filter continually changes, assembly from the same species pool is likely to result in multiple or alternate states as the physical regime alters rates of biotic processes. Rather than a single subset of species or succession to a single endpoint, many outcomes are possible depending on how physical and biotic processes interact. Although the sequence of states may appear random, knowledge of how specific interactions between physical and biological conditions result in particular states can allow prediction of critical transitions.

Implications of the model for the concept of assembly rules

Our study suggests that assembly of a stream fish system can be understood by quantifying a small number of factors in realistic combinations. Knowing how recruitment is affected by flow conditions and integrating these relationships across a series of years was the key to understanding community assembly. This is not to claim, however, that the system is simple and easily understood; our results also underline how important particular interactions between physical and biotic factors are in determining the outcome from a fixed pool of interacting species. Precise details matter a great deal.

How can this study be reconciled with others (e.g., Drake *et al.*, 1993) which suggest more dominant, higher-order interactions and greater complexity? One possibility is that the fishes of stream systems are not organized as patches which are readily and routinely invaded, as in Drake's model or in the case of the birds of the Bismarck Archipelago (Diamond, 1975). In streams, alien species can only be transplanted or invade from upstream or downstream. Thus, it is possible that the system itself has a major role in restricting and governing the success of an invader, whereas in terrestrial systems invasion is easier and biotic interactions are more problematic. On the other hand, Case (1996) has recently shown that simple systems like ours can exhibit very complex dynamical behavior because the outcomes of biotic interactions vary depending upon their magnitude – exactly what we have postulated for the role of seasonally variable stream flows.

Intensified study of community assembly is a very promising avenue in the study of ecological complexity, one of the central unsolved problems in ecology (Weiner, 1995). Our study shows that population interactions with the physical environment can be sufficient to understand patterns of relative abundance, local extinction, and invasion into a community. This implies that a limited number of variables can explain fluctuations in relative abundance in this system. Perhaps this is a very simple system, but the fact remains that we did not need to posit greater complexity due to emergent properties to understand the essential features of stream fish assembly dynamics. We concede that this does not mean that emergent properties are unimportant in the assembly of other ecosystems. One may question whether there are general assembly or response rules (Drake, 1990a, Keddy, 1992; Weiher & Keddy, 1995b) that apply across many ecosystems, or whether it is premature to speak of rules. Nevertheless, comparative study of assembly in a number of representative ecosystems is likely to be helpful in illuminating the study of complexity in ecology.

Acknowledgments

We are indebted to Dr Peter Moyle, director of the Martis Creek field studies, and the many students of the Department of Wildlife, Fish, and Conservation Biology at the University of California, Davis who assisted with field sampling. This work was supported in part by a Graduate Fellowship to Dr Elizabeth Strange from the Electric Power Research Institute. We appreciate the comments of Dr Christopher Dewees, Dr Peter Moyle, Dr Evan Weiher, and an anonymous reviewer on earlier drafts.

References

Allen, K.R. (1951). The Horokiwi stream: a study of a trout population. *New Zealand Marine Department Fisheries Bulletin* 10. 238 pp.

Baltz, D.M. & Moyle, P.B. (1993). Invasion resistance to introduced species by a native assemblage of California stream fishes. *Ecological Applications* 32:246–255.

Barkai, A. & McQuaid, C. (1988). Predator–prey role reversal in a marine benthic ecosystem. *Science* 242:62–64.

Baltz, D.M., Moyle, P.B. & Knight N.J. (1982). Competitive interactions between benthic stream fishes, riffle sculpin, *Cottus gulosus*, and speckled dace, *Rhinichthys osculus*. *Canadian Journal Fisheries and Aquatic Sciences* 39:1502–1511.

Begon, M. (1985). A general theory of life-history variation. In *Behavioural Ecology: Ecological Consequences of Adaptive Behaviour*. ed. R.M. Silby & R.H. Smith, pp. 91–97. London: Blackwell Sci. Pub.

Carpenter, S.R. & Kitchell, J.F. (ed.) (1993). *The Trophic Cascade in Lakes*. Cambridge: Cambridge University Press.

Case, T.J. (1991). Invasion resistance, species build-up and community collapse in metapopulation models with interspecies competition. *Biological Journal of the Linnaean Society* 42:239–266.

Case, T.J. (1996). Surprising behavior from a familiar model and implications for competition theory. *American Naturalist* 146:961–966.

Chesson, P.L. (1986). Environmental variation and the coexistence of species. In *Community Ecology*, ed. J.M. Diamond & T.J. Case, pp. 240–256. NY, USA: Harper and Row.

Cole, L.C. (1954). The population consequences of life history phenomena. *Quarterly Review Biology* 29:103–137.

Cooper, E.L. (1953). Mortality rates of brook trout and brown trout in the Pigeon River, Ostego County, Michigan. *Progressive Fish Culturist* 15:163–169.

DeAngelis, D.L., Svoboda, L.J., Christensen, S.W. & Vaughan, D.S. Stability and return times of Leslie matrices with density-dependent survival: applications to fish populations. *Ecological Modelling* 8:149–163.

Diamond, J.M. (1975). Assembly of species communities. In *Ecology and Evolution of Communities*, ed. M.L. Cody and J.M. Diamond, pp. 342–444. Cambridge: Harvard University Press.

Drake, J.A. (1988). Models of community assembly and the structure of ecological landscapes. In *Proceedings of the International Conference on Mathematical Ecology*, ed. L. Gross, T. Hallam, & S. Levin, pp. 585–604. Singapore: World Press.

Drake, J.A. (1990a). Communities as assembled structures: do rules govern pattern? *Trends in Ecology and Evolution* 5: 159–163.

Drake, J.A. (1990b). The mechanics of community assembly and succession. *Journal of Theoretical Biology* 147: 213–233.

Drake, J.A. (1991). Community-assembly mechanics and the structure of an experimental species ensemble. *American Naturalist* 137: 1–26.

Drake, J.A., Flum, T.E., Witteman, G.J., Voskuil, T., Hoylman, A.M., Creson, C., Kenny, D.A., Huxel, G.R., Larue, C.S. & Duncan, J.R. (1993). The construction and assembly of an ecological landscape. *Journal of Animal Ecology* 62: 117–130.

Erman, D.C., Andrews, E.D., & Yoder-Williams, M. (1988). Effects of winter floods on fishes in the Sierra Nevada. *Canadian Journal of Fisheries and Aquatics Science* 45: 2195–2200.

Gard, R. & Seegrist, D.W. (1972). Abundance and harvest of trout in Sagehen Creek, California. *Transactions of the American Fisheries Society* 101: 463–477.

Gard, R. & Flittner, G.A. (1974). Distribution and abundance of fishes in Sagehen Creek, California. *Journal of Wildlife Management* 38: 347–358.

Garman, G.C. & Nielsen, L.A. (1982). Piscivority by stocked brown trout (*Salmo trutta*) and its impact on the nongame fish community of Bottom Creek, Virginia. *Canadian Journal of Fisheries and Aquatic Sciences* 39: 862–869.

Gilpin, M., Carpenter, M.P., & Pomerantz, M.J. (1986). The assembly of a laboratory community: multispecies competition in *Drosophila*. In *Community Ecology*, ed. J.M. Diamond and T.J. Case, pp. 23–40. NY: Harper and Row.

Grime, J.P. (1974). Vegetation classification by reference to strategies. *Nature* 250: 26–31.

Grime, J.P. (1977). Evidence for the existence of three primary strategies in plants and its relevance to ecological and evolutionary theory. *American Naturalist* 111: 1169–1194.

Hunt, R.L. (1969). Overwinter survival of wild fingerling brook trout in Lawrence Creek, Wisconsin. *Journal of the Fisheries Research Board of Canada* 26: 1473–1483.

Keddy, P.A. (1992). Assembly and response rules: two goals for predictive community ecology. *Journal of Vegetation Science* 3: 157–164.

Latta, W.C. (1962). Periodicity of mortality of brook trout during first summer of life. *Transactions of the American Fisheries Society* 91: 408–411.

Levin, S.A. & Goodyear, C.P. (1980). Analysis of an age-structured fishery model. *Journal of Mathematical Biology* 9: 245–274.

MacArthur, R.H. (1962). Some generalized theorems of natural selection. *Proceedings of the National Academy of Sciences, USA* 48: 1893–1897.

MacArthur, R.H. & Wilson, E.O. (1967). *Theory of Island Biogeography.* Princeton: Princeton University Press. 203 pp.

McFadden, J.T. (1969). Dynamics and regulation of salmonid populations in streams. In *Symposium on Salmon and Trout in Streams*, ed. T.G. Northcote, pp. 313–329. Institute of Fisheries, University of British Colombia, Vancouver.

McFadden, J.T. Cooper, E.L. (1962). An ecological comparison of six populations of brown trout (*Salmo trutta*). *Transactions of the American Fisheries Society* 81: 202–217.

McFadden, J.T., Alexander, G.R. & Shetter, D.S. (1967). Numerical changes and population regulation in brook trout, *Salvelinus fontinalis. Journal of the Fisheries Research Board of Canada* 24: 1425–1459.

Matson, P.A. & Hunter, M.D. (ed.) (1992). Top-down and bottom-up forces in population and community ecology. *Ecology* 73:723–765.

Meffe, G.K. (1984). Effects of abiotic disturbance on coexistence of predator–prey fish species. *Ecology* 65: 1525–1534.

Meffe, G.K. (1985). Predation and species replacement in American southwestern fishes: a case study. *The Southwestern Naturalist* 30: 173–187.

Minckley, W.L., & Meffe G.F. (1987). Differential selection by flooding in stream-fish communities of the arid American southwest. In *Community and Evolutionary Ecology of North American Stream Fishes*, ed. W.J. Matthews & D.C. Heins, pp. 93–104. University Oklahoma Press, Norman.

Mortensen, E. (1977). Population, survival, growth and production of trout *Salmo trutta* in a small Danish stream. *Oikos* 28: 9–15.

Moyle, P.B. (1976). *Inland Fishes of California.* Berkeley, CA, USA: University of California Press. 405 pp.

Moyle, P.B. & Vondracek, B. (1985). Persistence and structure of the fish assemblage of a small California stream. *Ecology* 66: 1–13.

Moyle, P.B. & Herbold, B. (1987). Life-history patterns and community structure in stream fishes of western North America: comparisons with eastern North America and Europe. In *Community and Evolutionary Ecology of North American Stream Fishes*, ed. W.J. Matthews & D.C. Heins, pp. 25–34. Norman: University of Oklahoma Press.

Murphy, G.L. (1968). Pattern in life history and the environment. *American Naturalist* 102: 391–403.

Needham, P.R., Moffett, J.W. & Slater, D.W. (1945). Fluctuations in wild brown trout populations in Convict Creek, California. *Journal of Wildlife Management* 9: 9–25.

Nee, S. (1990). Community construction. *Trends in Ecology and Evolution* 5: 337–340.

Nehring, R.B. (1988). Stream fisheries investigations. Federal aid project F-51-R, Job No. 1, Fish flow investigations, Federal Aid in Fish and Wildlife Restoration, Job Final Report F-51-R, Colorado Div. Wildl., Fish Res. Sec., Fort Collins. 34 pp.

Pianka, E.R. (1970). On 'r' and 'K' selection. *American Naturalist* 104: 592–597.

Pimm, S.L. (1989). Theories of predicting success and impact of introduced species. In *SCOPE 37, Biological Invasions: A Global Perspective*, ed. J.A. Drake, H.A. Mooney, F. Di Castri, F. Kruger, R. Groves, M. Rejmanek, & M. Williamson, pp. 351–368. Chichester, UK: John Wiley.

Post, W.M. & Pimm, S.L. (1983). Community assembly and food web stability. *Mathematical Biosciences* 64: 169–192.

Robinson, J.V. & Dickerson, Jr. J.E. (1987). Does invasion sequence affect community structure? *Ecology* 68: 587–595.

Robinson, J.V. & Edgemon, M.A. (1988). An experimental evaluation of the effect of invasion history on community structure. *Ecology* 69: 1410–1417.

Rosenzweig, M.L. (1995). *Species Diversity in Space and Time.* Cambridge: Cambridge University Press.

Schaffer, W.M. (1974). Optimal reproductive effort in fluctuating environments. *American Naturalist* 108: 783–790.

Seegrist, D.W. & Gard, R. (1972). Effects of floods on trout in Sagehen Creek, California. *Transactions of the American Fisheries Society* 102: 478–482.

Shetter, D.S. (1961). Survival of brook trout from egg to fingerling stage in two Michigan trout streams. *Transactions of the American Fisheries Society* 90: 252–258.

Silby, R.M. & Calow, P. (1985). Classification of habitats by selection pressures: a synthesis of life-cycle and r/K theory. In *Behavioural Ecology: Ecological Consequences of Adaptive Behavior*, ed. R.M. Silby & R.H. Smith, pp. 75–90. London: Blackwell Science.

Southwood, T.R.E. (1977). Habitat, the templet for ecological strategies? *Journal of Animal Ecology* 46: 337–366.

Stearns, S.C. (1992). *The Evolution of Life Histories.* Oxford: Oxford University Press.

Strange, E.M. (1995). Pattern and process in stream fish community organization: field study and simulation modeling. Unpublished doctoral dissertation, University of California, Davis.

Strange, E.M., Moyle, P.B. & Foin, T.C. (1992). Interactions between stochastic and deterministic processes in stream fish community assembly. *Environmental Biology of Fishes* 36: 1–15.

van der Valk, A.G. (1981). Succession in wetlands: a Gleasonian approach. *Ecology* 62: 688–696.

Weiher, E. & Keddy, P.A. (1995a). The assembly of experimental wetland plant communities. *Oikos* 73: 323–335.

Weiher, E. & Keddy, P.A. (1995b). Assembly rules, null models, and trait dispersion: new questions from old patterns. *Oikos* 74: 159–164.

Weiner, J. (1995). On the practice of ecology. *Journal of Ecology* 83: 153–158.

Wilbur, H. & Alford, R.A. (1985). Priority effects in experimental pond communities: responses of *Hyla* to *Bufo* and *Rana*. *Ecology* 66: 1106–1114.

Winemiller, K.O. (1992). Life-history strategies and the effectiveness of sexual selection. *Oikos* 63: 318–327.

Winemiller, K.O. & Rose, K.A. (1992). Patterns of life-history diversification in North American fishes: implications for population regulation. *Canadian Journal of Fisheries and Aquatic Sciences* 49: 2196–2218.

Wootton, R.J. (1990). *Ecology of Teleost Fishes.* London: Chapman and Hall. 404 pp.

12

Functional implications of trait–environment linkages in plant communities

Sandra Díaz, Marcelo Cabido, and Fernando Casanoves

Introduction

In this chapter, three fundamental concepts in community ecology are dealt with: assembly rules, trait–environment linkages, and plant functional types. Although all of them have received considerable attention in the literature, they have been rarely discussed in an integrated way. The detection of general rules underlying observed patterns has been a major aim of community ecology as it grows into an integrated, predictive science (Keddy, 1989; Drake, 1990; Barbault & Stearns, 1991; Grime, 1993). The interest in identifying consistent and predictable associations between plant traits, types of plants, and environmental conditions is an integral part of this search for generalization. These aspects have received renewed interest in the last few years. In the face of the challenges of massive loss of biodiversity and global climate change, accurate predictions are not anymore simply desirable for the sake of 'good science'. They have become an urgent need.

This chapter is aimed at presenting an approach in which plant traits are used to construct functional types and to identify consistent trait–environment linkages. On this basis, present fundamental community/ecosystem processes can be predicted, as well as their likely shifts under changing climatic conditions. Conceptual issues are first analyzed and empirical studies in the literature summarized concerning assembly rules (in a broad sense), trait–environment linkages, and functional types. As an illustration of the approach, an example of the operation of environmental conditions as filters/assembly rules on a regional pool of plant traits along a steep climatic gradient is presented.

Assembly rules: interactions and filters

The issue of assembly rules is central to community ecology. Assembly rules are generalized restrictions to coexistence, and represent constraints on how communities are selected as subsets of a species pool (Diamond, 1975; Keddy, 1989; Wilson & Gitay, 1995). The concept of assembly rules was born in the context of animal ecology (Diamond, 1975) and most of the theory and empirical examples to date have remained in that context. Assembly rules have been suggested for plants as well (Lawton, 1987; Cody, 1989; Drake, 1990; Watkins & Wilson, 1992; Wilson *et al.*, 1995a). However, most of those authors emphasize interactions between organisms, rather than with other selective forces. Wilson & Gitay (1995) explicitly define an assembly rule as 'a restriction on species presence or abundance that is based on the presence or abundance of one or several other species, or types of species (not simply the response of individual species to the environment)'.

Other authors have defined assembly rules in a broader – or looser – sense (Fig. 12.1). According to Keddy (1989, 1992), filters of any kind imposed to the regional species pool can be regarded as assembly rules. The objective of assembly rules should then be the prediction of which subset of the total species pool for a given region will occur in a specified habitat. While Diamond (1975) associates the idea of assembly rules to 'forbidden combinations', thus emphasizing the importance of biotic interactions, Keddy (1992) links them to the idea of deletion, reinforcing his broader focus. He explicitly mentions climatic conditions, disturbance regime, and biotic interactions as examples of filters. Antecedents of these ideas may be found in the work of Woodward and Diament (1991), although these authors do not explicitly mention assembly rules. They proposed a conceptual model in which climate, fire (disturbance), and site productivity (interactions) act as successive 'filters', selecting certain traits and functions out of the regional species pool. These 'filters' fit into the definition of assembly rules in the broad sense mentioned above and adopted hereafter in this chapter.

Filters and trait–environment linkages at different scales

The concept of trait–environment linkages refers to sets of plant attributes consistently associated with certain environmental conditions, irrespective of the species involved (Keddy, 1992). It is plant traits (and therefore plant function) which is the subject of 'filtering' processes (Woodward & Diament, 1991; Keddy, 1992). Climatic, disturbance, and interaction filters tend to act at decreasing spatial (and to some degree temporal) scales (Fig. 12.1). At any

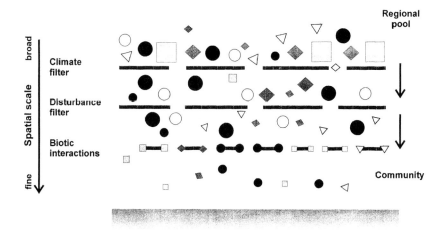

Fig. 12.1. Assembly rules in a broad sense are filters at different scales imposed to the regional pool of species, traits, or functional types (adapted from Woodward & Diament, 1991 and Keddy, 1992).

particular site, a hierarchy of filters can be found: only those traits/functions which can survive under the prevailing climatic conditions, and then under the predominant disturbance regime, have the opportunity to be 'filtered out' (or not) by the interactions with other organisms.

Most of the work on assembly rules for plants has concentrated on the interactions filter (e.g., Wilson & Roxburgh, 1994; Wilson et al., 1995b). On the other hand, there are uncountable examples documenting the selective action of climate and disturbance on plant communities, but they have not been presented within the context of assembly rules. However, the concept of 'filters' at different scales is present in some cases – either implicitly or explicitly. The works of van der Valk (1981) focused in topographic–edaphic conditions and interactions between plants, of Noble and Slatyer (1980) on disturbance and vital attributes, and of Box (1981) and Woodward (1987) on relations between vegetation and climate at regional to global scales are examples.

Climate change and changing filters

There is growing evidence that the Earth's climate and atmospheric composition may change to an unprecedented degree in human history during the next century (Houghton et al., 1996). As a consequence, many research groups around the world are involved in the prediction of likely responses of natural and semi-natural plant communities. The consideration of a hierarchy of environmental 'sieves' seems a useful approach to this problem. Climate change is expected to modify filters at all scales. Although it is regarded mainly as a change in climatic filters, it is also likely to modify disturbance filters through altered frequency of fires, flammability, flooding regime, or land use patterns (Houghton *et al.*, 1996). It may also change the biotic interactions filters. This is because species are expected to respond individualistically, as they have done in the past (Davis, 1981; Huntley, 1992; Bradshaw & McNeilly, 1991), and therefore communities will likely disassemble and reassemble in different ways, and thus plants may experience different neighborhoods.

Spatial gradients as proxies for temporal change

Present distribution of organisms in space – especially along steep environmental gradients – may be used as a model of possible changes in time. Paleoecological studies tend to support this view (Solomon & West, 1986; Delcourt & Delcourt, 1987, 1991). Richardson and Bond (1991) proposed a hierarchy of factors controlling pine invasion: climate, disturbance and biotic interactions with the resident biota. They concluded that the effects of predicted global warming on the distribution of pines are unlikely to be simple functions of temperature and precipitation, except at climatic extremes.

Using present distribution of vegetation in space as a proxy for temporal changes is liable to be inaccurate when important features of the expected environmental change have no clear equivalent at present. Examples of these are increased levels of atmospheric CO_2 or altered neighborhoods, including influx of potentially dominant alien species (Woodward & Diament, 1991). However, the analysis of present spatial patterns, together with paleoecological data, are arguably the only empirical support in investigating likely vegetation changes at a broad scale (region to globe).

Plant traits and ecosystem function

Within a framework of abiotic constraints and prevailing disturbance regime, the individual traits of the dominant living organisms strongly influence community/ecosystem processes (Hobbie, 1992; Schulze & Mooney, 1994; Jones *et al.*, 1994; Schulze & Zwölfer, 1994; Schulze, 1995). This is implicit in the concept of positive-feedback switches in plant communities, i.e., processes in which the dominant organisms modify the environment, making it more suitable for themselves (Wilson & Agnew, 1992). Whereas vegetative traits (e.g., size, leaf turnover, longevity, chemical composition) tend to

Table 12.1. *Examples of individual plant traits which strongly influence processes of the community/ecosystems in which they are dominant*

Individual traits	Community/ecosystem processes	Source
Relative growth rate	Productivity	Grime et al., 1988; Shaver et al., 1997
Leaf turnover rate	Nutrient cycling	Schulze & Chapin, 1987; Nadelhoffer et al., 1991; Reich et al., 1992; Shaver et al., 1997
	Production efficiency	Reich et al., 1992
Nutrient content	Nutrient cycling	Swift et al., 1979; Aber & Melillo, 1982; McClaugherty et al., 1985; Schulze & Chapin, 1987; Hobbie, 1992
	Carrying capacity for herbivores	McNaughton et al., 1989; Harris, 1991; Shaver et al., 1997
Biomass	Flammability	Christensen, 1985; Dublin et al., 1990
Lifespan	Inertia	Richardson & Bond, 1991; Chapin et al., 1993
Canopy structure	Aerodynamic conductance	Jarvis & McNaughton, 1986; Kelliher et al., 1993
	Interception, water relations, runoff	Calder, 1990; Woodward & Diament, 1991; Holling, 1992; Kelliher et al., 1993
	Roughness/albedo	Schulze & Zwölfer, 1994
	Temperature buffering	Schulze, 1982; Holling, 1992
	Soil stability	Holling, 1992
Secondary growth	Carbon sequestration	Larcher, 1995; Schulze, 1982
Ramification	Structural complexity	Lawton, 1983, 1987; Brown, 1991; Marone, 1991
Root architecture	Water uptake	Woodward & Diament, 1991; Kelliher et al., 1993; Sala et al., 1997
Reserve organs	Resilience	Grime, 1979; Noble & Slatyer, 1980
Pollination mode	Expansion over landscape	Faegri & Van der Pijl, 1979; Schulze & Zwölfer, 1994
Persistent seed bank	Resilience	Thompson & Grime, 1979; Thompson et al., 1993
Seed number	Expansion over landscape	Noble, 1989; Hodgson & Grime, 1990
Dispersal mode	Expansion over landscape	Howe & Smallwood, 1982; Noble, 1989; Hodgson & Grime, 1990; Leishman et al., 1992
Presence of root symbionts	Diversity	Grime et al., 1987; Gange et al., 1990
	Nutrient cycling	McNaughton & Oesterheld, 1990
	Rate of succession	Amaranthus & Perry, 1994

be associated to ecosystem processes *in situ* (productivity, nutrient cycling, carrying capacity), regeneration traits (e.g., seed output, dispersal mode, seed persistence) tend to determine stability, recolonization after major disturbances, and migration over the landscape (Table 12.1). This means that, given a certain set of abiotic and disturbance constraints, it may be possible to predict major community/ecosystem processes on the basis of consistent trait–environment linkages.

Plant functional types

As mentioned earlier, given a regional species pool, local conditions filter traits, rather than taxa (Woodward & Diament, 1991; Keddy, 1992). Traits are not filtered independently from each other, since selective pressures act on integrated individuals (Gould & Lewontin, 1979). There is evidence that plant traits tend to be associated in recurrent, predictable patterns. Plant design constraints (Grime, 1977; Grime *et al.*, 1988), tight integration among physiological processes (Chapin, 1980; Chapin *et al.*, 1993), allocational trade-offs and codependency between traits (Lambers & Poorter, 1992; Reich *et al.*, 1992) widely overlap with phylogenetic constraints in determining these consistent specialization patterns (Harvey & Pagel, 1991; Westoby *et al.*, 1995).

Because of the existence of these recurrent specialization patterns, the apparently unmanageable diversity of plant species in natural ecosystems can be summarized into fewer functional types (FTs), namely sets of plants exhibiting similar responses to environmental conditions and having similar effects on the dominant ecosystem processes (Walker, 1992; Gitay & Noble, 1997). The attempts to reduce communities to a reasonably low number of 'building blocks' are as old as ecology itself. They have given origin to a myriad of related, and not always clearly defined, concepts, such as guilds, functional types, growth forms, and strategies (see Simberloff & Dayan, 1991; Gitay & Noble, 1997 for critical review). More recently, and as accurate and swift predictions are urgently called for, these issues seem to have come back into the limelight of ecological research (e.g., Steffen *et al.*, 1992). The term 'functional types' is usually preferred, at least within the context of plant community ecology. This may be because its connotations are somewhat looser than the ones of 'guilds' or 'strategies'.

In the last years, several research groups have tried to construct plant functional types in a *a posteriori* way, on the basis of the screening of numerous traits organized into comparative databases. These initiatives vary in aims, geographical scope, and the range of species and traits considered. For example, based on vegetative and regeneration traits, Grime *et al.* (1988) classified more than 400 herbaceous and woody species within the British flora into strategies, with resource use and response to disturbance being the main criteria. Leishman and Westoby (1992) took a similar approach for 300 species of Australian semi-arid woodlands. Montalvo *et al.* (1991) and Fernández-Alés *et al.* (1993) analyzed vegetative and regeneration traits of 179 and 42 Mediterranean grassland species, respectively, classifying them into FTs with different responses to stress, grazing and ploughing, and discussing them within an evolutionary context. Diaz *et al.* (1992) classified herbaceous plant species and populations from central Argentina on the basis of morphological and life-history traits, and analyzed their role in community-level processes. McIntyre *et al.* (1995) classified herbaceous plants from Australia according to their responses to disturbance. Golluscio and Sala (1993) classified 24 forbs from the Patagonian steppe into groups with different strategies of soil water uptake. Boutin and Keddy (1993) classified 43 herbaceous species from wetlands across eastern North America into guilds, mostly on the basis of vegetative traits measured in the field and under controlled conditions. In all the cases in which both vegetative and regeneration traits were considered, there was only weak coupling between the two sets of traits. This suggests that a satisfactory classification of FTs on the basis of a single criterion is unlikely.

A case study: trait–environment linkages and FTs in central-western Argentina

Climatic variations along a regional gradient

Diaz and Cabido (1995, 1997) studied dominant plant traits across a steep environmental gradient in central-western Argentina, with a difference in annual rainfall of more than 800 mm, and a difference in altitude of more than 1500 m between extreme points (Table 12.2). The dominant vegetation types ranged between montane grasslands (highest and wettest extreme), woodlands (intermediate points), and open xerophytic shrublands (driest extreme). Open halophytic shrublands on lowlands with saline soils are scattered along the gradient, and were considered as points representing extreme and permanent water deficit. On the basis of published vegetation relevés and meteorological data, the gradients were divided into 13 climatically homogenous sectors (Table 12.2).

Table 12.2. *Climatically homogeneous sectors identified along a regional gradient in central-western Argentina (ca. 31° 25′–32°S, 64° 10′–68° 37′W).*

Sector	Dominant vegetation	Overall cover (%)	Altitude (m asl)	Mean temperature (°C)	Annual rainfall (mm)
1	Montane grasslands	100	2155	8.1	911.5
2	Montane grasslands	100	1850	8.9	840.4
3	Montane grasslands	100	1450	11.4	887.3
4	Montane grasslands	100	1000	13.1	996
5	Montane woodlands	100	900	13.1	996
6	Montane woodlands	90	750	15.6	826.4
7	Montane woodlands	90	600	17.5	662
8	Xerophytic woodlands	80	350	19.6	520
9	Xerophytic woodlands	70	368	19.6	520
10	Xerophytic woodlands	50	652	18.3	381
11	Xerophytic open shrublands	50	500	18.2	260
12	Xerophytic open shrublands	30	641	18	85

Cold dry winters and rainfall heavily concentrated to the warm season are characteristic of the climate over the whole region. Sector 13 (not shown in table) is composed of open halophytic shrublands under conditions corresponding to Sectors 9–12. See Fig. 12.2(*a*) for further details on selected Sectors.

The regional pool: plant species and functional types

Diaz and Cabido (1995, 1997) then selected the 100 most abundant plant species out of the regional pool. The species set comprised 30 families and numerous growth forms. There seems to be no obvious limitation to the dispersal of propagules of these species across the entire region. There is no substantial geographical barrier, the degree of habitat fragmentation is very low, and there are abundant animal dispersers. On the basis of the multivariate analysis of key plant traits, they classified the dominant species into eight functional types (Table 12.3). The traits that best discriminated between FTs were vegetative traits, which were also strongly correlated with each other. There was no consistent association between these attributes and regeneration traits (Table 12.3). Different FTs were dominant in different Sectors along the gradient (Fig. 12.2(*b*), for further details see also Diaz & Cabido, 1997).

Climatic conditions as filters

On the basis of these results, we formally tested whether different climatic conditions consistently 'filtered' certain categories of plant traits out of the regional pool, i.e., what values of specific leaf area, longevity, seed size, etc.

Table 12.3. *Plant functional types (FTs) produced by TWINSPAN and DCA on a set of the 100 most abundant species along two climatic gradients in central-western Argentina (Diaz & Cabido, 1997)*

Functional type	Vegetative traits	Regeneration traits	Examples
FT 1	Growth form: tussock grasses and large-leaved forbs (≤100 cm). C_3 pathway, high SLA and LWR, moderately long lived, moderate carbohydrate storage in reserve organs, and some short-term C immobilization into standing dead matter, moderate resistance to drought, shoot biomass peak spring to early autumn, concentrated to mid summer.	various seed sizes, shapes and numbers (<100 to >1000), various dispersal agents, wind pollination, reproductive peak in spring to early autumn, concentrated to mid summer.	*Poa stuckertii* *Stipa tenuissima* *Deyeuxia hieronymi*
FT 2.	Growth form: short graminoids (≤50 cm). C_4 and C_3 pathway, high SLA, and LWR, annual or short lived, very low carbohydrate storage and C immobilization into xylem, low resistance to drought, shoot biomass peak spring to early autumn.	various seed sizes and shapes, seed number low to intermediate (<1000), various dispersal agents, wind pollination, reproductive peak in spring to early autumn.	*Neobouteloua lophostachya* *Juncus uruguensis* *Briza subaristata*
FT 3	Growth form: short (≤50 cm) herbaceous and semi-woody erect, creeping, or rosette-like dicots. C_3 pathway, high LWR, intermediate SLA, lifespan short to moderate, moderate carbohydrate storage in reserve organs, very	seed number low to intermediate (<1000), various seed sizes, seed shapes, and dispersal agents, animal unspecialized and specialized pollination, reproductive peak highly variable, uncommon in mid summer.	*Nierembergia hippomanica* *Plantago myosurus* *Bidens triplinervia*

	low C immobilization into xylem and bark, moderate resistance to drought, shoot biomass peak highly variable, uncommon in mid summer.	
FT 4	Growth form: saxicolous or epiphytic rosettes (≤100 cm). CAM pathway, high LWR, very low SLA, evergreen, moderate carbohydrate storage in reserve organs, some short-term C immobilization into standing dead material, moderate resistance to drought, shoot biomass constant along the annual cycle or with a peak in spring.	seed size intermediate to large (2–10 mm), irregular seed shape, moderate seed number (100–1000), animal- and wind-assisted dispersal, animal specialised pollination, reproductive peak in spring.	*Tillandsia bryoides* *Deinacanthon urbanianum* *Dyckia floribunda*
FT 5	Growth form: trees (>300 cm). C₃ pathway, high to moderate SLA, low LWR, very long lived, deciduous or semi-deciduous, very high carbohydrate storage and C immobilization, very low ramification at ground level, high drought resistance, leaf biomass peak in late spring to early summer.	very numerous large flattened seeds (4–10 mm), various dispersal agents (vertebrate dispersal common), animal specialized pollination, reproductive peak in early spring to early autumn.	*Acacia caven* *Prosopis flexuosa* *Schinopsis haenkeana*
FT 6	Growth form: evergreen shrubs and small trees (≤300 cm). C₃ pathway, low SLA and LWR, very long lived, evergreen, high carbohydrate storage and C immobilization, ramified at the ground level, high resistance to drought, leaf biomass constant or very small peak in spring	numerous large seeds (4–10 cm), various seed shapes and dispersal agents, animal specialized pollination, reproductive peak in spring to early autumn.	*Larrea cuneifolia* *Zucagnia punctata* *Heterothalamus alienus*

Table 12.3. (*continued*)

Functional type	Vegetative traits	Regeneration traits	Examples
FT 7	Growth form: aphyllous or scale-leafed shrubs (≤200 cm). C_3 pathway, extremely low SLA and LWR, long lived, evergreen, high carbohydrate storage, some C immobilization, highly ramified at the ground level, high resistance to drought and salinity, green biomass constant throughout the year.	numerous small (≤2 mm) and spheroidal seeds, various dispersal and pollination agents, reproductive peak in spring to early autumn.	*Heterostachys ritteriana* *Senna aphylla* *Suaeda divaricata*
FT 8	Growth form: globular, cylindrical, and columnar-branched stem succulents of various sizes. CAM pathway, aphyllous, with green succulent stems, long to very long lived, evergreen, some carbohydrate storage and C immobilization, very low ramification at the ground level, very high resistance to drought, green biomass constant throughout the year.	numerous small- to intermediate-sized, spheroidal seeds (≤2–4 mm), dispersed by highly mobile animals, animal specialized pollination, reproductive peak in spring to early autumn.	*Opuntia sulphurea* *Cereus validus* *Gymnocalycium* spp.

See Table 12.4 for further explanation of traits and definition of scales of measurement.

were associated to different sectors along the gradient (see Table 12.4 for description of traits and categories). The null hypothesis of trait frequency distribution in local sectors being indistinguishable from the regional pool was rejected in 71% of the pair-wise comparisons (Table 12.5), strongly suggesting the existence of environmental filters. Vegetative traits were 'filtered' more often than regeneration traits (74% and 63% of the individual comparisons, respectively). Specific leaf area (SLA), lifespan, ramification, C immobilization into support tissue, canopy height, and pollination mode were the traits showing differences in the largest number of pair-wise comparisons.

The climatic factors which appear to have the strongest 'filtering effect' at the regional level were those related with the low temperatures (both means and extremes) predominating at high altitude (>750 m asl). The Sector most different from the regional pool were high mountain grasslands, with the differences becoming stronger as altitude increased (Sectors 1–4 in Table 12.5, Sector 1 in Fig. 12.2). The 'filtering effect' of water deficit became important at lower altitudes (<750 m asl). It was very strong in sitituations where water deficit is permanent throughout the year, as in the case of halophytic shrublands (Sector 13 in Table 12.5 and Fig. 12.2).

Semiarid to arid woodlands, woodland–shrublands, and shrublands (Sectors 8–12 in Table 12.5, Sectors 10 and 12 in Fig. 12.2) showed differences with the regional pool for a comparatively smaller number of traits. This is probably because water deficit disappears or is ameliorated during some months, allowing the survival of a relatively wide variety of trait combinations during the favorable season. There was a set of plants which remain active during the whole year, and tend to be evergreen, sometimes succulent, with CAM or more commonly C_3 photosynthesis. There is also a less persistent set of species which is only present or active during the growing season. This is represented by annual herbaceous plants (mainly C_4 grasses), and deciduous shrubs and trees. Finally, montane woodlands (Sector 6 in Fig. 12.2 and Sectors 5–7 in Table 12.5) were transitional between vegetation and climatic types. Accordingly, they presented a variety of categories for each plant trait, and thus showed comparatively the smallest differences with the regional pool.

The results summarized in Table 12.5 and Fig. 12.2 strongly suggest a filtering effect exerted by climatic conditions at the regional scale. This resulted in consistent trait–climate linkages, with possible implications for ecosystem processes (discussed in following sections). Moderately to highly disturbed areas were excluded from this study, and local biotic interactions were not analyzed. However, a similar trait-based multivariate approach can be taken to investigate their possible roles as filters.

Table 12.4. *Traits recorded on the 100 most abundant species along a climatic gradient in central-western Argentina.*

Trait	Description of classes
Photosynthetic pathway	CAM = 1; C_4 = 2; C_3 = 3
Specific leaf area (SLA) (surrogate for relative growth rate; Poorter & Bergkotte, 1992; Reich et al., 1992)	aphyllous = 0; >0–<100 cm² g⁻¹ = 1; 10–<100 = 2; >100 = 3
Leaf weight ratio (LWR)	LWR<1 = 0; LWR ≅ 1 = 1; LWR>1 = 2
Leaf succulence (indicator of resistance to drought and/or salinity)	non-succulent = 0; slightly succulent = 1; highly succulent = 2
Canopy height	≤ 20 cm =1; 20–60=2; >60–<100 = 3; 100–≤ 300 = 4; >300–<600 = 5, ≥ 600 = 6
Lifespan	annual = 1; biennial = 2; 3–10 years = 3; 11–50 = 4; >50 = 5
Carbon immobilization into support tissue in the form of compounds which cannot be used in further biosynthesis (xylem and bark)	herbaceous monocots = 0; herbaceous dicots = 1; semi-woody dicots = 2; woody dicots with trunk and bark = 3
Ramification at the ground level (adaptive under severe drought; Carlquist, 1988, Hargrave et al., 1994)	non-woody species = 0; 1 single trunk = 1; 2 – 10 = 2; >10 = 3
Drought resistance (taproot and/or highly succulent stem)	no evident drought-resisting organs = 0; taproot or highly succulent stem = 1
Thorniness	no thorns = 0; slightly thorny = 1; very thorny (e.g., Cactaceae) = 2
Shoot phenology (seasonality of maximum production of photosynthetic tissue)	no evident peak = 1; winter, autumn, early spring = 2; late spring, spring, spring-summer, late summer–early autumn = 3; late spring–summer, summer = 4
Seed size	<2 = 1; 2–<4 = 2; 4–10 = 3; >10 = 4
Number of seeds per plant	<100 seeds = 1; 100–999 = 2; 1000–5000 = 3; >5000 = 4
Seed dispersal mode	no obvious dispersal agent = 0; animals with relatively low mobility (ants, rodents) = 1; highly mobile animals (large mammals, bats, birds) = 2; wind = 3
Pollination mode	anemophyllous = 0; unspecialized zoophyllous = 1; specialized zoophyllous = 2

Ten randomly chosen, healthy-looking adult individuals were considered for traits measured in the field or in the laboratory.

Table 12.5. *Summary results of pair-wise comparisons (Chi-squared statistic) between trait frequency distributions in local sectors and in the region considered as a whole (regional pool)*

Traits													
	1	2	3	4	5	6	7	8	9	10	11	12	13
Vegetative traits													
Specific leaf area	*	*	*	*	_	*	*	*	*	*	*	*	*
Lifespan	*	*	*	*	*	_	*	*	*	*	*	*	*
Ramification	*	*	*	*	*	_	*	*	*	*	*	*	*
Leaf weight ratio	*	*	*	*	*	_	_	*	*	*	*	*	*
Carbon immobilization	*	*	*	*	_	*	*	_	*	*	*	*	*
Canopy height	*	*	*	_	*	_	*	*	*	*	*	*	*
Leaf succulence	*	*	*	*	*	*	*	*	*	_	_	_	_	*
Thorniness	*	*	*	*	*	_	*	*	*	_	_	_	_
Drought resistance	*	*	*	*	_	_	_	_	_	*	*	*	*
Photosynthetic pathway	*	*	*	*	_	*	*	_	_	_	_	_	*
Shoot phenology	*	*	*	_	_	*	_	_	_	_	_	_	*
Regeneration traits													
Pollination mode	*	*	*	*	*	*	*	*	_	*	*	*	*
Seed size	*	*	*	_	_	_	_	*	*	_	*	*	*
Seed number	*	*	*	*	*	*	_	_	_	_	_	_	*
Dispersal mode	*	*	_	_	_	*	_	_	*	*	*	_	_

* = difference with regional pool significant at $P < 0.05$; simultaneous inference; Bonferroni's correction (Agresti, 1990); – = no significant difference ($P \geq 0.05$). See Tables 12.2 and 12.4 for definition of sectors and traits, respectively.

Patterns in vegetative traits

The major patterns of association between different categories of traits and Sectors along the gradient are summarized in Fig. 12.2 (*c*)) (only five representative Sectors included for the sake of simplicity). Different sets of traits tended to be associated with each other and with particular Sectors. This is particularly obvious on comparison of the extreme situations (Sector 1 vs. Sectors 12 and 13 in Fig. 12.2 (*c*)). High relative growth rate (high specific leaf area), high investment in photosynthetic tissue (high leaf weight ratio), small stature, and short lifespan were distinctive plant traits under comparatively cold and moist conditions. In contrast, low relative growth rate, high investment in support tissue, intermediate stature, and high persistence in time were associated with severe water deficit. These patterns were reflected in the dominant FTs in each Sector (Fig. 12.2 (*b*)), and were consistent with trade-offs repeatedly mentioned in the literature (Grime, 1977; Chapin, 1980; Lambers & Poorter, 1992; Reich *et al.*, 1992). Woodlands and woodland–shrublands

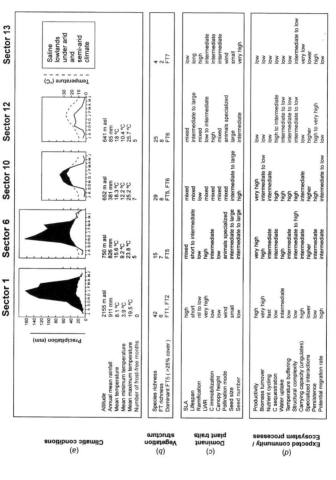

Fig. 12.2. Summary of climatic conditions (*a*), vegetation structure (*b*), dominant plant traits (*c*), and expected community/ecosystem processes (*d*) in representative Sectors along a regional gradient in central-western Argentina. See Table 12.2 for climatic conditions in the rest of the Sectors, and Table 12.4 for definition and scales of measurement of traits mentioned in (*c*). Frequency distribution of different trait categories in all the Sectors are available upon request.

(Sectors 5–11 in Table 12.5, Sectors 6 and 10 in Fig. 12.2) showed not only intermediate values for plant traits, but also the widest variety of values. This, again, was in agreement with the relative abundance of different FTs in them (Fig. 12.2 (*b*)).

Patterns in regeneration traits

In the case of regeneration traits, anemophylly was the most frequent pollination syndrome in grasslands and halophytic shrublands (Sectors 1 and 13 in Fig. 12.2 and Sectors 1–4 and 13 in Table 12.5), whereas specialized zoophyllous syndromes (mammal-, bird-, and hymenopteran-pollination) were the most frequent in woodlands and xerophytic shrublands (Sectors 6 to 12 in Fig. 12.2, Sectors 5 to 12 in Table 12.5). Small seeds predominated in grasslands and halophytic shrublands. In the first case, this was associated with low seed production per plant. In the second case, in contrast, seeds were produced in large numbers. The predominant pattern in woodlands and xerophytic shrublands was the production of numerous big seeds. As in the case of analysis by FTs (Díaz & Cabido, 1997), these traits were only loosely associated with each other and with vegetative traits.

Trait–environment linkages, FTs, and ecosystem function

Assuming that (a) different local environmental conditions are filtering certain traits (and therefore FTs) out of the regional pool; and (b) predominant plant traits have a considerable influence at the community and ecosystem levels (Table 12.1), then the next step in this approach is the prediction of the relative magnitude of major community/ecosystem processes on the basis of consistent trait–environment linkages (Fig. 12.2 (*d*)).

Woodlands and woodland–shrublands (Sectors 6 and 10 in Fig. 12.2, Sectors 5–10 in Table 12.5) should exhibit maximal plant biomass accumulation, carbon sequestration in biomass, temperature buffering, and soil water retention, as compared with high-altitude grasslands and xerophytic and halophytic vegetation (Sectors 1 and 12, and 13 in Fig. 12.2, Sectors 1–4 and 11–13 in Table 12.5). They should also have maximum structural complexity (spatial arrangement of plant structures, *sensu* Brown, 1991), given both by the convergence of several FTs and by the complex architecture of most trees and shrubs. Structural complexity has strong influence on higher trophic levels (Lawton, 1983, 1987; Marone, 1991) and its positive association with animal diversity in these systems, in particular, has been documented by Gardner *et al.* (1995). The predominance of specialized animal pollination also suggests that specialized plant–animal interactions may play a more important role (or at least may be more common) in woodlands and woodland–shrublands than in other systems.

In the case of montane grasslands (Sector 1 in Fig. 12.2, Sectors 1–4 in Table 12.5), maximum biomass turnover, productivity, and nutrient cycling, and only moderate capacity for C sequestration in biomass and water uptake by the rhizosphere are some of the expected community/ecosystem processes. These systems should also exhibit maximum carrying capacity for ungulates. Their dominant functional types have high leaf/support tissue ratio (McNaugton *et al.*, 1989). They should also show comparatively high nutritional quality: RGR tends to be associated positively with N concentration in plant tissue (Lambers & Poorter, 1992; Poorter & Bergkotte, 1992; Reich *et al.*, 1992) and negatively with the concentration of defensive compounds (Bryant *et al.*, 1983). From the structural viewpoint, these communities are far less complex than woodlands and shrublands (Díaz *et al.*, 1992, 1994; Gardner *et al.*, 1995).

Minimum biomass accumulation, biomass turnover, productivity, nutrient cycling, C sequestration in biomass, and carrying capacity for ungulates were predicted for open shrublands (Sectors 12 and 13 in Fig. 12.2, Sectors 11–13 in Table 12.5). The low overall vegetation cover per unit area (Table 12.2) also supports these predictions. The dominant FTs have various root architectures, but the overall root web is less developed than those predominating in woodlands and woodland–shrublands. This, together with the low vegetation cover, should lead to a comparatively low water uptake by vegetation. Structural complexity is higher than in grasslands, but lower than in woodlands and woodland–shrublands.

Woodlands, woodlands–shrublands, and short open shrublands should show maximum persistence if environmental conditions have become unsuitable for regeneration. This is because the dominant FTs are likely to survive as adults for very long periods under climatic conditions unfavourable for regeneration. On the other hand, their capacity for expansion over the landscape is expected to be moderate to low, mainly because their dominant FTs take a long time to establish and reach maturity. Grasslands should show a very low persistence *in situ* under unfavourable conditions, but very fast expansion rates, mainly due to the faster establishment and development of their constituent FTs, and the predominance of wind pollination.

Further studies of the actual persistence of seeds in soil banks are clearly needed in order to improve these predictions. At first inspection, however, the predominance of large seeds in woodlands, woodland–shrublands, and open xerophytic shrublands, in contrast to smaller seeds in grasslands and especially in halophytic shrublands (associated with high seed numbers) (Fig 12.2 (*c*)), suggests differences among Sectors in capacity for regeneration *in situ* following disturbances, including extreme climatic events, such as untimely or unusually severe droughts or frosts.

Predicting ecosystem function under changing climatic conditions

It is possible to apply the same rationale explained above to the prediction of major community/ecosystem processes in the face of climate change. Under changing environmental conditions, there would be shifts in predominant plant traits and thus ecosystem functions along regional gradients. One thing that should be considered is the persistence *in situ* (inertia) of the established vegetation and the establishment rate of the invading populations. The interplay between these two aspects could lead to very different time lags between environmental and vegetation changes. Although it is widely accepted that species will show individualistic responses as they have done in the past, it is reasonable to expect that those species that show similarities in essential functional characteristics will behave in relatively similar ways in their migration over the landscape. Therefore, on the assumption of consistent trait–environment linkages, one should be able to predict expansions and retreats of different functional types, future predominant traits, and future ecosystem processes, at different points along regional gradients. This rationale is valid provided at least one member of each functional type has the capacity to migrate over the landscape fast enough to keep pace with changing environmental conditions (Schulze & Swölfer, 1994; Chapin *et al.*, 1996). This requisite seems to be met by the plants we analyzed, since there was no strong correlation between FTs and dispersal ability.

We used the climatic predictions given by the GISS global circulation model (Hansen *et al.*, 1983) for central-western Argentina, in order to hypothesize changes in plant distribution over the study area. Current GCMs have many inadequacies and significant errors on regional scales (Rochefort & Woodward, 1992). Therefore, we stress the fact that this is simply an exercise to illustrate the approach.

By the time atmospheric CO_2 concentration reaches 700 ppm, mean monthly temperature and rainfall values obtained from GISS suggest a shift towards hotter and drier conditions in the case of xerophytic woodlands, shrublands and halophytic ecosystems (Sectors 10–13 in Fig. 12.2, Sectors 8–13 in Table 12.5). Therefore, a progressive predominance of plant traits may be expected typical of very dry conditions (Sectors 12 and 13 in Fig. 12.2, Sectors 11–13 in Table 12.5) over more mesic shrubland–woodlands and woodlands (Sector 10 in Fig. 12.2, Sectors 9 and 10 in Table 12.5). A direct consequence would be a shift from ecosystem processes listed in the central column of

Fig. 12.2(d) towards those in the two columns to the right. But these changes are likely to be very slow, since the present vegetation is expected to be highly persistent in the established phase once environmental conditions have become unfavorable, and the invading FTs need a long time to establish and reach maturity. The time lag between climate and vegetation changes may be of the order of centuries.

Hotter and wetter conditions are predicted in the case of montane grasslands (Sector 1 in Fig. 12.2, Sectors 1–4 in Table 12.5). A shift in altitudinal vegetation belts may thus be expected, with plant traits now typical of lower montane vegetation (Fig. 12.2(c), second column to the left) predominating over those now typical of high montane grasslands (Fig. 12.2(c), left column). Ecosystem processes listed in Fig. 12.2(d) should change accordingly. The rates of change should be faster in this case. FTs predominating in high-mountain grasslands show low persistence as adults once environmental conditions have become unfavourable for establishment, and the potentially invading FTs can establish and reach maturity at fast to moderate rates. Changes in plant distribution along this gradient might be evident within decades.

Concluding remarks

Efforts aimed at identifying assembly rules (Keddy, 1992) and at predicting ecosystem function (Schulze & Zwölfer, 1994) at the level of functional types are seemingly more fruitful than those focused on plant species. Functional diversity reduces the number of variables to be dealt with, and it is probably a better predictor of ecosystem processes. This is not to underestimate the importance of species richness within FTs. This may provide redundancy to the system, and therefore enhance stability/resilience (Schulze & Zwölfer, 1994).

Because of the occurrence of recurrent patterns of specialization among plants, it is possible to identify FTs and trait–environment linkages on the basis of selected individual traits. However, vegetative traits influencing *in situ* resource acquisition and storage, and regeneration traits influencing recolonization after disturbance and expansion over the landscape are only loosely coupled and cannot be predicted from each other. Therefore, predictions of vegetation responses under changing climatic conditions (Nemani & Running, 1989; Chapin, 1993) or disturbance regime (McIntyre *et al.*, 1995) simply on the basis of a unique, or a highly reduced set of either vegetative or regeneration traits are likely to be inaccurate.

Using plant traits and information about the fundamental environmental conditions operating in an area as inputs allows the identification of plant functional types and of consistent trait–environment linkages both at the local and regional scale. On these bases, predictions of present and future community/ecosystem processes can be made. In taking this last step, two essential points should be taken into account, in order to minimize the risks involved in scaling-up. First, when choosing individual plant traits, not only should short-term physiological attributes, such as relative growth rate, be considered, but morphogenetic traits, reproduction and dispersal in time and space, and relationships with other trophic levels (e.g., root symbionts, herbivores, pollinators) should also be taken into account. Although the prediction of ecosystem function on the basis of individual plant traits involves a scaling-up process, this approach is also a top-down one, since the community/ecosystem level perspective guides the selection of variables to be measured at the individual level, rather than the reverse (Keddy, 1992). Secondly, empirical testing of predictions at the relevant scale (by means of ecosystem experiments and medium to long-term monitoring across the landscape) is the next obvious and indispensable step.

Acknowledgments

We are grateful to M. Balzarini, E. Pucheta, N. Fernández, and R. Rodríguez for useful comments, to G. Funes and V. Falczuk, for their help in data collection, and to D. Abal-Solís and G. Milano for preparing graphs and figures. This work was supported by Universidad Nacional de Córdoba, Fundación Antorchas – The British Council and the International Scientific Cooperation Program of the European Union.

References

Aber, J.D. & Melillo, J.M. (1982). Nitrogen immobilization in decaying hardwood leaf litter as a function of initial nitrogen and lignin content. *Canadian Journal of Botany* 60: 2263–2269.

Agresti, A. (1990). *Categorical data analysis*. Chichester, UK: John Wiley.

Amaranthus, M.P. & Perry, D.A. (1994). The functioning of ectomycorrhizal fungi in the field: linkages in space and time. *Plant and Soil* 159: 133–140.

Barbault, R. & Stearns, S. (1991). Towards an evolutionary ecology linking species interactions, life-history strategies and community dynamics: an introduction. *Acta Oecologica* 12: 3–10.

Boutin, C. & Keddy, P.A. (1993). A functional classification of wetland plants. *Journal of Vegetation Science* 4: 591–600.

Box, E.O. (1981). *Macroclimate and Plant forms: An Introduction to Predictive Modelling in Phytogeography*. The Hague: Junk.

Bradshaw, A.D. & McNeilly, T. (1991). Evolutionary response to global climatic change. *Annals of Botany* 67: 5–14.

Brown, V.K. (1991). The effects of changes in habitat structure during succession in terrestrial communities. In *Habitat Structure, The Physical Arrangement of Objects in Space*, ed. S.S. Bell, E.D. McCoy & H.R. Mushinsky, pp. 141–168. London: Chapman Hall.

Bryant, J.P., Chapin III, F.S., & Klein, D.R. (1983). Carbon/nutrient balance of boreal plants in relation to vertebrate herbivory. *Oikos* 40: 357–368.

Calder, I.R. (1990). *Evaporation in the Uplands*. Chichester, UK: John Wiley.

Carlquist, S. (1988). *Comparative Wood Anatomy*. Berlin: Springer-Verlag.

Chapin, F.S.III (1980). The mineral nutrition of wild plants. *Annual Review of Ecology and Systematics* 11: 233–260.

Chapin, F.S.III (1993). Functional role of growth forms in ecosystem and global processes. In *Scaling Physiological Proceses: Leaf to Globe*, ed. J.R. Ehleringer & C.B. Field, pp. 287–312. San Diego: Academic Press.

Chapin, F.S.III, Autumn, K., & Pugnaire, F. (1993). Evolution of suites of traits in response to environmental stress. *The American Naturalist* 142: S78–S92.

Chapin, F.S.III, Bret-Harte, M.S., Hobbie, S. & Zhong, H. (1996). Plant functional types as predictors of the transient response of arctic vegetation to global change. *Journal of Vegetaion Science* 7: 347–357.

Christensen, N.L. (1985). Shurbland fire regimes and their evolutionary consequences. In *The Ecology of Natural Disturbance and Patch Dynamics*, ed. S.T.A.S. Pickett & P.S. White, pp. 85–100. Orlando, FL: Academic Press.

Cody, M.L. (1989). Discussion: structure and assembly of communities. In *Perspectives in Ecological Theory*, J. Roughgarden, R.M. May, & S.A. Levin, pp. 227–241. Princeton, NJ: Princeton University Press.

Davis, M.B. (1981). Quaternary history and the stability of forest communities. In *Forest Succession: Concepts and Applications*, ed. D.C. West, H. Shugart & D.B. Botkin, pp. 132–153. NY: Springer-Verlag.

Delcourt, H.R. & Delcourt, P.A. (1991). *Quaternary Ecology – A Paleoecological Perspective*. London: Chapman & Hall.

Delcourt, P.A. & Delcourt, H.R. (1987). *Long Term Forest Dynamics of the Temperate Zone. Ecological Studies 63*. New York: Springer-Verlag.

Diamond, J.M. (1975). Assembly of species communities. In *Ecology and Evolution of Communities*, ed. M.L. Cody & J.M. Diamond pp. 342–444. Cambridge, MA, USA: Hardvard University Press.

Díaz, S. & Cabido, M. (1995). Plant functional types and trait–environment linkages: a multiscale approach. *Bulletin of the ESA (Suppl.)* 76: 64.

Díaz, S. & Cabido, M. (1997). Plant functional types and ecosystem function in relation to global change: a multiscale approach. *Journal of Vegetation Science* 8: 463–474.

Díaz, S., Acosta, A., & Cabido, M. (1992). Morphological analysis of herbaceous communities under different grazing regimes. *Journal of Vegetation Science* 3: 689–696.

Díaz, S., Acosta, A., & Cabido, M. (1994). Community structure in montane grasslands of central Argentina in relation to land use. *Journal of Vegetation Science* 5: 483–488.

Drake, J.A. (1990). Communities as assembled structures: do rules govern pattern? *Trends in Ecology and Evolution* 5: 159–164.

Dublin, H.T., Slinclair, A.R.E., & McGlade, J. (1990). Elephants and fire as causes of multiple states in the Srengeti-Mara woodlands. *Journal of Animal Ecology*: 1147–1164.

Faegri, K. & van der Pijl, L. (1979). *The Principles of Pollination Ecology*. Oxford: Pergamon.

Fernández-Alés, R., Laffarga, J.M., & Ortega, F. (1993). Strategies in Mediterranean grassland annuals in relation to stress and disturbance. *Journal of Vegetation Science* 3: 313–322.

Gange, A.C., Brown, B.V., & Farmer, L.M. (1990). A test of mycorrhizal benefit in an early successional plant community. *New Phytologist* 115: 85–91.

Gardner, S.M., Cabido, M., Valladares, G.R., & Díaz, S. (1995). The influence of habitat structure on arthropod diversity in Argentine semi-arid Chaco. *Journal of Vegetation Science* 6: 349–356.

Gitay, H. & Noble, I.R. (1997). What are plant functional types and how should we seek them? In *Plant Functional Types*, ed. T.M. Smith, H.H. Shugart, & F.I. Woodward, pp. 3–19. Cambridge: Cambridge University Press.

Golluscio, R. & Sala, O.E. (1993). Plant functional types and ecological strategies in Patagonian forbs. *Journal of Vegetation Science* 4: 839–846.

Gould, S.J. & Lewontin, R.C. (1979). The spandrels of San Marco and the Panglossian paradigm: a critique of the adaptationist programme. *Proceedings of the Royal Society of London* 205: 581–598.

Grime, J.P. (1977). Evidence for the existence of three primary strategies in plants and its relevance to ecological and evolutionary theory. *The American Naturalist* 111: 1169.

Grime, J.P. (1979). *Plant Strategies and Vegetation Processes*. Chichester, UK: John Wiley.

Grime, J.P. (1993). Ecology sans frontières. *Oikos* 68: 385–392.

Grime, J.P., Hodgson, J.G., & Hunt, R. (1988). *Comparative Plant Ecology: A Functional Approach to Common British Species*. London: Unwin Hyman.

Grime, J.P., Mackey J.M.L., Hillier, S.H., & Read, D.J. (1987). Floristic diversity in a model system using experimental microcosms. *Nature* 328: 420–422.

Hansen, J., Russell, G., Rind, D., Stone, P., Lacis, A., Lebedef, S., Ruedy, R., & Travis, L. (1983). Efficient three-dimensional global models for climate studies. Models I and II. *Monthly Weather Review* 111: 609–662.

Hargrave, K.R., Kolb, K.J., Ewers, F.W., & Davis, S.D. (1994). Conduit diameter and drouht-induced embolism in *Salvia mellifera* Greene (Labiatae). *New Phytologist* 126: 695–705.

Harris, P. (1991). Classical biocontrol of weeds: its definition, selection of effective agents, and administrative-political problems. *Canadian Entomologist* 123: 827–849.

Harvey, P.H. & Pagel, M.D. (1991). *The Comparative Methods in Evolutionary Biology*. Oxford: Oxford University Press.

Hobbie, S.E. (1992). Effects of plant species on nutrient cycling. *Trends in Ecology and Evolution* 7: 336–339.

Hodgson, J.G. & Grime, J.P. (1990). The role of dispersal mechanisms, regenerative strategies and seed banks in the vegetation dynamics of the British landscape. In *Species Dispersal in Agricultural Habitats*, R.G.H. Bunce & D.C. Howard, pp. 61–81. London: Belhaven.

Holling, C.S. (1991). Cross-scale morphology, geometry, and dynamics of ecosystems. *Ecological Monographs* 62: 447–502.

Houghton, J.J., Meiro Filho, L.G., Callander, B.A., Harris, N., Kattenberg, A. & Maskell, K. (1996). *Climate Change 1995: The Science of Climate Change. Contribution of Working Group I to the Second Assessment Report of the Intergovernmental Panel on Climate Change*. Cambridge: Cambridge University Press.

Howe, H.F. & Smallwood, J. (1982). Ecology of seed dispersal. *Annual Review of Ecology and Systematics* 13: 201–228.

Huntley, B. (1992). Rates of change in the European palynological record of the last 13 000 and their climatic interpretation. *Climate Dynamics* 6: 185–191.

Jarvis, P.G. & McNaughton, K.G. (1986). Stomatal control of transpiration: scaling up from leaf to region. *Advances in Ecological Research* 15: 1–49.

Jones, C.G., Lawton, J.H., & Shachak, M. (1994). Organisms as ecosystem engineers. *Oikos* 69: 373–386.

Keddy, P.A. (1989). *Competition*. London: Chapman & Hall.

Keddy, P.A. (1992). Assembly and response rules: two goals for predictive community ecology. *Journal of Vegetation Science* 3: 157–164.

Kelliher, F.M., Leuning, R., & Schulze, E-D. (1993). Evaporation and canopy characteristics of coniferous forests and grasslands. *Oecologia* 95: 153–163.

Lambers, H. & Poorter, H. (1992). Inherent variation in growth rate between higher plants: a searchj for physiological causes and ecological consequences. *Advances in Ecological Research* 23: 187–261.

Larcher, W. (1995). *Physiological Plant Ecology*. Berlin: Springer-Verlag.

Lawton, J.H. (1983). Plant architecture and the diversity of phytophagous insects. *Proceedings Sixth Annual Tall Timbers Fire Ecology Conference*. 28: 23–39.

Lawton, J.H. (1987). Are there assembly rules for successional communities? In *Colonization, Succession and Stability*, ed. A.J. Gray, M.J. Crawley, & P.J. Edwards, pp. 225–244. Oxford: Blackwell Scientific Publications.

Leishman, M.R. & Westoby, M. (1992). Classifying plants into groups on the basis of associations of individual traits–evidence from Australian semi-arid woodlands. *Journal of Ecology* 80: 417–424.

Leishman, M.R., Hughes, L., French, K., Amstrong, D., & Westoby, M. (1992). Seed and seedling biology in relation to modelling vegetation dynamics under global climate change. *Australian Journal of Botany* 40: 599–613.

McClaugherty, C.A., Pastor, J., Aber, J.D. & Melillo, J.M. (1985). Forest litter decomposition in relation to soil nitrogen dynamics and litter quality. *Ecology* 66: 266–275.

McIntyre, S., Lavorel, S., & Tremont, R.M. (1995). Plant life-history attributes: their relationship to disturbance response in herbaceous vegetation. *Journal of Ecology* 83: 31–44.

McNaughton, S.J. & Oesterheld, M. (1990). Extramatrical abundance and grass nutrition in a tropical grazing ecosystem, the Serengeti National Park, Tanzania. *Oikos* 59, 92–96.

McNaughton, S.J., Oesterheld, M., Frank, D.A., & Williams, K.J. (1989). Ecosystem-level patterns of primary productivity and herbivory in terrestrial habitats. *Nature* 341: 142–144.

Marone, L. (1991). Habitat features affecting bird spatial distribution in the Monte desert, Argentina. *Ecologia Australis* 1: 77–86.

Montalvo, J., Casado, M.A., Levassor, C., & Pineda, F.D. (1991). Adaptation of ecological systems: compositional patterns of species and morphological and functional traits. *Journal of Vegetation Science* 2: 655–666.

Nadelhoffer, K.J., Giblin, A.E., Shaver, G.R., & Laundre, J.A. (1991). Effects of temperature and organic matter quality on C, N, and P mineralization in soils from six arctic ecosystems. *Ecology* 72: 242–253.

Nemani, R.R. & Running, S.W. (1989). Testing a theoretical climate–soil–leaf area hydrologic equilibrium of forests using satellite data and ecosystem simulations. *Agricultural and Forest Meteorology* 44: 245–260.

Noble, I.R. (1989). Attributes of invaders and the invading process: terrestrial and vascular plants. In *Biological Invasions: A Global Perspective*, ed. J.D. Drake *et al.*, pp. 301–313. Chichester, UK: John Wiley.

Noble, I.R. & Slatyer, R.O. (1980). The use of vital attributes to predict successional changes in plant communities subject to recurrent distrubances. *Vegetatio* 43: 5–21.

Poorter, H. & Bergkotte, A. (1992). Chemical composition of 24 wild species differing in relative growth rate. *Plant, Cell and Environment* 15: 221–229.

Reich, P.B., Walters, M.B., & Ellsworth, D.S. (1992). Leaf life-span in relation to leaf, plant and stand characteristics among diverse ecosystems. *Ecological Monographs* 62: 365–392.

Richardson, D.M. & Bond, W.J. (1991). Determinants of plant distribution: evidence from pine invasions. *The American Naturalist* 137: 639–668.

Rochefort, L. & Woodward, F.I. (1992). Effects of climate change and a doubling of CO_2 on vegetation diversity. *Journal of Experimental Botany* 43: 1169–1180.

Sala, O.E., Lauenroth, W.K., & Golluscio, R.A. (1997). Plant functional types in temperate arid regions. In *Plant Functional Types*, ed. T.M. Smith, H.H. Shugart, & F.I. Woodward, pp. 217–233. Cambridge: Cambridge University Press.

Schulze, E-D. & Chapin, F.S. (1987). Plant specialization to environments of different resource availability. *Ecological Studies* 61: 120–148.

Schulze, E-D. & Mooney, H.A. (1994). *Biodiversity and Ecosystem Function.* Berlin: Springer-Verlag.

Schulze, E-D. & Zwölfer, H. (1994). Fluxes in ecosystems. In *Flux Control in Biological Systems*, ed. E-D. Schulze pp. 421–445. New York: Academic Press.

Schulze, E-D. (1982). Plant life forms and their carbon, water and nutrient relations. *Encyclopedia of Plant Physiology, New Series* 12B: 615–676.

Schulze, E-D. (1995). Flux control at the ecosystem level. *Trends in Ecology and Evolution* 10: 40–43.

Shaver, G.R., Giblin, A.E., Nadelhoffer, K.J., & Rastetter, E.B. (1997). Plant functional types and ecosystem change in Arctic tundras. In *Plant Functional Types*, ed. T.M. Smith, H.H. Shugart, & F.I. Woodward, pp. 153–173. Cambridge: Cambridge University Press.

Simberloff, D. & Dayan, T. (1991). The guild concept and the structure of ecological communities. *Annual Review of Ecology and Systematics* 22: 115–143.

Solomon, A.E. & West, D.C. (1986). Atmospheric carbon dioxide change: agent of future forest growth or decline. In *Effects of Changes in Stratospheric Ozone and Global Climate*, ed. J.G. Titus, pp. 23–38. Washington DC: US Environmental Protection Agency.

Steffen, W.L., Walker, B.H., Ingram, J.S., & Koch, G.W. (1992). *Global Change and Terrestrial Ecosystems: The Operational Plan*. The Royal Swedish Academy of Sciences: IGBP-Secretariat.

Swift, M.J., Heal, O.W., & Anderson J.M. (1979). *Decomposition in Terrestrial Ecosystems.* Berkeley, CA: University of California Press.

Thompson, K. & Grime, J.P. (1979). Seasonal variations in the seed banks of herbceous species in ten contrasting habitats. *Journal of Ecology* 67: 893–921.

Thompson, K., Band, S.R., & Hodgson, J.G. (1993). Seed size and shape predict persistence in soil. *Functional Ecology* 7: 236–241.

van der Valk, A.G. (1981). Succession in wetlands: a Gleasonian approach. *Ecology* 62: 688–696.

Walker, B.H. (1992). Biodiversity and ecological redundancy. *Conservation Biology* 6: 18–23.

Watkins, A.J. & Wilson, J.B. (1992). Fine-scale community structure of lawns. *Journal of Ecology* 80: 15–24.

Westoby, M., Leishman, M.R., & Lord, J.M. (1995). On misinterpreting the 'phylogenetic correction'. *Journal of Ecology* 83: 531–534.

Wilson, J.B. & Agnew, A.D.Q. (1992). Positive-feedback switches in plant communities. *Advances in Ecological Research* 23: 263–336.

Wilson, J.B. & Roxburgh, S.H. (1994). A demonstration of guild-based assembly rules for a plant community and determination of intrinsic guilds. *Oikos* 69: 267–276.

Wilson, J.B. & Gitay, H. (1995). Community structure and assembly rules in a dune slack: Variance in richness, guild proportionality, biomass constancy and dominance/diversity relations. *Vegetatio*, 116: 93–106.

Wilson, J.B., Allen, R.B., & Lee, W.G. (1995a). An assembly rule in the ground and herbaceous strata of a New Zealand rain forest. *Functional Ecology* 9: 61–64.

Wilson, J.B., Sykes, M.T., & Peet, R.K. (1995b). Time and space in the community structure of a species-rich limestone grassland. *Journal of Vegetation Science* 6: 729–740.

Woodward, F.I. (1987). *Climate and Plant Distribution*. Cambridge: Cambridge University Press.

Woodward, F.I. & Diament, A.D. (1991). Functional approaches to predicting the ecological effects of global change. *Functional Ecology* 5: 202–212.

13

When does restoration succeed?

Julie L. Lockwood and Stuart L. Pimm

Introduction

How effective have efforts been to restore nature? Has anyone ever success-fully restored a degraded site, species for species, or function for function? We attempt to answer these key questions of restoration ecology by review-ing 87 published studies. We also try to relate our results to ecological theory, which may strike the reader as overly ambitious. Our ambition stems from prior experience of assembling other scattered and highly subjective descrip-tions of nature – food webs. On reading these early compilations, many field workers felt (correctly) that they could collect much better data. As a conse-quence, the quality and quantity of new studies improved (Pimm *et al.*, 1991) and served to refine food web theory considerably.

Obviously, all restoration ecologists wish for more and better data. It is hoped our ideas will suggest specific choices from the bewildering array of observations for them to report. The ecological theories presented here will surely need improvement; indeed, it may be necessary to abandon them com-pletely. Theories improve most when confronted with data, even data as ill-defined as subjective accounts of restoration's successes and failures.

The simplest division in the expressed goals of restoration is between func-tion and structure. By 'function' we mean ecosystem processes, such as pri-mary productivity, water purification, soil erosion, and the loss or retention of nutrients. By 'structure' we mean measures of species composition. These measures range from a complete list of an area's original inhabitants, to such qualitative terms as 'wildlife habitat'. Function and structure are unlikely to be independent, of course. What does ecological theory suggest about the likelihood of restoring function or structure?

On getting function

Large-scale surveys of ecosystem processes reveal consistent patterns. It is possible to predict terrestrial primary productivity from evapotranspiration (Rosenzweig, 1968), secondary productivity from primary (McNaughton *et al.*, 1989; Cyr & Pace, 1993), the stocks of soil carbon and nitrogen from a site's Holdridge life-zone (Post & Pasteur, 1985) and many other features (see Schulze & Mooney, 1993). These patterns are broadly independent of the species the ecosystems contain. This does not mean that species composition does not matter, for much scatter remains about these statistical relationships. It does suggest, however, that nature achieves these relationships using a vast array of possible species combinations. Thus, in order to restore ecosystem functions, it may not be necessary to restore any special species combination.

To restore ecosystem functions, it may not even be necessary to have the original number of species. The connections between the number of species an area supports and its ecosystem processes are tenuous. Clearly, with no species there can be no functions. Yet, except for systems with unrealistically few species, there is, as yet, little evidence for a relationship between structure and function (Naeem *et al.*, 1994; Lockwood & Pimm, 1994). Finally, restoration of ecosystem function does not even require native species (Aronson *et al.*, 1993).

Particular ecosystem functions simply appear to be consistent with a range of species numbers and broad arrays of species compositions including native and introduced species. Metaphorically, the restoration of function should be an easy target to hit.

On getting structure

In contrast, the restoration of the original species composition, or any particular species composition, must surely require a close approximation to the original ecosystem processes. For example, without fire some prairies will become forests. Although restoration of this ecosystem function may be necessary to restore structure, surely it is not sufficient. Simply burning an area may still not restore the original prairie's composition. Differences in species composition in prairies may be determined in part by biotic interactions superimposed on the template of ecosystem processes. Metaphorically, structure will be a hard target to hit, because more constraints must be satisfied.

Critics will argue that this case has been overstated: subtle differences in environmental condition may drive all differences in species composition.

After all, not all prairie fires are the same. Restore these exact conditions and perhaps the original species composition will return.

The experience of restoration provides a test of these competing hypotheses. If nature is so sloppy as to accommodate a wide variety of species compositions within a narrow range of ecosystem conditions, then function should be much easier to restore than structure. There will be many pathways to the former and few to the latter.

Partial or complete structure

Critics will now (rightly) notice that we have applied a double standard. 'Structure' is meant to be some precise measure of species composition – often the original! 'Function' is intended to mean some rough measures of productivity, nutrient retention, and so on. Society often imposes these different standards upon restoration. There are circumstances when it does not. Some partial structure may be the goal; by this we mean any loose grouping of native species but not a particular sub-set of that group – for example, hardwood trees, prairie plants, ground cover, 'habitat for wildlife' and so on are considered partial structure.

Following earlier arguments, function and partial structure should be equally easy to restore: there are multiple ways to achieve both. There are a variety of species composition that will provide suitable 'habitat for wildlife' or include a group of prairie plants. One might define, say, a prairie as easily by its partial structure (grasses, forbs, but no trees) as by its processes (short hydroperiod, episodic fires). Ease of definition may also equate to ease of restoration: the more precise an object's definition, the more difficult it will be to create.

Secondary succession or community assembly?

So far, a crude theory of the likely patterns in the success of a restoration has been derived. It is based on the number of acceptable endpoints and thus the number of pathways that will lead us there. Our analogy begs the question: what if there are few pathways, but they are easy, well-worn ones? Perhaps familiarity is more important than number. Long-held ideas on secondary succession suggest that community composition moves along pathways towards a final end-point with goal-driven determination.

Communities assemble by the addition of species through invasions and by the loss of species through extinctions. In this sense, a pathway is the sequence of species invasions and extinctions that leads to a particular species composition. Succession, by its very name, implies that the pathways are traditional and commonly traveled. Some restorations may be able to mimic these traditional roles, and so incorporate natural processes of disturbance and recovery.

Unfortunately, there is no guarantee that desired species A need follow the introduction of species B. As a society, neither 'every cog and wheel' (Leopold, 1948), – all the species, nor the 'assembly instructions' – the ecological history of the site have been kept. In ecological succession, species follow a time-worn path through community assembly. In many restorations, these pathways may be lost due to excessive habitat fragmentation, lingering pollution, and so on.

What does the theory of community assembly tell us?

If the familiar pathway toward community regeneration is lost, what does ecological theory tell us to expect? Computer simulations and small-scale experiments in community assembly now recognize two endpoints along a continuum of possible behaviors (Drake, 1988; Luh & Pimm, 1993; Lockwood *et al.* 1997). The rate at which species are allowed to attempt colonization determines the rate at which the species composition of that community turns over. Communities in which species are allowed to attempt invasion often, churn through those species that establish regularly and quickly. Virtually all species in the species pool (i.e., the group of species that potentially could be community members) become members of the community at some time during the assembly process. Each species persists for only a very short period of time; no species come to dominate (Lockwood *et al.*, 1997).

Communities in which species are allowed to attempt invasion much less often will show a greater persistence, and may, in fact, become completely invasion resistant. By 'persistent' it is typically meant that no (or, in practice, little) turnover in species composition occurs (Drake, 1990; Pimm, 1991). A small set of all species in the pool will become community members and they will persist for long periods. Given that different places will likely experience different histories, each species has its chance somewhere, rather than sometime (Drake *et al.*, 1993).

Interestingly, theory also suggests that from any initial set of species, several states (each with different species compositions) are possible (Drake, 1988, 1990; Luh & Pimm, 1993). The presence of a species early in the development of the community may have profound effects on the species composition at any selected point in the assembly history. Communities are not just the result of processes that operate today, but are also the products of a long, capricious history. The historical sequence in which species are added to and removed from a community determine its extant structure.

Within the practical realm of restoration, assembly theory then suggests that it may be possible to re-establish a set of species, but not always the desired set. If species attempt invasion often, any resultant community will likely be ephemeral. Indeed, the repeat of any exact mix of species may be statistically improbable. If species attempt invasion occasionally, when composition is highly dependent on history, restoring a community from its extant, constituent species may be impossible (Drake, 1988, 1990; Pimm, 1991; Luh & Pimm, 1993). This Humpty-Dumpty effect (Pimm, 1991) holds that species important in determining structure may not be present in the community composition we see today. Thus, there is no guarantee that practical restoration can exactly or sufficiently repeat a community's history, even if all the species of the extant community are available for reintroduction, if the reintroduction attempts are sufficiently separated in time and abiotic conditions are sufficiently replicated.

In sum, and based on results from community assembly experiments, it is first argued that restoring ecosystem functions or partial structure should be relatively easy to achieve. The restoration of a particular species composition will not be easy or it may be impossible. Second, it is argued that communities in which functional or structural goals are successfully restored may be transitory and may quickly revert back to an 'unsuccessful' state. The only exceptions to these two 'rules' come when the restorationist can employ the natural patterns of secondary succession.

Restoration as ecological experimentation

The most difficult bridge to span between the disciplines of ecological theory and restoration practice will likely remain the similarity of their terms and the contrariety in their meanings (Pimm, 1984). We now try to translate our theoreticians' lexicon (e.g., persistence and community composition) into the vocabulary of practical restoration. In so doing, our method of evaluating restoration projects will be outlined within the context of community assembly theory.

Restoration often involves the manipulation of the physical and chemical environment. These may involve major changes and we do not underrate their difficulty. A restoration attempt is also an uncontrolled experiment in assembling a community (Pimm, 1991; Luh & Pimm, 1993). Species are added ('seeding') to the community either by planting or inoculating the restoration site with a desired set of species or by natural colonization of species from adjacent sources. Species are removed ('weeding') either through active removal (often actual weeding) of unwanted species or from the unintended loss of all individuals of a desired species that is purposefully introduced.

Recall that assembly theory suggests two 'rules', one concerning the achievement of persistence and the other the achievement of a particular species composition. Our translations between the theoretical and practical meanings of these two terms, given below, provide our two criteria for evaluating success and failure. A restored community is considered to be persistent when the turnover in species composition falls to a sufficiently low level (i.e., something resembling the natural turnover rate for a particular system). Ideally, those responsible for management of the restoration will cease to 'weed and seed' when this happens. Thus, when management stops (no more weeding and seeding), we consider persistence achieved.

The appropriate species composition is considered to be successfully restored when those responsible for evaluating a project judge that a biological goal has been successfully attained. Thus, it is assumed that, if the restorationist has been successful at restoring a goal, it is because he has re-established an appropriate species composition. From our theory derived above, function and partial structure goals are expected to be successfully attained using a variety of 'correct' species compositions. Structural goals can only be attained using a limited set of species compositions. By further dividing goals into function, full structure and partial structure categories, the rate at which each is successfully attained can be assessed and thus the applicability of our theory can be tested.

Persistence may not be tightly linked with the restoration of species composition. Thus, management may cease even though several goals may not have been attained (e.g., successfully establishing a persistent set of prairie plants which do not necessarily include those rare species most desired). Similarly, goals may be achieved but management may not stop (e.g., improving lake water clarity through continued 'weeding and seeding' of desired species). Thus, these two criteria are considered relatively independent: one criteria may be met without having to necessarily accomplish the other.

Restoration projects can be reviewed and identified as either 'successes' or 'failures' in terms of these two major features: achievement of persistence and an appropriate species composition. First, restorationists are either 'weeding and seeding' or else they have achieved a sufficiently low level of species turnover that this is not deemed necessary. Second, restorationists either achieve or fail to achieve their other goals. A project is deemed completely successful only if it satisfies all projected biological goals and achieves persistence as defined by the cessation of management.

By using these translations of practical to theoretical terms, a project can be judged as successful or failed based simply on knowledge gained from published reports. Some of these reports appear in peer-reviewed journals. Others do not, and may not be as carefully scrutinized. Still others are secondhand reports from compilations of case studies. Wherever possible, the primary source was obtained along with other corroborating publications. It is recognized, however, that not all publications have passed through equally stringent publication guidelines and that relying on secondhand interpretations may misrepresent results. The results from peer-reviewed and not-peer-reviewed plus compilation publications will be compared to assay the effects of data quality.

It is understood that the terms 'success' and 'failure' are loaded with meaning. Here it is assumed that practitioners hope to restore some attribute(s) of the original community and have these attributes persist for long periods of time. Under other assumptions, a project classified as 'failed' may be quite successful (e.g., the project successfully revitalizes a fishing industry or improves the quality of life for nearby residents).

Our criteria of success and failure incorporate several biases associated with the sociological setting of restoration efforts. The authors' judgment of whether or not a specific goal was met is relied upon. There is likely a bias toward optimistic judgments on the part of some authors. This bias is difficult to assess, but will likely inflate the ease with which goals were attained. When management ceases, it will be assumed this is because changes in species composition are minimal enough to require no further intervention. This is likely not always true. Individuals responsible for an expensive, legally mandated restoration efforts, for example, may 'declare victory and go home', ceasing to assume responsibility regardless of the site's changing species composition. Although some projects that potentially fit this class can be identified, it is not possible to reliably estimate the size of this bias nor correct for it. Thus, it is also likely that the ease at which persistence is achieved is overestimated. It is acknowledged that our criteria will consistently err on the side of optimism and therefore our conclusions must be carefully tempered.

The literature base

Literature was gathered using computer databases, citation indexes, and bibliographies. To include a project, it had to satisfy three constraints. First, each project must have intended goals. Second, projects in which the damaged ecosystem was assigned a new use, through exotic species introduction or physical manipulation were excluded (Aronson *et al.*, 1993; Bradshaw 1988). By doing this, the scope of the review has been narrowed to include only those projects that seek to restore some portion of the native ecosystem. Third, each project was subjected to at least some initial management intervention. This may include simply the cessation of pollution or some other source of degradation, but most often there was some physical manipulation of the site.

Eighty-seven restoration projects met our criteria. Projects were incorporated as we found them, so our sample represents how easily they could be attained. Thus, if projects describing wetland restorations in the continental United States were easier to find than those for forests in central Siberia, they were more likely to be included in our analyses. Our review does include a variety of community types, however, and they are located around the world. The compilation of restoration efforts is representative, not exhaustive.

All the projects are listed in the Appendix. It includes the author(s) of the publications, the location of the project, the success category (see below), and the author(s) judgment of whether or not each goal was met. Three descriptive variables are also included: duration, system type, and size. The duration of each project was calculated by subtracting the date of initial management from the date of latest publication. Thus, each project is judged during the time period for which we have information and not beyond. For many recent projects, this information is necessarily preliminary. Projects range in duration from 1 to 53 yr with the average being 6.3 yr. Duration is categorized as short-term (1–5 yr), medium-term (6–20 yr), and long-term (> 20 yr). Each project is classified broadly according to system type. 'Freshwater' includes rivers, streams and lakes (15 studies), 'marine' includes seagrass beds, rocky subtidal areas, and artificial reefs (18 studies), 'terrestrial' includes prairies and woodlands (23 studies) and 'wetland' includes any area which is partially or periodically flooded (31 studies). The size of the site where the restoration takes place is grouped into three categories: small (1–10 ha), medium (11–50 ha), and large (> 51 ha). There are 59 small, 20 medium, and only 8 large sites.

Classifying the goals

Structural goals include anything that pertains to the return of species. The success rates of returning a community species-for-species (full structure) v. limited sets of appropriate species (partial structure) is of interest, so we report these goals separately.

For example, consider two projects intended to restore a freshwater wetland. One may list the establishment of a 'sufficient diversity of freshwater plants' as its goal. The second lists the return of each of the species observed at the site prior to disturbance. The first desires the return of diversity *per se* and thus pays less attention to the original species composition nor even to the use of only original species (although not exotics). Some of the original species may never return, nor are they encouraged to do so, but the finished product still resembles a native freshwater wetland. The return of partial structure is sufficient for a successful judgment. The second project demands the return not only of native species but most of the original species composition. Here, those responsible are very concerned about which species are present and which are not. Full structure is desired and required for a successful judgment.

Functional goals comprise variables that ecologists consider under the broad umbrella of ecosystem processes. Functional goals, like partial structure, do not require the return of a specific set of species. Functional goals may call for similar intervention as for structural goals. For example, a functional goal might be achieved by the planting of *Spartina* to control erosion but the goal is that process and not *Spartina* and its associated biota.

Classifying the results

Unsuccessful

In 17 projects, management neither ceased nor were all the biological goals achieved. Commonly, projects under this category faced physical restraints. For example, improper turbidity in a seagrass restoration prevented the long-term establishment of underwater vegetation (Teas, 1977), or previously unknown contaminants uncovered during restoration operations caused the die-off of large portions of the newly planted marsh vegetation (Josselyn *et al.*, 1990).

Partially successful

This class includes the majority of the projects considered (53). There are two subclasses of partial success:

All goals met but management continues

In 11 studies, management continues in efforts to control for unwanted species or species are continually added to the site, but the other goals are met.

For example, in many prairie restorations when other than a desired species enters, it is actively removed. If a desired species fails or becomes only poorly established it is repeatedly introduced. It is possible to achieve intended goals while this management continues. In one case, the goals of establishing a few dominant prairie grasses and controlling erosion were both met (Kramer, 1980). Although those responsible were concerned about species composition and continued to manage the site to direct which species established, the goals stated at the outset of the project were apparently met by the current, if changing, species composition.

Also included in this class are restorations where management continues in efforts to correct problems with physical characteristics of the site but not necessarily to control for unwanted species. These projects often face the difficulties associated with restoring water quality, soil properties or elevation characteristics. For example, when large river systems or lakes are considered for restoration those responsible must improve water quality by stopping non-point pollution sources (Berger, 1992; Markarewicz & Bertram, 1991). Often, such improvements are achieved allowing for the return of many native fish. However, the continuation of this function (i.e., providing fish habitat) is dependent on the continued improvements in water quality through constant management. Here functions are restored without much regard to the exact species composition that supports that function (i.e., as long as fish have something to feed on they will survive; it is not necessary to return the set of species that the fish originally fed on).

Management stops but not all goals are met
This type of project is the most common in our review, with 42 case studies. Typically, some of the desired species persist and some, but not all, functions are restored. For example, the Salmon River salt marshes were restored through a series of projects initiated in the early 1980s (Frenckel & Morlan, 1990). The project was successful in restoring a diversity of plants typical of salt marshes and biomass production. The native species richness of the site was not restored. Since all three goals were mentioned and only two achieved, we classify this project as partially successful.

There is a subset of 13 projects in which management ceased but the reasons for the cessation are not clearly detailed or expected. Most often, management ceased despite none of the stated goals having been met. An unusual example involves a wetland that, after years of neglect, was set for restoration in mitigation for construction on a nearby site (Josselyn *et al.*, 1990). As their goals, the authors listed restoring a diversity of wetland plants and the use of the area by birds. Neither goal was met. Inadvertently, the area did provide habitat for an endangered plant, the Humboldt Bay Owl's clover (*Orthocarpus castillejoides* var. *humboldtiensis*) causing those responsible to cease management (at least temporarily).

Table 13.1. *Under our success criterion, the achievement of persistence and a species composition that will meet the projected goals of the project determine success*

Characteristics of the project	Total number of projects that fit characterization	Success category	Type of goals listed
Persistent and met all biological goal	17	Successful	13 functional goals 18 partial structure 1 complete structure
Persistent and not all goals met	42	Partially Successful	41 functional goals 32 partial structure 25 complete structure
Not persistent and all goals met	11	Partially Successful	13 functional goals 10 partial structure 0 complete structure
Not persistent and not all goals met	17	Unsuccessful	13 functional goals 10 partial structure 7 complete structure

Using this criterion, only 17 (20%) of the projects in our literature base we considered completely successful. Within those 17 projects, 32 goals were given. Only one of those was the restoration of the native community species-for-species.

Complete success

In this class are 17 projects that satisfied all prescribed goals and management ceased. For example, a project in Florida to mitigate marsh habitat destruction due to phosphate mining listed as its goals the establishment of a diversity of trees and shrubs common to freshwater wetlands in the area (Erwin, 1985). Both of these goals were met and management was determined no longer necessary, thus we consider this project completely successful.

Results

(a) An equal percentage (20%) of all projects are completely successful or unsuccessful; most projects are only partially successful (60%: Table 13.1).

(b) There is no difference in the success rates reported from publications from peer reviewed journals and publications from non-reviewed or compiled sources. In 24 peer reviewed publication, 4 projects were judged complete successes, 4 failed, and 16 achieved partial success. In 63 non-reviewed sources, 13 were considered complete successes, 13 failed and 37 were partially successful. These proportions are almost identical ($\chi^2 = 0.46$, 2df, $P = 0.79$) and we henceforth discount biases between sources.

Table 13.2. *Biological goals listed by all projects*

Function	Successful	Failed	Total
Water quality	11 (92%)	1	12
Sustaining habitat	16 (73%)	6	22
Productivity	13 (38%)	21	34
Erosion control	9 (75%)	3	12
Total	49 (61%)	31	180
Structure			
Partial	46 (66%)	24	70
Complete	2 (6%)	32	34
Total	48 (46%)	56	104

There are two types of goals: functional and structural. We divide the structural goals into those that are restoring only partial structure (e.g., diversity *per se*) and those that are restoring complete structure (i.e., portions of the community species-for-species). The number of successes and failures are listed beside each goal type. Function and partial structure are considered restored more often than complete structure.

(c) Of the 184 goals given, 97 (52%) were achieved (Table 13.2). Functional goals were listed a total of 80 times and considered restored by the author(s) in 49 of those attempts (61%). Structural goals were listed 104 times with success occurring in 48 of those attempts (46%).

(d) Most of the successes under structure, however, came from restoring diversity *per se* or establishing the dominant species (46 out of the 48 successes) rather than a specific species composition (Table 13.2). Restoring a specific assemblage species-for-species was listed as a goal 34 times but only two attempts were considered successful (6%). Thus, function and partial structure are restored at least 10 times more frequently than species-for-species restoration ($\chi^2 = 17.76$, $P < 0.001$).

(e) When we consider the subset of projects deemed 'completely successful' only one listed among its goals the return of full structure (Southward, 1979; Hawkins & Southward, 1992). The other goals, function and partial structure, were mentioned 13 and 18 times, respectively. The restoration of complete structure is considerably underrepresented among this group of completely successful projects ($\chi^2 = 7.1$, $P < 0.05$). Those projects that wished to restore a diversity of the appropriate species are at least as likely to be included in our complete success category as those who wished to restore function. However, those projects that attempted to restore the original community species-for-species are far less frequently considered completely successful.

Table 13.3. *No association existed between complete success and such descriptive variables as duration, size or type of system*

	Duration	Size	System type
Complete success	$\chi^2 = 1.565$ df = 2 $P = 0.457$	$\chi^2 = 2.346$ df = 2 $P = 0.3094$	$\chi^2 = 0.153$ df = 3 $P = 0.9849$

(f) We found no χ^2 association between projects deemed completely successful and any descriptive variables (Table 13.3). This is likely the result of combining a large number of projects that incorporate many temporal and spatial scales.

(g) It is considered that 59 of 87 attempts (61%) to have achieved persistence. This rate falls to 48% if we exclude a set of 17 projects in which it is unclear why management ceased.

Discussion

A broad-scale view of the practice of restoration has purposely been taken to search for patterns. Two have been found. First, those who practise restoration (or who evaluate restoration projects) are frequently pleased with efforts to restore such features as water quality, habitat structure suitable for deer, a substantial biomass of grasses or trees, or reduce the rate of soil loss. In contrast, only a very few are pleased with efforts to restore a particular species composition. Thus, function and partial structure are indeed considered restored more often than species composition.

Second, in a total of 34 deliberate attempts, only two projects restored original species composition. In these projects, those responsible were likely utilizing secondary successional pathways and the spread of adjacent extant communities (Gore & Johnson, 1981; Southward, 1979; Hawkins & Southward, 1992). In one project a river was diverted along a 2.5 km stretch while coal was mined from under the original bed. Restoration then proceeded by re-inforcing the natural bed, controlling erosion and forcing water flow back along its original course. The upstream portion of the river served as the source pool for colonizing invertebrates and eventually that set of characteristic species established (Gore & Johnson, 1981). The second successful attempt was comparable. After an oilspill and its clean-up along a rocky tidal coastline, the invertebrate sessile community was destroyed. Eventually, the species typical of the area returned after the remaining oil and solvent disappeared. Again, the source for the colonizing species was presumably an undisturbed adjacent coastline since no reintroduction of species was attempted (Southward, 1979; Hawkins & Southward, 1992). Thus, it may be that original species richness will only be restored when secondary successional pathways are still available for exploitation.

Two possible theoretical explanations have been outlined above for why these patterns exist. The first emphasizes the necessity of exactly replicating original physical conditions. The second is the nature of community assembly dynamics: even given identical physical conditions, restoration of a particular species composition may be impossible, at best improbable, and even then, only transitory. When restorationists can use the historically familiar, frequently explored pathways of secondary succession, it may be possible to escape such a pessimistic fate.

We report a relatively high rate of persistence, which we equate with the cessation of management. This may be the product of the relatively short average duration of the projects (6.3 yr). It remains to be seen whether persistence is indeed achieved despite the cessation of management. The level of persistence achieved by most restoration projects is important beyond satisfying political and economic requirements. Collecting long-term community turnover data from restoration projects may be the only way to distinguish which of the two assembly scenarios described above (i.e., constant species turnover or achieving alternative persistent states) best describe most restoration attempts.

No matter which of these two assembly models proves the best descriptor of most restoration attempts, incorporating several separate restoration protocols within any one site may be most efficacious at restoring lost species. Restoring original composition is a worthy goal, of course. Those who practise restoration should accept that the communities they build might be functional replicates, but – as our models predict – they will not be structural replicates. Any attempt at hitting that one 'target' will almost surely fail for the reasons given above.

Varying the restoration protocol across the restoration site will increase the numbers and variety of species the site will support. For example, some areas may be devoted to repeated and frequent attempts at re-establishing a set of desired native species. In adjacent areas, however, these same species are given only a few scattered opportunities at establishment. These variations in protocol effectively provide different colonization rates and sequences across the restoration site. These differences will likely produce a series of variant communities which, taken together, may hold all of the desired native species while also providing a complete set of original functions. In their totality, these variant communities (together called a meta-communities by Wilson, 1992) may be the best way to ensure that most desired species are restored. To our knowledge, this suggestion has not yet been explored by any restoration.

References

Aronson, J, Floret, C. Le Floch E. Ovalle C., & Pontanier, R. (1993). Restoration and rehabilitation of degraded ecosystems in arid and semi-arid lands. I. A view from the south. *Restoration Ecology* 1: 8–17.

Berger, J.J. (1992). Restoring attributes of the Willamette River. In *Restoration of Aquatic Ecosystems*. National Resource Council, Washington, DC: National Academic Press.

Bradshaw, A.D. (1998). Alternative endpoints for reclamation. In *Rehabilitating Damaged Ecosystems* vol II, ed. John Cairns Jr. Boca Raton, FA: CRC Press, Inc.

Cyr, H. & Pace, M.L. (1993). Magnitude and patterns of herbivory in aquatic and terrestrial ecosystems. *Nature* 361 (6488): 148–149.

Drake, J.A. (1988). Models of community assembly and the structure of ecological landscapes. In *Mathematical Ecology*, ed. T. Hallam, L. Gross, & S. Levin, Singapore: World Press.

Drake, J.A. (1990). The mechanics of community assembly and succession. *Journal of Theoretical Biology*. 147: 213–233.

Drake, J.A., Flum, T.E., Witteman, G.J., Voskuil, T., Hoylman, A.M., Creson, C., Kenny, D.A., Huxel, G.R., LaRue, C.S. & Duncan, J.R. (1993). The construction and assembly of an ecological landscape. *Journal of Animal Ecology* 62: 117–130.

Erwin, K.L., Best, G.R., Dunn, W.J. & Wallace, P.M. (1985). Marsh and forested wetland reclamation of a central Florida phosphate mine. *Wetlands* 4: 87–104.

Frenkel, R.E. & Morlan, J.C. (1990). Restoration of the Salmon River salt marshes: retrospect and prospect. Final Report to the US Environmental Protection Agency, Seattle, Washington.

Gore, J.A. & Johnson, L.S. (1981). Strip-mined river restoration. *Water Spectrum* 13(1): 31–38.

Hawkins, S.J. & Southward, A.J. (1992). The *Torrey Canyon* oil spill: recovery of rocky shore communities. In *Restoring the Nation's Marine Environment*, ed. G.W. Thayer, Maryland Sea Grant College, College Park, MD.

Josselyn, M., Zedler, J., & Griswold, T. (1990). Lincoln street marsh wetland mitigation (Appendix II). In *Wetland Creation and Restoration: Status of the Science*, ed. J.A. Kusler & M.E. Kentula. Washington, DC: Island Press.

Kramer, G.L. (1980). Prairie studies at caterpillar tractor co. Peoria proving ground. In *Proceedings of the Seventh North American Prairie Conference*, ed. C.L. Kucera. University of Missouri, Columbia, MO.

Leopold, A. (1948). *A Sand County Almanac*, Oxford: Oxford University Press.

Lockwood, J.L. & Pimm, S.L. (1994). Species: would any of them be missed? *Current Biology* 4(5): 455–457.

Lockwood, J.L., Powell, R., Nott, M.P., & Pimm, S.L. (1997). Assembling ecological communities in time and space. *Oikos* 80: 549–553.

Luh, H-K. & Pimm, S.L. (1993). The assembly of ecological communities: a minimalist approach. *Journal of Animal Ecology* 62: 749–765.

McNaughton, S.J., Oesterheld, M., Frank, D.A., & Williams, K.J. (1989). Ecosystem productivity and herbivory in terrestrial habitats. *Nature* 341: 142–144.

Makarewicz, J.C. & Bertram, P. (1991). Evidence for the restoration of the Lake Erie ecosystem. *BioScience* 41(4): 216–223.

Naeem, S., Thompson, L.J., Lawlor, S.P., Lawton, J.H., & Woodfin, R.M. (1994). Declining biodiversity can alter the performance of ecosystems. *Nature* 368: 734–737.

Pimm, S.L. (1984). The complexity and stability of ecosystems. *Nature* 307: 321–326.

Pimm, S.L. (1991). *The Balance of Nature? Ecological issues In the Conservation of Species and Communities.* Chicago, IL: University of Chicago Press.

Pimm, S.L., Lawton, J.H., & Cohen, J.E. (1991). Food web patterns and their consequences. *Nature* 350: 669–674.

Post, W.M. & Pasteur. A. (1985). Global patterns in carbon cycling. *Nature* 317: 613–616.

Rosenzweig, M.L. (1968). Net primary productivity of terrestrial environments: predictions from climatological data. *American Naturalist* 102: 67–84.

Schulze, E-D. & Mooney, H.A. (1993). Biodiversity and ecosystem function. Berlin: Springer-Verlag.

Southward, A.J. (1979). Cyclic fluctuations in population density during eleven years recolonization of rocky shores in West Cornwall following the 'Torrey Canyon' oil-spill in 1967. In *Cyclic Phenomena in Marine Plants and Animals*, ed. E. Naylor & A.G. Harnoll. Oxford: Pergamon Press.

Teas, H.J. (1977). Ecology and restoration of mangrove shorelines in Florida. *Environmental Conservation* 4(1): 51–58.

Wilson, D.S. (1992). Complex interactions in metacommunities, with implications for biodiversity and higher levels of selection. *Ecology* 73: 1984–2000.

Appendix

All projects used in this study are listed by author. Also included are location, achievement of persistence and all goals, success according to both criteria, and the intended goals. Each project is categorized under three descriptive variables: duration, system type, and size.

Table 13.4 *Project attributes*

Author(s)	Peer	Location of project	Persistence?	All goals?	Complete success?	Type of goal	Goal attained?	Duration	System type	Size
Berger, J.J.	No	Williamette River, OR	No	No	No	Functional Partial	(1) yes (2) no	1 1	Freshwater	2
Berger, J.J.	No	Mattole River Watershed, CA	No	No	No	Functional Functional	(1) yes (2) no	2 2	Freshwater	3
Berger, J.J.	No	Blanco River, CO	Yes	Yes	Yes	Partial	(1) yes	1	Freshwater	2
Buckner, D.L. & Wheeler, R.L.	No	Freshwater marsh, Front Range, CO	Yes	Yes	Yes	Partial Functional	(1) yes (2) yes	1 1	Wetland	1
Erwin, K.L.	Yes	Agrico phosphate , FL mine restoration	Yes	No	Partial	Partial Partial	(1) no (2) yes	1 1	Wetland	2
Fonseca, M.S. et al.	Yes	Eelgrass bed, NC	Yes	No	Partial	Complete Complete	(1) no (2) no	1 1	Marine	
Hutchinson, D.E.	No	Prairie restoration at Homestead Monument	No	Yes	Partial	Partial	(1) yes	3	Terrestrial	2
Pritchett, D.A.	No	Vernal Pools in Santa Barbara, CA	Yes	No	Partial	Complete	(1) no	1	Freshwater	1
Josselyn et al.	No	Vernal Pools in San Diego County, CA	Yes	No	Partial	Complete	(1) no	1	Freshwater	1
Josselyn, M. et al.	No	Hayward Regional Shoreline, CA	Yes	No	Partial	Functional Partial Functional	(1) no (2) yes (3) no	1 1 1	Wetland	1
Josselyn, M. et al.	No	Bracut Marsh, CA	Yes	No	Partial	Partial Functional Partial	(1) no (2) no (3) no	2 2 2	Wetland	1

Table 13.4 (*continued*)

Author(s)	Peer	Location of project	Persistence?	All goals?	Complete success?	Type of goal	Goal attained?	Duration	System type	Size
Josselyn, M. et al.	No	Lincoln Street Marsh, CA	No	Yes	Partial	Functional Functional Functional Functional Functional Partial	(1) yes (2) yes (3) yes (4) yes (5) yes (6) yes	1 1 1 1 1 1	Wetland	1
Levin D.A. & Willard, D.E.	No	The Harbor Project, Sandusky Bay, OH	Yes	Yes	Yes	Partial Functional	(1) yes (2) yes	2 2	Wetland	1
Levin D.A. & Willard, D.E.	No	Lack Puckaway, WI	Yes	No	Partial	Functional Partial Functional Functional	(1) yes (2) yes (3) yes (4) no	1 1 1 1	Freshwater	2
Moss, B. et al.	Yes	Norfolk Broadland Lakes, UK	Yes	Yes	Yes	Functional Partial	(1) yes (2) yes	1 1	Freshwater	
Butler, R.S. et al.	Yes	Lake Tohopekaliga, FL	Yes	No	Partial	Complete	(1) yes	1	Freshwater	2
Josselyn, M. et al. & Roberts, L.	No	Sweetwater Marsh, CA	Yes	No	Partial	Partial Functional Functional Partial	(1) yes (2) no (3) no (4) no	2 2 2 2	Wetland	1
Thorhaug, A.	Yes	Florida Keys seagrass bed, FL	Yes	Yes	Yes	Partial	(1) yes	1	Marine	1
Erwin, K.L.	Yes	Agrico phosphate mine second site, FL	Yes	Yes	Yes	Partial Partial	(1) yes (2) yes	1 1	Wetland	1
Mclaughlin, P.A. et al.	Yes	Turkey Point seagrass bed, FL	Yes	No	Partial	Functional Complete	(1) yes (2) no	2 2	Marine	1

Asuquo Obot, E.	Yes	Nigerian freshwater wetland	Yes	No	Partial	Complete	(1) no	Wetland	2
Smith, I. *et al.*	Yes	Seagrass bed, NC	Yes	No	Partial	Functional	(1) no	Marine	1
Anderson, B.W. & Ohmart, R.D.	No	Lower Colorado River Test plots	Yes	No	Partial	Partial Complete	(1) yes (2) no	Terrestrial	1 1
Childress, D.A. & Eng, R.L.	No	MT	Yes	Yes	Yes	Partial	(1) yes	Wetland	2
Hueckel, G.J. *et al.*	Yes	Elliott Bay, Puget Sound rocky intertidal	Yes	No	Partial	Functional	(1) yes	Marine	1
Jessee, W.N. *et al.*, Carter *et al.*	Yes	Pendelton Artificial Reef, CA	Yes	No	Partial	Functional Functional	(1) no (2) yes	Marine	1 1
Teas, H.J.	Yes	Mangrove shorelines, FL	No	No	No	Functional Partial	(1) no (2) no	Wetland	1 1
Cammen, L.	Yes	Salt Marsh, NC	Yes	No	Partial	Functional Functional Functional	(1) no (2) no (3) yes	Wetland	2 1 1
Gore, J.A. & Johnson, L.S	Yes	Tongue River, WY	Yes	No	Partial	Complete Complete Functional	(1) yes (2) no (3) no	Freshwater	1 1 1
Brunori, C.	No	Freshwater marsh along Elk River, MA	No	Yes	Partial	Functional	(1) yes	Wetland	1
Brunori, C.	No	Freshwater marsh on Bohemia River, MA	No	Yes	Partial	Functional	(1) yes	Wetland	1
Bontje, M.P.	No	Salt Marsh, NJ	Yes	No	Partial	Partial Partial Partial Partial Partial	(1) no (2) yes (3) yes (4) no (5) no	Wetland	2 1 1 1 1
Brown, V.K. & Gibson, C.W.D.	Yes	Grassland in UK	Yes	No	Partial	Complete	(1) no	Terrestrial	2

Table 13.4 (*continued*)

Author(s)	Peer	Location of project	Persistence?	All goals?	Complete success?	Type of goal	Goal attained?	Duration	System type	Size
Broome, S.W.	No	Saltwater Marsh at Pamlico estuarary, NC	No	No	No	Functional	(1) no	1	Wetland	1
Carangelo, P.D.	No	Hog Island mitigation, TX	No	No	No	Partial	(1) no	1	Marine	1
Carangelo P.D.	No	Huffco Project, TX	Yes	No	Partial	Functional Functional	(1) no (2) no	1 1	Marine	1
Carangelo, P.D.	No	TX	No	No	No	Functional	(1) no	1	Marine	1
Carangelo, P.D.	No	Clark Island Project, TX	No	No	No	Functional	(1) no	1	Marine	1
Bragg, T.B.	No	Allwine Prairie Transplants, NE	No	No	No	Complete	(1) no	1	Terrestrial	1
Bragg, T.B.	No	NE	No	Yes	Partial	Partial	(1) yes	2	Terrestrial	2
Steigman, K.L. & Ovenden, L.	No	Heard Wildlife Sanctuary transplants,	No	No	No	Complete	(1) no	1	Terrestrial	1
Nichols, O. *et al.*	No	Jarrah forest of Western Australia	Yes	No	Partial	Functional Complete Functional	(1) no (2) no (3) yes	1 1 1	Terrestrial	3
Hayes, T.D. *et al.*	No	Rolling plains prairie of north-cental Texas	No	Yes	Partial	Partial Partial	(1) yes (2) yes	2 2	Terrestrial	2
Southward, A.J.	No	Rocky shores of West Cornwall, UK	Yes	Yes	Yes	Complete	(1) yes	2	Marine	2
Pickart, A.J.	No	Dune revegetation at Buhne Point, CA	Yes	No	Partial	Complete Functional Functional	(1) no (2) yes (3) no	1 1 1	Terrestrial	1
Kramer, G.L.	No	Prairie establishment, IA	No	Yes	Partial	Functional Partial	(1) yes (2) yes	1 1	Terrestrial	1
Drake, L.D.	No	Prairie establishment, IA	Yes	Yes	Yes	Functional Partial	(1) yes (2) yes	1 1	Terrestrial	1

Author		Description								
Holler J.R.	No	Microfungla community in Prairie, WI	No	No	No	Complete / Functional	(1) no / (2) no	1 / 1	Terrestrial	1
Woehler, E.E. & Martin, M.A.	No	Prairie, WI	Yes	No	Partial	Complete	(1) no	1	Terrestrial	1
Dale, E.E. Jr. & Smith, T.C.	No	Prairie at Pea Ridge National Military Park, AR	No	No	No	Partial	(1) no	1	Terrestrial	1
Kirt, R.R.	No	Prairie at College of DuPage, IL	Yes	No	Partial	Complete	(1) no	1	Terrestrial	1
Webb, J.W. and Newling, C.J.	Yes	Salt Marshes Galveston Bay, TX	Yes	No	Partial	Functional / Functional / Functional	(1) yes / (2) yes / (3) no	1 / 1 / 1	Wetland	1
LaGrange, T.G. & Dinsmore, J.J	Yes	Freshwater wetlands, IA	Yes	No	Partial	Partial / Functional / Partial	(1) yes / (2) yes / (3) no	2 / 2 / 2	Wetland	1
Shisler, J.K. & Charette, D.J.	No	Salt Marshes, N.J.	Yes	No	Partial	Partial / Partial / Functional	(1) no / (2) no / (3) yes	1 / 1 / 1	Wetland	1
Beeman, S.	No	Salt Marsh, FL	No	Yes	Partial	Functional / Partial	(1) yes / (2) yes	1 / 1	Wetland	1
Gillo, J.L.	No	Marshland, FL	No	No	No	Functional / Partial / Functional	(1) yes / (2) no / (3) yes	1 / 1 / 1	Wetland	1
Goforth, H.W. Jr. & Williams M	No	Mangrove shorelines, FL	Yes	No	Partial	Functional / Partial	(1) yes / (2) no	1 / 1	Wetland	1
Newling, C.J. et al.	No	Saltwater marsh in L Apalachicola Bay, F	Yes	Yes	Yes	Functional / Functional / Partial / Functional	(1) yes / (2) yes / (3) yes / (4) yes	1 / 1 / 1 / 1	Wetland	1
Lewis, R.R. III & K.C. Haines	No	Mangrove restoration St Croix, Virgin Islands	Yes	No	Partial	Functional / Partial	(1) no / (2) no	1 / 1	Wetland	2

Table 13.4 (continued)

Author(s)	Peer	Location of project	Persistence?	All goals?	Complete success?	Type of goal	Goal attained?	Duration	System type	Size
Lewis, R.R. III & R.C. Phillips	No	Seagrass bed at Craig Key, FL	Yes	No	Partial	Partial / Functional	(1) no / (2) no	1 / 1	Marine	1
Kenworthy, W.J. et al.	No	Seagrass bed in Back Sound, NC	Yes	No	Partial	Functional / Partial / Functional	(1) yes / (2) yes / (3) no	1 / 1 / 1	Marine	1
George, D.H.	No	Lagoon at Cape Canaveral, FL	Yes	Yes	Yes	Partial / Partial	(1) yes / (2) yes	1 / 1	Wetland	1
Hoffman, R.S.	No	Eelgrass bed at Mission Bay, CA	Yes	No	Partial	Complete / Functional	(1) no / (2) no	1 / 1	Marine	1
Harrison, P.G.	No	Eelgrass bed southwestern British	Yes	Yes	Yes	Functional / Partial	(1) yes / (2) yes	1 / 1	Marine	1
Nitsos, R.	No	Eelgrass transplants in Morro Bay, CA	Yes	Yes	Yes	Partial / Functional	(1) yes / (2) yes	1 / 1	Marine	1
Merkel, K.W.	No	Eelgrass bed in Chula Vista Wildlife Reserve	Yes	No	Partial	Partial	(1) no	1	Marine	1
Frenkel, R.E. & Morian, J.C.	No	Salmon River Salt Marshes, OR	Yes	No	Partial	Complete / Functional / Partial	(1) no / (2) yes / (3) yes	2 / 2 / 2	Wetland	2
Allen, H.H. & Webb J.W.	Yes	Saltmarsh in Mobile Bay, AL	No	No	No	Partial / Functional	(1) no / (2) no	1 / 1	Wetland	1
Lewis, R.R. III	No	Needlerush marsh, FL	Yes	No	Partial	Partial / Functional	(1) yes / (2) no	1 / 1	Wetland	1
Berger, J.J. & Kent, M. et al.	No	Kissimmee River demonstration project, FL	Yes	No	Partial	Complete / Complete / Complete / Complete	(1) no / (2) no / (3) no / (4) no	1 / 1 / 1 / 1	Wetland	3

Reference	Yes	Site	Yes	No	Partial	Complete	(1) no		Ecosystem	
Sperry, T.M./ Zimmerman J.H. J.A. Schwarzmeier/ Greene, H.C. & J.T. Curtis	Yes	University of Wisconsion Arboretum Prairie, WI	Yes	No	Partial	Partial	(1) no (2) yes	3 3	Terrestrial	2
Woehler, E.E. & Martin, M.A.	No	Prairies on public land, WI	Yes	No	Partial	Parital Partial	(1) no (2) yes	1 1	Terrestrial	1
Wombacher, J & R. Garay/ Burton, P.J. et al.	No	Prairie at Knox College, IL	No	No	No	Complete Complete	(1) no (2) no (3) no	3 3 3	Terrestrial	1
Burton, P.J. et al./ Schulenberg, R.	No	Morton Aboretum, IL	No	No	No	Partial Complete Partial	(1) no (2) no (3) no	3 3 3	Terrestrial	1
Makarewicz, J.C. & Bertram P. Charlton M.N. et al.	Yes	Lake Erie	No	No	No	Functional Partial	(1) no (2) yes	2 2	Freshwater	3
Urabe, J.	Yes	Japanese Pond	Yes	Yes	Yes	Functional Functional	(1) yes (2) yes	2 2	Freshwater	1
Lutz, K.A.	No	Prairie, OH	Yes	No	Partial	Complete Partial	(1) no (2) yes	2 2	Terrestrial	1
Edwards, R.D. & Woodhouse, W.W.	No	Saltwater Marsh in Pamlico Esturary, NC	Yes	No	Partial	Complete Partial Functional	(1) no (2) yes (3) yes	1 1 1	Wetland	1
Haynes, R.J. & Moore, L.	No	Bottomland forest in the Southeastern US	Yes	Yes	Yes	Partial Functional Partial	(1) yes (2) yes (3) yes	2 2 2	Terrestrial	2
Vassar, J.W. et al.	No	Prairie in Mark Twain National Forest, MO	Yes	Yes	Yes	Partial	(1) yes	1	Terrestrial	2
Kilburn, P.D.	No	Prairies, IL	Yes	Yes	Yes	Partial	(1) yes	1	Terrestrial	1

Table 13.4 (*continued*)

Author(s)	Peer	Location of project	Persistence?	All goals?	Complete success?	Type of goal	Goal attained?	Duration	System type	Size
Keller, W. & . Yan, N.D	Yes	Sudbury Lakes, Ontario, Canada	No	No	No	Functional Complete	(1) yes (2) no	2 2	Freshwater	3
Gameson, A.L.H. . & Wheeler, A	No	Thames Estuary, UK	No	Yes	Partial	Functional Partial	(1) yes (2) yes	3 3	Wetland	3
Sparks, R.	No	Illinois River	No	Yes	Partial	Functional Functional Partial	(1) yes (2) no (3) yes	3 3 3	Freshwater	3
Edmondson, W.T.	No	Lake Washington, WA	Yes	No	Partial	Functional Functional Partial Partial	(1) yes (2) no (3) no (4) yes	2 2 2 2	Freshwater	3
Moore. D.C. & Rodger, G.K.	Yes	Garroch Head, Isle of Bute,Scotland	No	Yes	Partial	Partial Functional	(1) yes (2) yes	2 2	Marine	2
Olmstead, L.I. & . Cloutman, D.G	Yes	Mud Creek, AR	Yes	No	Partial	Complete Functional Functional	(1) no (2) yes (3) no	1 1 1	Freshwater	1

References

Allen, H.H. & Webb, J.W. (1983). Influence of breakwaters on artificial saltmarsh establishment on dredged material. In *Proceedings of the Ninth Annual Conference on Wetlands Restoration and Creation*, ed. F.J. Webb Jr. Hillsborough Community College, Tampa, FA.

Anderson, M.R. & Cottam, G. (1970). Vegetational change on the Greene prairie in relation to soil characteristics. In *Proceedings of a Symposium on Prairies and Prairie Restoration*, ed. P. Schramm, Knox College, Galesburg, IL.

Anderson, B.W. & Ohmart, R.D. (1979). In *The Mitigation* Symposium: a national workshop in mitigating losses of fish and wildlife habitats. General technical report RM-65. USDA Forest Service.

Asuquo Obot, E., Chinda, A., & Braid, S. (1992). Vegetation recovery and herbaceous production in a freshwater wetland 19 years after a major oil spill. *African Journal of Ecology* 30: 149–156.

Beeman, S. (1983). Techniques for the creation and maintenance of intertidal saltmarsh wetlands for landscaping and shoreline protection. In *Proceedings of the Tenth Annual Conference on Wetland Restoration and Creation*, ed. F.J. Webb, Hillsborough Community College, Tampa, FA.

Berger, J.J. (1992a). Restoring attributes of the Willamette River. In *Restoration of Aquatic Ecosystems*. National Resource Council, Washington, DC: National Academic Press.

Berger, J.J. (1992b). Citizen restoration efforts in the Mattole river watershed. In *Restoration of Aquatic Ecosystems*. National Resource Council. Washington, DC: National Academic Press.

Berger, J.J. (1992c). The Blanco River. In *Restoration of Aquatic Ecosystems*. National Resource Council. Washington, DC: National Academic Press.

Bontje, M.P. (1987). The application of science and engineering to restore a salt marsh, 1987. In *Proceedings of a Conference: Increasing Our Wetland Resources*, ed. J. Zelany & J.S. Feierbend. Washington, DC: National Wildlife Federation.

Bragg, T.B. (1976). Allwine prairie preserve: a reestablished bluestem grassland research area. In *Proceedings of the 5th North American Prairie Conference*, Iowa State University, IA.

Bragg, T.B. (1988). Prairie transplants: preserving ecological diversity. In *The Prairie: Roots of Our culture; Foundation of our economy. Proceedings of the Tenth North American Prairie Conference*, ed. A. Davis & G. Stanford. Dallas, TX: Native Prairie Association of Texas.

Broome, S.W. (1987). Creation and development of brackish-water marsh habitat. In *Proceedings of a Conference: Increasing our Wetland Resources*, ed. J. Zelany & J.S. Feierbend. Washington, DC: National Wildlife Federation.

Brown, V.K. & Gibson, C.W.D. (1994). Re-creation of species-rich calcichlorous grassland communities. In *Grassland Management and Nature Conservation*, ed. R. Haggar.

Brunori, C.R. (1987). Examples of wetland creation and enhancement in Maryland. In *Proceedings of a Conference: Increasing our Wetland Resources*, ed. J. Zelany & J.S. Feierbend. Washington, DC: National Wildlife Federation.

Buckner, D.L. & Wheeler, R.L. (1990). Construction of cattail wetlands along the east slope of the Front Range in Colorado. In *Environmental Restoration*, ed. J.J. Berger. Washington, DC: Island Press.

Burton, P.J., Robertson, K.R., Iverson, L.R. & Risser, P.G. (1988). Use of resource partitioning and disturbance regimes in the designing and management of restored prairies. In *Reconstruction of Disturbed Arid Lands: An Ecological Approach*, ed. E.B. Allen. Boulder, CO: Westview Press, Inc.

Butler, R.S., Hulon, M.W., Moyer, E.J. & Williams, V.P. (1992). Littoral zone invertebrate communities as affected by a habitat restoration project on Lake Tohopekaliga, Florida. *Journal of Freshwater Ecology* 7(3): 317–328.

Cammen, L.M. (1976a). Macroinvertebrate colonization of *Spartina* marshes artificially established on dredge spoil. *Estuarine and Coastal Marine Science* 4: 357–372.

Cammen, L.M. (1976b). Abundance and production of macroinvertebrates from natural and artificially established salt marshes in North Carolina. *The American Midland Naturalists* 96(2): 487–493.

Carangelo, P.D. (1987). Creation of seagrass habitat in Texas: results of research investigations and applied programs. In *Proceedings of a Conference: Increasing our Wetland Resources*, ed. J. Zelany & J.S. Feierbend, Washington, DC: National Wildlife Federation.

Carter, J.W., Jessee, W.N. Foster, M.S. & Carpenter, A.L. (1985). Management of artificial reefs designed to support natural communities. *Bulletin of Marine Science* 37(1): 114–128.

Charlton, M.N., Milne J.E. Booth, W.G. & Chiocchio, F. (1993). Lake Erie offshore in 1990: restoration and resilience in the Central Basin. *Journal of Great Lakes Research* 19(2): 291–309.

Childress, D.A. & Eng, R.L. (1979). Dust abatement project with wildlife enhancement on Canyon Ferry Reservoir, Montana. In The mitigation symposium: a national workshop in mitigating losses of fish and wildlife habitats. General technical report RM–65. USDA Forest Service.

Dale, E.E. Jr. & Smith, T.C. (1980). Changes in vegetation on a restored prairie at Pea Ridge National Military Park, Arkansas. In *Proceedings of the Seventh North American Prairie Conference*, ed. C.L. Kucera. University of Missouri, Columbia.

Edmondson, W.T. (1977). Recovery of Lake Washington from eutrophication. In *Recovery and Restoration of Damaged Ecosystems*, ed. J. Cairns Jr., K.L. Dickson, & E.E. Herricks. Charlottesville, VA: University of Virginia Press.

Edwards R.D. & Woodhouse Jr. W.W. (1982). Brackish marsh development. In *Proceedings of the Ninth Annual Conference on Wetlands Restoration and Creation*, ed. F.J. Webb Jr. Hillsborough Community College, Tampa, FL.

Erwin, K.L. (1990). Agrico 8.4 acre wetland (Appendix II). In: *Wetland Creation and Restoration: Status of the Science* ed. J.A. Kusler, & M.E. Kentula. Washington, DC: Island Press.

Erwin, K.L., Best, G.R. Dunn, W.J. & Wallace, P.M. (1985). Marsh and forested wetland reclamation of a central Florida phosphate mine. *Wetlands* 4: 87–104.

Fonseca, M.S., Kenworthy, W.J. Colby D.R. Rittmaster, K.A. & Thayer, G. (1990). Comparison of fauna among natural and transplanted eelgrass *Zostera marina* meadows: criteria for mitigation. *Marine Ecology Progress Series* 65: 251–264.

Frenkel, R.E. & Morland, J.C. (1990). Restoration of the Salmon River salt marshes: retrospect and prospect. Final Report to the US Environmental Protection Agency, Seattle, WA.

Gameson, A.L.H. & Wheeler, A. (1977). Restoration and recovery of the Thames Estuary. In: *Recovery and Restoration of Damaged Ecosystems*, ed. J. Cairns Jr., K.L. Dickson, & E.E. Herricks. Charlottesville, VA: University of Virginia Press.

George, D.H. (1982). Lagoon restoration at Cape Canaveral Air Force Station,

Florida. In *Proceedings of the Ninth Annual Conference on Wetlands Restoration and Creation*, ed. F.J. Webb Jr. Hillsborough Community College, Tampa, FL.

Gilio, J.L. (1983). Conversion of an impacted freshwater wet prairie into a functional aesthetic marshland. In *Proceedings of the Tenth Annual Conference on Wetland Restoration and Creation*, ed. F.J. Webb. Hillsborough Community College, Tampa, FL.

Goforth, H.W., Jr. & Williams, M. (1983). Survival and growth of red mangroves (*Rhizophora manle* L.) planted upon marl shorelines in the Florida Keys (A five-year study). In *Proceedings of the Tenth Annual Conference on Wetland Restoration and Creation*, ed. F.J. Webb, Hillsborough Community College, Tampa, FL.

Gore, J.A. & Johnson, L.S. (1981). Strip-mined river restoration. *Water Spectrum* 13(1): 31–38.

Green, H.C & Curtis, J.T. (1953). The re-establishment of prairie in the University of Wisconsin arboretum. *Wildflower* 29: 77–88.

Harrison, P.G. (1988). Experimental eelgrass transplants in southwestern British Columbia, Canada. In *Proceedings of the California Eelgrass Symposium: Chula Vista, California*, ed. K.W. Merckel & J.L. Stuckrath, National City, CA: Sweetwater River Press.

Hayes, T.D., Riskind, D.H. & Pace III, W.L. (1987). Patch-within-patch restoration of man-modified landscapes within Texas state parks. In *Landscape Heterogeneity and Disturbance*, ed. M.G. Turner. NY: Springer-Verlag.

Haynes, R.J. & Moore, L. (1987). Re-establishment of bottomland hardwoods within National Wildlife Refuges in the southeast. In *Proceedings of a Conference: Increasing our Wetland Resources*, ed. J. Zelany & J.S. Feierbend Washington, DC: National Wildlife Federation.

Hoffman, R.S. (1988a). Fishery utilization of natural versus transplanted eelgrass beds in Mission Bay, San Diego, California. In *Proceeding of the California Eelgrass Symposium: Chula Vista, California*, ed. K.W. Merckel & J.L. Stuckrath. National City, Sweetwater River Press.

Hoffman, R.S. (1988b) Recovery of eelgrass beds in Mission Bay, San Diego, California following beach restoration work. In *Proceedings of the California Eelgrass Symposium: Chula Vista, California*, ed. K.W. Merckel & J.L. Stuckrath. National City, CA: Sweetwater River Press.

Holler, J.R. (1981). Studies of soil microfungal populations of a prairie restoration project. *Proceedings of the Seventh North American Prairie Conference*, ed., C.L. Kucera. University of Missouri, Columbia.

Hueckel, G.J., Buckley, R.M., & Benson, B.L. (1989). Mitigating rocky habitat loss using artificial reefs. *Bulletin of Marine Science* 44(2): 913–922.

Hutchison, D.E. (1992). Restoration of the Tall Grass prairie at Homestead Monument, Beatrice, Nebraska. *Rangelands* 14(3): 174.

Jessee, W.N., Carpenter, A.L., & Carter, J.W. (1985). Distribution patterns and density estimates of fishes on a southern California artificial reef with comparisons to natural reef-kelp habitats. *Bulletin of Marine Science* 37(1): 214–226.

Josselyn, M. Zedler, J. & Griswold, T. (1990a). Vernal pool project in San Diego (Appendix II). In *Wetland Creation and Restoration: Status of the Science*, ed. J.A. Kusler & M.E. Kentula. Washington, DC: Island Press.

Josselyn, M. Zedler, J. & Griswold, T. (1990b). Sweetwater Marsh mitigation in San Diego County (Appendix II). In *Wetland Creation and Restoration: Status of the Science*, ed. J.A. Kusler & M.E. Kentula. Washington, DC: Island Press.

Josselyn, M. Zedler, J. & Griswold, T. (1990c). Hayward Regional Shoreline, Alameda county (Appendix II). In *Wetland Creation and Restoration: Status of the Science*, ed. J.A. Kusler & M.E. Kentula. Washington, DC: Island Press.

Josselyn, M. Zedler, J. & Griswold, T. (1990d). Bracut marsh mitigation bank. In *Wetland Creation and Restoration: Status of the Science*, ed. J.A. Kusler & M.E. Kentula. Washington, DC: Island Press.

Josselyn, M. Zedler, J. & Griswold, T. (1990e). Lincoln street marsh wetland mitigation (Appendix II). In *Wetland Creation and Restoration: Status of the Science*, ed. J.A. Kusler & M.E. Kentula. Washington, DC: Island Press.

Keller W. & Yan, N.D. (1991). Recovery of crustacean zooplankton species richness in Sudbury area lakes following water quality improvements. *Canadian Journal of Fisheries and Aquatic Sciences* 48: 1635–1644.

Kenworthy, W.J. Fonseca, M.S. Homziak, J., & Thayer, G.W. (1980). Development of a transplanted seagreas (*Zostera marina* L.) meadow in Back Sound, Carteret county, North Carolina. In *Proceedings of the Seventh Annual Conference on Wetland Restoration and Creation*, ed. D.P. Cole. Hillsborough Community College, Tampa, FA.

Kilburn, P.D. (1970). Hill prairie restoration. In *Proceedings of a Symposium on Prairies and Prairie Restoration*, ed. P. Schramm. Knox College, Galesburg, IL.

Kirt, R.R. (1990). Quantitative trends in the progression toward a prairie state by seed broadcasting and seedling transplant methods. In *Proceedings of the 12th North American Prairie Conference.*

Kramer, G.L. (1980). Prairie studies at caterpillar tractor co. Peoria proving ground. In *Proceedings of the Seventh North American Prairie Conference*, ed. C.L. Kucera. University of Missouri, DC.

LaGrange, T.G. & Dinsmore, J.J. (1989). Plant and animal community responses to restored Iowa wetlands. *Prairie Naturalist* 21(1): 39–48.

Levin, D.A. & Willard, D.E. (1990a). Lake Puckaway, Wisconsin (Appendix II). In *Wetland Creation and Restoration: Status of the Science*, ed. J.A. Kusler & M.E. Kentula. Washington, DC: Island Press.

Levin, D.A. & Willard, D.E. (1990b). The Harbour Project, Sandusky Bay, Ohio (Appendix II). In *Wetland Creation and Restoration: Status of the Science*, ed. J.A. Kusler & M.E. Kentula. Washington, DC: Island Press.

Lewis R.R. III. (1982). Restoration of a needlerush (*Juncus roemerianus* Scheele) following interstate highway construction II. Results after 22 months. In *Proceedings of the Ninth Annual Conference on Wetlands Restoration and Creation*, ed. F.J. Webb Jr. Hillsborough Community College, Tampa, FA.

Lewis R.R. III. & Haines, K.C. (1980a). Large scale mangrove restoration on St Croix, US Virgin Islands – II. Second Year. In *Proceedings of the Seventh Annual Conference on Wetland Restoration and Creation*, ed. D.P. Cole, editor. Hillsborough Community College, Tampa, FL.

Lewis R.R. III. & Phillips, R.C. (1980b). Experimental seagrass mitigation in the Florida Keys. In *Proceedings of the Seventh Annual Conference on Wetland Restoration and Creation*, ed. D.P. Cole. Hillsborough Community College, Tampa, FL.

Loftin, K.M., Toth, L.A., & Obeysekeka, J.T.B. (1988). *Proceedings: Kissimmee River Restoration Symposium*, Orlando, FL.

Lutz, K.A. (1988). Prairie establishment in southwestern Ohio. In *Proceedings of the Eleventh North American Prairie Conference: Prairie Pioneers: Ecology, History and Culture*, ed. T.B. Bragg & J. Stubbendreck. Lincoln, NE: University of Nebraska Printing.

McLaughlin, P.A., Treat, S.F., & Thorhaug, A. (1983). A restored seagrass (*Thalassia*) bed and its animal community. *Environmental Conservation* 10(3): 247–254.

Makarewicz, J.C. & Bertram, P. (1991). Evidence for the restoration of the Lake Erie ecosystem. *BioScience* 41(4): 216–223.

Merkel, K.W. (1988). Eelgrass transplanting in South San Diego Bay, California. In *Proceedings of the California Eelgrass Symposium: Chula Vista, California*, ed. K.W. Merckel & J.L. Stuckrath. National City, CA: Sweetwater River Press.

Moore, D.C. & Rodger, G.K. (1991). Recovery of a sewage sludge dumping ground. II. Macrobenthic community. *Marine Ecology Progress Series* 75: 301–308.

Morrison, D. (1987). Landscape restoration in response to previous disturbance. In *Landscape Heterogeneity and Disturbance*, ed. M.G. Turner. NY: Springer-Verlag.

Moss, B., Balls, H., Irvine, K., & Stansfield, J. (1986). Restoration of two lowland lakes by isolation from nutrient-rich water sources with and without removal of sediment. *Journal of Applied Ecology* 23: 391–414.

Newling, C.J., Landin, M.C. & Parris, S.D. (1983). Long-term monitoring of the Apalachicola Bay wetland habitat development site. In *Proceedings of the Tenth Annual Conference on Wetland Restoration and Creation*, ed. F.J. Webb. Hillsborough Community College, Tampa, FL.

Nichols, O., Wykes, B.J. & Majer, J.D. (1988). The return of vertebrate fauna and invertebrate fauna to bauxite mined areas in south-western Australia. In *Animals in Primary Succession: The Role of Fauna in Reclaimed Lands*, ed. J.D. Majer. Cambridge: Cambridge University Press.

Nitsos, R. (1988). Morro Bay eelgrass transplant. In *Proceedings of the California Eelgrass Symposium: Chula Vista, California*, ed. K.W. Merckel & J.L. Stuckrath. National City, CA: Sweetwater River Press.

Olmstead, L.I. & Cloutman, D.G. (1974). Repopulation after a fish kill in Mud Creek, Washington county, Arkansas following pesticide pollution. *Transactions of the American Fisheries Society* 1: 79–87.

Pickart A.J. (1990). Dune revegetation at Buhne Point, King Salmon, California. In *Environmental Restoration*, ed. J.J. Berger. Washington, DC: Island Press.

Pritchett, D.A. (1990). Creation and monitoring of vernal pools at Santa Barbara, California. In *Environmental Restoration*, ed. J.J. Berger Washington, DC: Island Press.

Schulenberg R. (1970). Summary of Morton Arboretum prairie restoration work, 1963 to 1968. In *Proceedings of a Symposium on Prairies and Prairie Restoration*, ed. P. Schramm. Knox College, Galesburg, IL.

Shisler, J.K. & Charette, D.J. (1984). Evaluation of artificial salt marshes in New Jersey. *New Jersey Agricultural Experiment Station* publication number P–40502–01–84.

Smith, I. Fonseca, M.S. Rivera, J.A. & Rittmaster, K.A. (1988). Habitat value of natural versus recently transplanted eelgrass, *Zostera marina*, for the Bay scallop, *Argopecten irradians*. *US Fishery Bulletin* 87: 189–196.

Sparks, R. (1992). The Illinois River–floodplain ecosystem. In *Restoration of Aquatic Ecosystems. National Resource Council*. Washington, DC: National Academic Press.

Sperry, T.M. (1983). Analysis of the University of Wisconsin–Madison prairie restoration project. In *Proceedings of the Eighth North American Prairie Conference*, ed. R. Brewer. Western Michigan University, Kalamazoo, MI:

Southward, A.J. (1979). Cyclic fluctuations in population density during eleven
    years recolonization of rocky shores in West Cornwall following the 'Torrey
    Canyon' oil-spill in 1967. In Cyclic Phenomena in Marine Plants and
    Animals, ed. E. Naylor & A.G. Harnoll. Oxford: Pergamon Press.
Steigman, K.L. & Ovenden, L. (1988). Transplanting tallgrass prairie with a
    sodcutter. In The Prairie: Roots of our Culture; Foundation of our Economy.
    Proceedings of the Tenth North American Prairie Conference. ed. A. Davis &
    G. Stanford. Native Prairie Association of Texas, Dallas, TX.
Teas, H.J. (1977). Ecology and restoration of mangrove shorelines in Florida.
    Environmental Conservation 4(1): 51–58.
Thorhaug, A. (1983). Habitat restoration after pipeline construction in a tropical
    estuary: seagrasses. Marine Pollution Bulletin 14(11): 422–425.
Toth, L.A. (1991). Environmental responses to the Kissimmee River demonstration
    project. Environmental Sciences Division, South Florida Water Management
    District Technical Publication 91–02.
Urabe, J. (1994). Effect of zooplankton community on seston elimination in a
    restored pond in Japan. Restoration Ecology 2(1): 61–70.
Vassar, J.W., Henke, G.A. & Blakely, C. (1978). Prairie restoration in North-
    Central Missouri. In Proceedings of the Sixth North American Prairie
    Conference, The Prairie Peninsula – In the Shadow of Transition, ed. R.L.
    Stuckey & K.J. Reese. College of Biological Sciences, The Ohio State
    University, OH.
Webb, J.W. & Newling, C.J. (1985). Comparison of natural and man-made salt
    marshes in Galveston Bay Complex, Texas. Wetlands 4: 75–86.
Woehler, E.E. & Martin, M.A. (1980). Annual vegetation changes in a
    reconstructed prairie. In Proceedings of the Seventh North American Prairie
    Conference, ed. C.L. Kucera. University of Missouri, MO.
Wombacher, J. & Garay R. (1973). Insect diversity and associations in a restored
    prairie. In Proceedings of the Third Midwest Prairie Conference, Kansas State
    University, Manhattan, KS.
Zimmerman, J.H., & Schwartz, J.A. (1976). Experimental prairie restoration at
    Wingra overlook. In Proceedings of the Seventh North American Prairie
    Conference, ed. C.L. Kucera. University of Missouri, MO.

14

Epilogue: From global exploration to community assembly

Paul Keddy

In accordance with the plan laid down, we proceed to the consideration of follies into which men have been led by their eager desire to pierce the thick darkness of futurity.

Charles Mackay (1841)

Introduction

The editors of a symposium volume are allowed to frame the question, choose the participants, contribute their own chapter, and offer opinions on drafts of each chapter. In this sense, they are like hosts at a party: they already own the house, and so should attempt to remain in the background, introduce the guests as they arrive, serve drinks and snacks unobtrusively, occasionally intervene if there are awkward periods of silence, and ensure that, in the end, the guests find their own coats and car keys and leave in a good humor. In general, then, it is probably unwise for editors to also presume to add commentaries, conclusions, or, as in some cases, multiple chapters. They risk becoming like the unfortunate host, who, having indulged too freely in his own liquor and become intoxicated with his own importance, proceeds to bore everyone with loud opinions on every topic that arises. In assuming the role of hosts, our intention was to remain sober, restricting ourselves to an introduction and a single contributed chapter. The purpose of the volume, after all, was to present a variety of views, not, like the over-refreshed host, to force readers to accept ours. Indeed, if our own views could not be adequately expressed in two sections, a third or fourth would probably be equally unsatisfactory, resembling the swaying host, who assumes that pushing his face closer and increasing the volume of his voice will make him more compelling. The summary and conclusions of this volume were therefore to be contributed by a guest, Jared Diamond, the author whose work so obviously inspired many of the chapters. Alas, Dr Diamond's field work conflicted with

393

publication deadlines, and so the task of summing up has fallen back upon editors. I am, to shift analogies, in the unenviable position of the opening act at a rock concert, who, finding the concert star delayed in traffic, must try to entertain (or at least distract) the audience to avoid damage to the stadium.

Since this volume will appear near the end to the twentieth century, it seems reasonable to begin by putting the topic of assembly rules within the context of broader developments in ecology, beginning with voyages of exploration and passing through the concept of habitat templates for communities. I then propose to discuss several aspects of assembly rules that challenged all participants, and offer a few brief observations on the chapters themselves.

The end of a great era: global exploration

We have now reached the end of a great era in biology: the era of global exploration, map-making, large collections of new species, and classification (e.g., Morris, 1973; Morison, 1978; Edmonds, 1997). This was an essential first step in ecology, revealing, as it did, the diversity of life forms on earth, and documenting the pool of species from which communities are assembled. While there remain new discoveries to be made, particularly in poorly known groups such as the arthropods and fungi, and under the oceans or on tropical tepui, the great period of explorers and collectors in sailing ships and steamers has passed. Morris (1973, vol. 1, pp. 232–249) recounts how only a little over a hundred years ago, 16 September 1864, the British Association for the Advancement of Science met in Bath, England. Among the celebrities were the two most controversial figures of African exploration, Richard Burton and John Speke. *The Times* called their impending confrontation a gladiatorial exhibition. The topic of debate? The source of the Nile.

From the comfort of our offices near the end of the twentieth century, it may be difficult to imagine the trials and tribulations of early explorers and collectors. In the above case, the joint expedition to Africa in 1858 ended one phase at Lake Tanganyika with Speke nearly blind from trachoma and Burton half-paralyzed by malaria; Speke's solitary reconnaissance trip 25 days later brought him to the shore of Lake Victoria. On Livingstone's last trip, still on the hunt for the Nile headwaters, he was 'delayed by tribal wars, constantly sick, losing his teeth one by one' when he reached the Arab slaver's village of Ujiji and languished near death. On 10 November 1871, he was discovered by Henry Stanley of the *New York Herald*, and greeted with the now famous 'Dr Livingstone, I presume?' Such stories of the era of exploration are delightfully recounted by James Morris (1973) in his three-volume history of the British Empire, *Pax Britannica*. Other examples include:

Ferdinand Magellan (*c.* 1480–1521) who endured conspiracy, mutiny, cheating by provisioners, and scurvy, eventually dying in the Philippine Islands while his ship, the Victoria, went on to be the first to circumnavigate the world.

James Cook (1728–1779) who charted much of the Pacific ocean including New Zealand and Australia, but who was killed by natives in Hawaii in a dispute over a stolen boat.

Alexander von Humboldt (1769–1859) who explored the northern areas of South America, set the world altitude limit for mountain climbing while ascending Mount Chimborazo, prepared a treatise on the political economy of Mexico, encouraged young scientists including Charles Darwin and Louis Agassiz, and died at 90 while working on the fifth volume of *Kosmos*, an overview of the structure and behaviour of the known universe.

Augustin de Candolle (1778–1841) and his son Alphonse (1806–1893) who laid many of the foundations of plant geography and ecology.

Alfred Russel Wallace (1823–1913), explorer of the Amazon Basin and Malay Archipelago, co-discoverer of the evolution by natural selection, author of more than ten books ranging in topics from *Palm Trees of the Amazon* (1853) to *Man's Place in the Universe* (1903), and founding father of zoogeography.

One of the most remarkable discoveries of the early explorers was the sheer diversity of life forms on Earth. New climates and new lands yielded a fantastic, even unbelievable, array of new species. Within each of the major zoogeographic realms, biomes and ecosystem types that they discovered, there were further scales of variation. Moisture gradients, elevation gradients, fires and herds of wandering herbivores, all generated patterns at more local scales. As the results of map-making, systematics, phytogeography and zoogeography coalesced, attention naturally turned to more fine-scale descriptions and the search for causation. This we may regard as the beginning of the modern era of ecology.

The beginning of a new era: the search for causation

In the current era, ecologists have increasingly turned their attention from tabulating the species pool, towards understanding the environmental factors that cause these hierarchically nested ecological patterns. The outcome of this last century of effort might be roughly summarized with three principles.

The first principle might state that *any particular community or ecosystem*

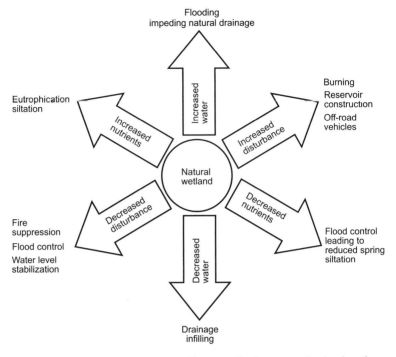

Fig. 14.1. Any specified community, in this case a freshwater wetland, arises from an array of opposing factors that together produce the local environmental conditions.

is produced by multiple environmental factors acting simultaneously. Any specified community (and that includes its species and functions) could be viewed as being the product of the pushing and pulling by opposing environmental factors. Since the concluding chapter is being written in November 1998 while my co-editor Dr Weiher drives us through wetlands along the Gulf Coast of Mississippi, I will use a wetland example (Fig. 14.1), but clearly the same principles will apply to every other ecosystem, be it birds on islands or fish in streams. It may be useful to consider this set of physical factors as a kind of *habitat template* (Southwood, 1977) which both guides and constrains the evolution of species, the formation of biological communities, and the functions that the community performs. For example, along most watercourses, the movement of water, and particularly the movement of floodwaters, creates gravel bars, eroding banks, sand bars, ox bows and deltas (Fig. 14.2). In these circumstances, the three most important processes producing the habitat template are (i) flooding, (ii) erosion and (iii) deposition. These are not entirely independent, of course, but this co-variation of factors is one of the realities of community ecology. The characteristics of species, the composition

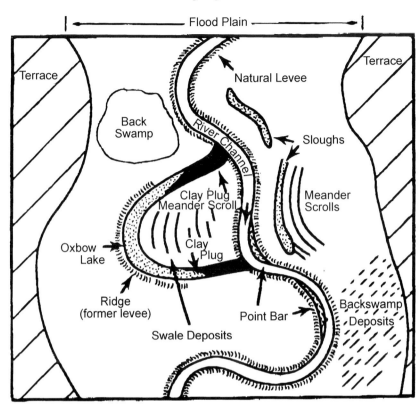

Fig. 14.2. Dynamic processes can create a wide variety of environmental conditions that act as a habitat template. In this case, a river creates habitats including terrace edges, sand bars, oxbows, sloughs and levees (from Mitsch & Gosselink, 1986).

of the communities, and the processes of ecosystems are all related to this template.

Nested within the first principle are a series of more specific relationships between communities and environmental factors. The second principle might therefore state that *there are quantitative relationships between environmental factors and the properties of communities*. The study of each factor in the habitat template provides an opportunity for exploration of such quantitative relationships. Examples might include (i) the production of tropical fish as determined by floodplain area, (ii) the diversity of plants as controlled by substrate fertility, or (iii) the zonation of invertebrates produced by different rates of burial by sediment. These relationships summarize the state of human knowledge about the factors that create and control ecological communities. The challenge for the ecologist is to unravel these factors, discover their consequences for communities, and determine their relative importance. The challenge for managers and conservationists is to understand these relationships, and then, if necessary, manipulate or regulate one or more of them to maintain or produce the desired characteristics of a landscape.

These challenges are made difficult by the many kinds of communities, and the many factors at work in them. The difficulty is compounded by a third principle, *the multiple factors that produce a community or ecosystem will change through time*. Disturbance from factors such as fire, storms, landslides and floods can influence the communities and species found at any site (Sousa, 1984; Pickett & White, 1985; Botkin, 1990). The habitat template along water courses in Fig. 14.2 is constantly changing as moving water reshapes the environment. Similarly, if humans change the factors acting on the community in Fig. 14.1, say, by decreasing spring flooding or increasing fertility, the balance of forces will shift and the composition or function of the community will begin to shift as well. It is far too easy, and therefore far too common, for humans to study small fragments of this habitat complex (say one species or one vegetation type in one oxbow lake), losing track of the fact that the particular species and community types are but transitory occurrences at any location. To understand communities, and to manage them wisely, it is essential to appreciate their multifactorial and dynamic nature.

Even if the species pool is determined by evolution and biogeography, and the type of species have been set by the habitat template, an important question remains: how much of the residual variance in composition can be accounted for by other constraints upon species composition? Is the site-to-site variation within one habitat type merely a result of stochastic factors, is it a consequence of subtle changes in the template, or are there other biological constraints upon the combinations of species that can occur at any location? From this perspective, assembly rules are nested within the habitat template and biogeographic realms as a further set of constraints upon species composition. All of the foregoing chapters in this book could be treated as elaborations of these principles, with varying degrees of emphasis upon biogeography, habitat templates and biological constraints on composition.

The search for assembly rules

The intractable number of components still remains a central problem in the causal analysis of ecological communities. To borrow from realm of general systems theory (Weinberg, 1975), one can recognize small, medium and large number systems. These have very different properties, and therefore require different approaches in analysis and modelling. *Small number systems* have few components and few interactions, and these systems are amenable to precise mathematical description; the trajectory of a bullet can be predicted with reasonable confidence. While science has in general been successful in analyzing small number systems, in ecology this can usually be done only by artificially isolating a set of populations from the many connections they have with other populations, an isolation achieved only by careful judgement (Starfield & Bleloch, 1991) or by use of containers (Fraser & Keddy, 1997). Desert rodents (Chapter 3) and stream fish (Chapter 12) might be workable examples of small number systems. At the other extreme are *large number systems* where there are so many components that the average behavior becomes a useful description of the system. The ideal gas laws provide one example; the position and velocity of a particular gas molecule are not of interest, but the properties of volume, temperature and pressure are. Large number systems may not occur in ecology. Regional floras having several thousand species (Chapter 5, Chapter 6) may be as large as we normally encounter, and even here the traits of the individuals probably differ enough that estimates of average behavior are misleading. Even congeneric species of animals such as frogs in the genus Rana cannot be easily combined: diet, size, physiology and behavior change with life history stage, and even adults of species differ in trait such as body size, temperature preferences and desiccation tolerance (Moore, 1949; Goin & Goin, 1971). A frog is neither a molecule nor a billiard ball.

Ecology is therefore inordinately difficult, precisely because communities and ecosystems are neither large nor small number systems (Lane, 1985). As *medium number systems*, they contain too many components to be treated analytically, and too few for statistical analysis. This may be why some ecologists are attracted to the study of complexity itself (Chapters 8 and 9). A rhinoceros, a frog and a grass plant cannot be averaged like ideal gas molecules, nor are their behaviors and population dynamics equivalent to random events. As the number of components increase arithmetically, the number of interactions increase geometrically. Some method of simplification is therefore necessary in order to solve problems involving medium number systems (Lane, 1985; Starfield & Bleloch, 1991). Functional groups, guilds and traits all have this potential to simplify medium number systems to small number systems, which may explain the popularity of such approaches (e.g., Chapters 1, 5, 9 and 12). Simplification requires carefully preserving critical interactions and components while excising or ignoring others. The inherent difficulty in doing this probably explains why, at present, medium number systems require analytical approaches that are as much an art as a science.

Rising to the challenge

Our introductory chapter ended the same way as our opening remarks at the symposium, with the challenge of the maze in Fig. 9 on page 18. At this point, it might be worthwhile, although hardly tactful, to evaluate the authors' responses to our challenge. How many authors suggested useful tactics to tra verse the maze? How many were successful in reaching the goal? Answering these two questions for each chapter, and summarizing the results in a table, might be a useful challenge for graduate courses. Students could read chapters, and, for each one, an assessor could have the task of evaluating both the tactics offered and the success demonstrated in attaining the goal. The graduate class as a whole could then discuss each assessor's judgement. By the end of the course, a chapter by chapter table rating of (i) value of tactics and (ii) degree of achievement of goals, could be constructed. The only further guideline that might assist such discussions is a reminder of the opening view on page 8 that:

> Contrary to common practice, merely documenting a pattern is not the study of community assembly . . . Asking if there is pattern in nature is akin to asking if bears shit in the woods. Null models provide a valuable and more rigorous way of demonstrating pattern, but they still do not specify assembly rules.

Booth and Larson (Chapter 7) may have provided the most controversial contribution. In reading their chapter, one is reminded of Andy Warhol's assertion that in the future everyone will have their 15 minutes of fame. For this to be possible, collective amnesia, a fascination with the ephemeral, and self-delusion are all required. How many ecological ideas trumpeted as new discoveries, they ask us, are simply reworded versions of ideas from the turn of the century? The lack of computer access to early texts, and the degree to which they are physically deteriorating in storage, suggests that it will be increasingly easy to pretend that re-statements of old ideas are novel contributions. Larson and Booth have challenged us all to be aware of the history of our discipline and to give credit where credit is due.

The final chapter in the book, Lockwood and Pimm's assessment of 87 published reports on restoration projects provides a realistic assessment of the practical limitations to re-assembling natural communities. Seventeen projects were rated unsuccessful, and 53 were rated partially successful, leaving only 17 (20%) as successful in attaining stated goals. These included 13 functional goals and 19 structural goals. It therefore appears that much of the rhetoric around assembly rules cannot yet be translated into practical success on the ground.

And so, corrected proofs in hand, we now excuse ourselves from our duties as hosts, and return to our rainy trip through the Longleaf Pine forests and Pitcher Plant savannas of the Gulf coast, a region with several dozen species of carnivorous plants and orchids, and a seemingly endless array of Asteraceae, Cyperaceae and Poaceae. The existence of this Longleaf Pine ecosystem (Peet & Allard, 1993) can only be explained by invoking processes over very different time scales. At the geological scale, over tens of millions of years, there were the factors of erosion, deposition, sea level change and leaching that produced these rolling hills and sand plains. At the historical scale of centuries, there were the recurring effects of fire, and the increasingly strong effects of human settlement. Current factors appear to include fire, seepage, infertility, summer drought, competition, and disturbance by burrowing crawfish. The coexistence of up to 40 species in a 50 by 50 cm quadrat requires consideration of all these different time scales. By invoking these processes, one can offer an entertaining explanation for why some areas have so many plants. Superimposed upon such factors, however, remains the tantalizing possibility that only some mixtures are biologically possible, or ecologically favored. The finer the scale becomes, the more difficult explanation seems to become. Sitting here at the Western Inn, in Stone County, southern Mississippi at 9.30 on a Saturday night, filled with catfish and chicken-fried steak (but no beer), the original questions posed in the introduction remain to tantalize the naturalist and the reader: just how do these multivariate factors select the flora of a pitcher plant savannah from the available pool, how deterministic is the habitat template, and why are there five species of pitcher plant in a savanna as opposed to any other number?

Acknowledgments

I thank my co-editor Evan Weiher for his many hours of hard work and endless attention to detail. We are both grateful to the contributors for their effort, enthusiasm, patience and general good humor, as well as to Maria Murphy and Alan Crowden at Cambridge University Press for piloting us though the publishing process. Dennis Whigham was helpful in the early stages of organising the symposium through the Ecological Society of America. NSERC provided the continuity of funding that made the project possible.

References

Botkin, D.B. (1990). *Discordant Harmonies. A New Ecology for the Twenty-first Century*. New York: Oxford University Press.

Edmonds, J. ed. (1997). *Oxford Atlas of Exploration*. New York: Oxford University Press.

Fraser, L.H. & Keddy, P. (1997). The role of experimental microcosms in ecological research. *Trends in Ecology and Evolution* 12: 478–481.

Goin, C.J. & Goin, O.B. (1971). *Introduction to Herpetology*. 2nd edn. San Francisco: W. H.Freeman.

Lane, P.A. (1985). A food web approach to mutualism in lake communities. In *The Biology of Mutualism, Ecology and Evolution*, ed. D.H. Boucher, pp. 344–374. New York: Oxford University Press.

Mackay, C. (1841). *Memoirs of Extraordinary Popular Delusions*. Reprinted in 1980 as *Extraordinary Popular Delusions and the Madness of Crowds*, with a foreword by A. Tobias. NY: Harmony Books.

Mitsch, W.J. & Gosselink, J.G. (1986). *Wetlands*. NY: Van Nostrand Reinhold.

Moore, J.A. (1949). Patterns of evolution in the genus *Rana*. In *Genetics, Paleontology and Evolution*, ed. G.L. Jepsen, G.G. Simpson, & E. Mayr, pp. 315–338. Princeton: Princeton University Press. Reprinted in 1963 by NY: Atheneum.

Morison, S.E. (1978). *The Great Explorers. The European Discovery of America*. NY: Oxford University Press.

Morris, J. (1973). *Pax Britannica. 3 Vols.* Faber & Faber Ltd. Reprinted 1992, London: Folio Society.

Peet, R.K. & Allard, D.J. (1993). Longleaf pine vegetation of the southern Atlantic and eastern Gulf Coast regions: a preliminary analysis. In *Proceedings of the Tall Timbers Fire Ecology Conference No. 18, The Longleaf Pine Ecosystem: Ecology, Restoration and Management*, ed. S.M. Hermann, pp. 45–104. Tallahassee, FL: Tall Timbers Research Station.

Pickett, S.T.A & White, P.S. (1985). *The Ecology of Natural Disturbance and Patch Dynamics*. Orlando, FL: Academic Press.

Sousa, W.P. (1984). The role of disturbance in natural communities. *Annual Review of Ecology and Systematics* 15: 353–391.

Southwood, T.R.E. 1977. Habitat, the template for ecological strategies? *Journal of Animal Ecology* 46: 337–365.

Starfield, A.M. & Bleloch, A.L. (1991). *Building Models for Conservation and Wildlife Management*, 2nd edn. Edina, MN: Burgers International Group.

Weinberg, G.M. (1975). *An Introduction of General Systems Thinking*. New York: John Wiley.

Index

403